GEOPHYSICAL APPLICATIONS OF ARTIFICIAL NEURAL NETWORKS AND FUZZY LOGIC

MODERN APPROACHES IN GEOPHYSICS

VOLUME 21

The titles published in this series are listed at the end of this volume.

GEOPHYSICAL APPLICATIONS OF ARTIFICIAL NEURAL NETWORKS AND FUZZY LOGIC

edited by

WILLIAM SANDHAM

University of Strathclyde,
Glasgow, Scotland, U.K.

and

MILES LEGGETT

Fugro-Jason,
Rotterdam, The Netherlands

SPRINGER-SCIENCE+BUSINESS MEDIA, B.V.

A C.I.P. Catalogue record for this book is available from the Library of Congress.

ISBN 978-90-481-6476-9 ISBN 978-94-017-0271-3 (eBook)
DOI 10.1007/978-94-017-0271-3

Printed on acid-free paper

Additional material to this book can be downloaded from http://extras.springer.com.

Originally published by Kluwer Academic Publishers in 2003
Softcover reprint of the hardcover 1st edition 2003

To my brother John for the many contributions he's made in my life

- WAS

To my wife Audrey and my family for their support

- ML

CONTENTS

SECTION B

LITHOLOGY, WELL LOGS, PROSPECTIVITY MAPPING AND RESERVOIR CHARACTERISATION

SECTION C

ELECTROMAGNETIC EXPLORATION

SECTION D

OTHER GEOPHYSICAL APPLICATIONS

LIST OF CONTRIBUTORS

ANDREAS AHL	*Institute of Meteorology and Geophysics, Austria.*
FRED AMINZADEH	*DGB-USA & FACT, USA.*
PING AN	*Schlumberger-Geoquest, USA.*
STUART BLUNDELL	*Integrated Geoscience Inc., USA.*
WARICK M. BROWN	*University of Western Australia, Australia.*
HENGCHANG DAI	*British Geological Survey, Scotland.*
TARIQ S. DURRANI	*University of Strathclyde, Scotland.*
ROBERT ESSENREITER	*PGS Seres AS, Norway.*
LI-YUN FU	*CSIRO Petroleum, Australia.*
TAMÁS D. GEDEON	*Murdoch University, Australia.*
DAVID I. GROVES	*University of Western Australia, Australia.*
DOUGLAS I. HART	*Western Geophysical, USA.*
QIAODENG HE	*Changchun University of Science and Technology, Peoples Republic of China.*
PHILLIP A. HIMMER	*Montana State University, USA.*
KOU-YUAN HUANG	*National Chiao Tung University, Taiwan.*
FOTIS KALANTZIS	*PGeoph., Canada.*
MARTIN KARRENBACH	*University of Karlsruhe, Germany.*
MILES LEGGETT	*Jason Geosystems bv, The Netherlands.*
CURTIS A. LINK	*University of Montana, USA.*
MICHAEL D. McCORMACK	*Optimization Associates, Inc., USA.*
WOOIL M. MOON	*University of Manitoba, Canada.*
MARY POULTON	*University of Arizona, USA.*
ENDERS A. ROBINSON	*Columbia University, USA.*
WILLIAM A. SANDHAM	*University of Strathclyde, Scotland.*
WOLFGANG SEIBERL	*Institute of Meteorology and Geophysics, Austria.*
ANTOINE TOUMANI	*DMT GmbH, Germany.*
SVEN TREITEL	*TriDekon, Inc., USA.*
EDMUND WINKLER	*Geological Survey of Austria, Austria.*
LIN ZHANG	*Chevron Petroleum Technology Company, USA.*
HUI ZHOU	*Changchun University of Science and Technology, Peoples Republic of China.*

PREFACE

The past fifteen years has witnessed an explosive growth in the fundamental research and applications of artificial neural networks (ANNs) and fuzzy logic (FL). The main impetus behind this growth has been the ability of such methods to offer solutions not amenable to conventional techniques, particularly in application domains involving pattern recognition, prediction and control. Although the origins of ANNs and FL may be traced back to the 1940s and 1960s, respectively, the most rapid progress has only been achieved in the last fifteen years. This has been due to significant theoretical advances in our understanding of ANNs and FL, complemented by major technological developments in high-speed computing.

In geophysics, ANNs and FL have enjoyed significant success and are now employed routinely in the following areas (amongst others):

1. Exploration Seismology. (a) Seismic data processing (trace editing; first break picking; deconvolution and multiple suppression; wavelet estimation; velocity analysis; noise identification/reduction; statics analysis; dataset matching/prediction, attenuation), (b) AVO analysis, (c) Chimneys, (d) Compression / dimensionality reduction, (e) Shear-wave analysis, (f) Interpretation (event tracking; lithology prediction and well-log analysis; prospect appraisal; hydrocarbon prediction; inversion; reservoir characterisation; quality assessment; tomography).
2. Earthquake Seismology and Subterranean Nuclear Explosions.
3. Mineral Exploration.
4. Electromagnetic / Potential Field Exploration. (a) Electromagnetic methods, (b) Potential field methods, (c) Ground penetrating radar, (d) Remote sensing, (e) inversion.

This research monograph was originally conceived at an EAGE pre-conference workshop which we convened in 1997, entitled "Geophysical Applications of Artificial Neural Networks and Fuzzy Logic". This one-day workshop was held on 26th May 1997, at the Palexpo Conference Centre, Geneva, Switzerland. The general aim of the workshop was to address the wide diversity of ANN and FL applications within geophysics, and had the following remit:

- To focus awareness of ANNs and FL in geophysics, with the possibility of establishing an EAGE ANN/FL Special Interest Group.
- To review the current status of ANNs and FL in geophysics.
- To identify the usefulness and pitfalls of ANNs and FL in geophysics.
- To identify commercial ANN and FL products in general, and geophysical implementations in particular.
- To establish the leading individuals and institutions working in the area.
- To address future applications for ANNs and FL in geophysics, through evaluation of new algorithms, methods and technology.

The general format of the Workshop consisted of a number of formal oral presentations, arranged in application category, followed by open discussion sessions. Based on the enthusiastic feedback and favourable responses resulting from the Workshop, we requested presenters to modify their presentations for suitable publication as the present chapters. We also solicited additional chapters from other established geophysical experts in ANNs and FL.

This book reflects a healthy diverse of papers by topic, contributor background (50% industry and 50% academia), and country of origin (10). Chapters have been organised into four major sections; (i) exploration seismology, (ii) lithology, well logs, prospectivity mapping and reservoir characterisation, (iii) electromagnetic exploration, and (iv) other geophysical applications (including earthquake seismology).

We have been very fortunate to include a Special Preface by Professor Fred Aminzadeh, whose experience and indeed international reputation in the general area of geophysics and artificial intelligence, is highly regarded by colleagues in industry and academia. His observations, comments and outlook on ANNs and FL in geophysics, make stimulating reading.

A two-part comprehensive bibliography has also been included at the end of the book. Part A lists introductory and advanced texts and review papers concerning the theory of ANNs and FL, and should prove a useful reference for those readers requiring knowledge of fundamentals or advanced issues regarding ANNs or FL. Part B is a compilation of all the major publications concerning geophysical applications of ANNs and FL, from the earliest works in the 1980s to the present day. It is broken down into references dealing with overviews of the subject, collective pattern recognition publications, and compilations of papers dealing with exploration seismology, earthquake seismology and subterranean nuclear explosions, mineral exploration, and electromagnetic / potential field exploration. These major subject headings are then broken down further into their respective sub-headings.

It is our hope that this text will be of use to those geophysicists already involved in the application of ANNs and FL within their discipline, and that it will contribute to the education of *future practitioners* of these very powerful techniques.

Acknowledgements

We wish to express our sincere appreciation to all of the delegates at the original EAGE pre-conference workshop we convened in 1997, for their participation, enthusiasm and positive feedback. This proved to be a major catalyst for the present text.

We also wish to record our deep thanks to Petra van Steenbergen and her staff at Kluwer Academic for their support, assistance, and continuous encouragement throughout the preparation of this manuscript.

WILLIAM SANDHAM

Signal Processing Group,
Institute of Communications and Signal Processing,
Dept. of Electronic and Electrical Engineering,
University of Strathclyde,
Glasgow G1 1XW,
SCOTLAND
`E-Mail: w.sandham@eee.strath.ac.uk`

MILES LEGGETT

Jason Geosystems bv,
Plaza Building,
Weena 598,
P.O. Box 1573, 3000BN,
Rotterdam,
THE NETHERLANDS
`E-Mail: miles@jason.nl`

SPECIAL PREFACE

"As complexity rises, precise statements lose meaning, and meaningful statements lose precision"

Lotfi A. Zadeh, Father of Fuzzy Logic.

Preamble

The last decade has witnessed significant advances in transforming geosciences and well data into drillable prospects, generating accurate structural models and creating reservoir models with associated properties. This has been made possible through improvements in data integration, quantification of uncertainties, effective use of geophysical modeling for better describing the relationship between input data and reservoir properties, and use of unconventional statistical methods. Soft computing techniques such as neural networks and fuzzy logic, and their appropriate usage in many geophysical and geological problems, have played a key role in the progress made in recent years.

However, there is a consensus of opinion that we have only begun to scratch the surface in realizing the full benefits of soft computing technology. Many challenges remain, in particular when we are faced with characterization of reservoirs having substantial heterogeneity and fracturing, exploration in areas with thin-bedded stacked reservoirs, and regions with poor data quality or limited well control and seismic coverage, and quantifying uncertainty and confidence intervals of the estimates. Inherent problems we need to overcome include inadequate and uneven well data sampling, non-uniqueness in cause and effect in subsurface properties versus geoscience data response, different scales of seismic, log and core data and finally, how to accommodate reservoir changes as the characterization progresses.

William Sandham and Miles Leggett should be commended for a fine editing job. This research monograph is an excellent compilation of articles dealing with the use of artificial neural networks and fuzzy logic in different geophysical applications.

These applications are divided into the following categories:

(a) Seismic exploration.
(b) Lithology, well logs, prospectivity mapping and reservoir characterisation.
(c) Electromagnetic exploration.
(d) Other geophysical applications.

In this Special Preface, I would like to highlight soft computing applications with most impact in the oil industry, in a somewhat different manner as follows:

(a) Automation in data editing and data mining.
(b) Non-linear signal (geophysical and log data) processing.
(c) Wave-equations with random or fuzzy coefficients.
(d) Data integration and reservoir property estimation.
(e) Quantification of data uncertainty, prediction error and confidence interval.

Automation in Data Editing and Data Mining

In recent years we have witnessed a massive explosion in the data volumes we have to deal with. This is a consequence of increased sampling rate, more data channels, larger offset and longer record acquisition, multi-component data, 4-D seismic and, most recently, the possibility of continuous recording in "instrumented oil fields". We therefore require efficient techniques to process such large data volumes. Automated techniques to refine the data (trace editing and filtering), selecting the desired event types (first break picking) or automated interpretation (horizon tracking) are necessities for large data volumes.

Fuzzy logic and neural networks have proven to be effective tools for such applications. To make use of large volumes of field data and the multitude of associated data volumes (e.g. different attribute volumes or partial stack gathers), effective data compression methods will be of increasing significance, both for fast data transmission and efficient data storage. Yet, the greatest impact of advances in data compression techniques will be realized when geoscientists have the ability to fully process and analyze data in the compressed domain. This will make it possible to carry out computer-intensive processing of large volumes of data in a fraction of the time, resulting in tremendous cost reductions. Data mining is another alternative to assist identification of the most information-rich part of such large data volumes. Again, in many recent reports, it has been demonstrated that neural networks and fuzzy logic, *in combination* with some of the more conventional methods such as eigenvalue or principal components analysis, are very useful.

Non-Linear Signal (Geophysical and Log Data) Processing

Although seismic signal processing has advanced significantly over the last four decades, the fundamental assumption of a "convolution model" is violated in many practical settings. In 1995, Sven Treitel posed the question: *"What if mother earth refuses to convolve?"*. Situations which do not adhere to the convolution model include highly heterogeneous environments, very absorptive media (such as unconsolidated sand and young sediments), fractured reservoirs, mud volcano, and karst and gas chimneys. In such cases we must consider non-linear processing and interpretation methods. Neural networks, fuzzy logic, genetic algorithms, fractals, chaos and complexity theory, are among such non-linear processing and analysis techniques that have proven to be effective. The highly heterogeneous earth model that geophysics attempts to quantify is an ideal place for applying these concepts. The subsurface lives in a high-dimensional space (the properties can be considered as the additional space components), but its actual response to external stimuli initiates an internal coarse-grain and self-organization that results in a low-dimensional structured behavior. Fuzzy logic and other non-linear methods can describe shapes and structures generated by chaos. These techniques will push the boundaries of seismic resolution, allowing smaller-scale anomalies to be characterized.

Wave-Equations with Random or Fuzzy Coefficients

Many of our geophysical analysis techniques such as migration, DMO, wave-equation modeling as well as the potential methods (gravity, magnetic, electrical), use conventional partial differential equations (PDEs) with deterministic coefficients. The same is true for the partial differential equations used in reservoir simulation. For many practical and physical reasons, deterministic parameters for the coefficients of these PDEs is unrealistic (for example medium velocities for seismic wave propagation or fluid flow for the Darcy equation). Stochastic parameters in these cases can provide us with a more practical characterization. Fuzzy coefficients for PDEs may prove to be even more realistic and simpler to parameterize.

Today's deterministic processing and interpretation ideas will give way to stochastic methods, even if the industry has to rewrite the book on geophysics. That is, using wave-equations with random and fuzzy coefficients, subsurface velocities and densities may be described in statistical and membership grade terms, thereby enabling a better description of wave propagation in the subsurface, particularly when a substantial amount of heterogeneity is present. More generalized applications of geostatistical techniques will emerge, making it possible to introduce risk and uncertainty at the early stages of the seismic data processing and interpretation loop.

Data Integration and Reservoir Property Estimation

Historically, the link between reservoir properties and seismic and log data have been established either through "statistics-based" or "physics-based" approaches. The latter, also known as model-based approaches, attempt to exploit the changes in seismic character or seismic attribute to a given reservoir property, based on physical phenomena. Here, the key issues are sensitivity and uniqueness. Statistics-based methods attempt to establish a heuristic relationship between seismic measurements and prediction values from examination of data only. It can be argued that a hybrid method, combining the strength of both statistics and physics-based methods would be most effective. Figure 1 shows the concept, schematically.

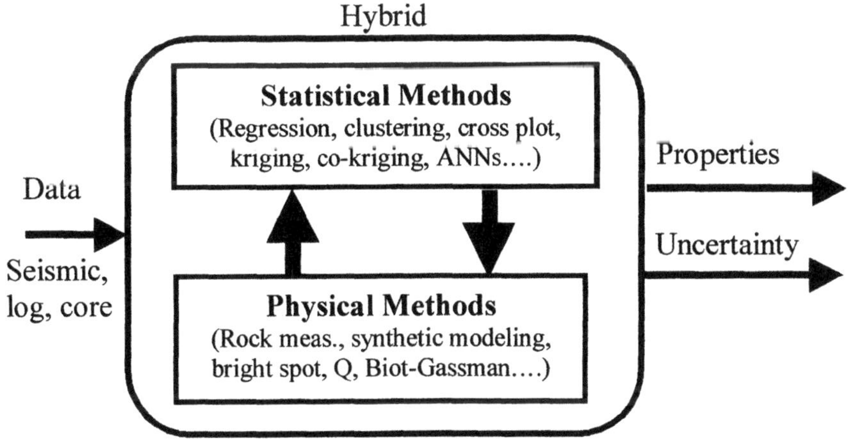

Figure 1. Schematic description of physics-based, statistics-based and hybrid methods for linking reservoir properties and seismic and log data.

Many geophysical analysis methods, and consequently seismic attributes, are based on physical phenomena i.e. based on theoretical physics (wave propagation, Biot-Gassman equation, Zoeppritz equation, tuning thickness, shear wave splitting, etc). Certain attributes may be more sensitive to changes in certain reservoir properties. In the absence of a theory, using experimental physics (for example rock property measurements in a laboratory environment such as the one described in the last section) and/or numerical modeling, one can identify or validate suspected relationships. Although physics-based methods and direct measurements (the ground truth) are the ideal and reliable way to establish such correlations, for various reasons it is not always practical. These reasons range from lack of known theories, differences between the laboratory environment and field environment (noise, scale, etc.), and the cost for conducting elaborate physical experiments.

Statistics-based methods aim at deriving an explicit or implicit heuristic relationship between measured values and properties to be predicted. Neural networks and fuzzy-neural networks-based methods are ideally suitable to establish such implicit relationships through proper training. We all attempt to establish a relationship between different seismic attributes, petrophysical measurements, laboratory measurements and different reservoir properties. In such statistics-based methods, one has to keep in mind the impact of noise on the data, the data population used for statistical analysis, scale, geologic environment, and the correlation between different attributes when performing clustering or regressions. The statistics-based conclusions have to be re-examined and their physical significance explored.

Quantification of Data Uncertainty, Prediction Error and Confidence Interval

One of the main problems we face is how to handle the non-uniqueness issue, and quantify uncertainty and confidence intervals in our analysis. We also need to understand that incremental improvements in prediction error and confidence stem from the introduction of new data or a new analysis scheme. Methods such as evidential reasoning and fuzzy logic are most suited for this purpose. Figure 2 shows the distinction between conventional probability and these techniques. "Point probability," describes the probability of an event, for example, having a commercial reservoir. The implication is we know exactly what this probability is. Evidential reasoning, provides an upper bound (plausibility) and lower bound (credibility) for the event. The difference between the two bounds is considered as the ignorance range. Our objective is to reduce this range through use of all the new information. Given the fact that in real life we may have non-rigid boundaries for the upper and lower bounds and we ramp up or ramp down our confidence for an event at some point, we introduce fuzzy logic to handle it and we refer to it as "membership grade". Next-generation earth modeling will incorporate quantitative representations of geological processes and stratigraphic / structural variability. Uncertainty will be quantified and built into the models.

On the issue of non-uniqueness, the more sensitive the particular seismic character is to a given change in reservoir property, the easier it is to predict it. The more unique the influence of the change in seismic character is to changes in a specific reservoir property, the higher the confidence level in such predictions. Fuzzy logic can handle subtle changes in the impact of different reservoir properties on the wavelet response. The multitude of wavelet responses (for example near, mid and far-offset wavelets) may

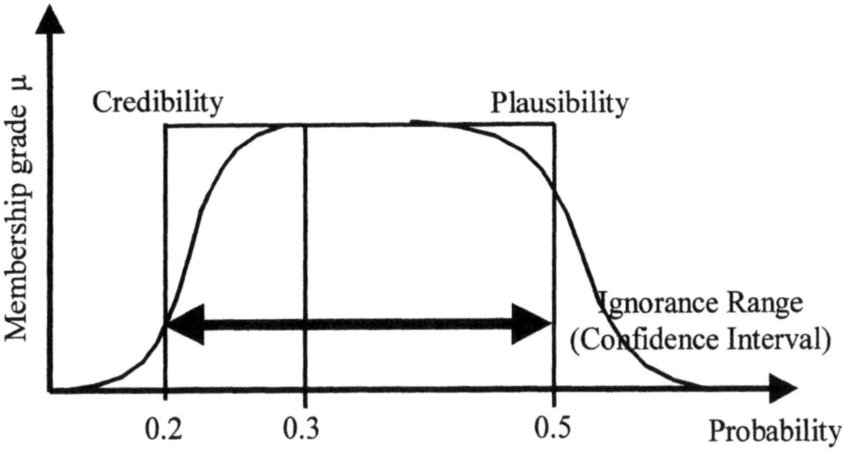

Figure 2. Point probability, evidential reasoning and fuzzy logic.

be compared more effectively through the use of neural networks. Let us assume that seismic patterns for three different lithologies (sand, shally sand and shale) are compared for different well information and seismic responses (both model and field data); the respective seismic character within the time window or the reservoir interval involves four "classes" of wavelets, (*w1*, *w2*, *w3* and *w4*). These 4 wavelets (basis wavelets) serve as a segmentation vehicle.

The histograms in Figure 3a show what wavelet classes are likely to be present for the given lithologies. In the extreme positive (EP) case we would have one wavelet uniquely representing one lithology. In the extreme negative case (EN) we would have a uniform distribution of all wavelets for all lithologies. Unfortunately, in most cases we are closer to NP than to EP. The question is how best we can move these distributions from the EN side to the EP side, thus improving our prediction capability and increasing our confidence level. The common sense solution is to add / enhance information content of the input data.

What if we use wavelet vectors comprised of pre-stack data (in the simple case, mid, near, far-offset data) as the input to a neural network, in order to perform the classification? Intuitively, this should lead to a better separation of different lithologies (or other reservoir properties). Likewise, including three component data as the input to the classification process would further improve the confidence level. Naturally, this requires introduction of a new "metric" measuring "the similarity" of these "wavelet vectors". This can be done using the new basis wavelet vectors as input to a neural network, and applying different weights to mid, near and far-offset traces. This is demonstrated conceptually in Figure 3 to predict lithology. Compare the sharper histograms of the vector wavelet classification (in this case, mid, near, and far-offset gathers) in Figure 3b, against those of Figure 3a based on scalar wavelet classification

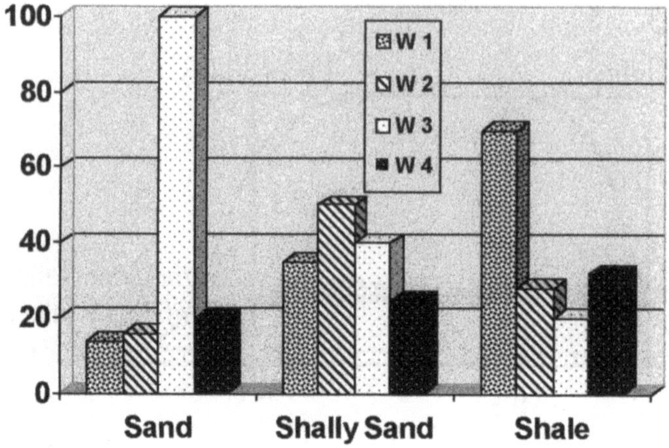

(a)

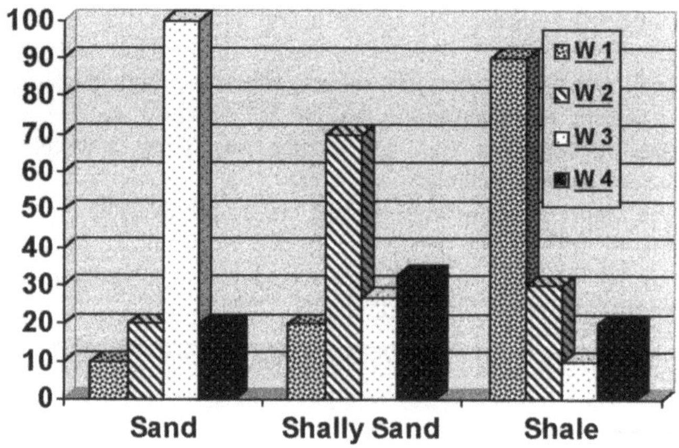

(b)

Figure 3. Statistical distribution of different wavelet types versus lithologies for (a) pre-stack data, (b) stacked data.

Concluding Remarks

We have discussed the main areas where fuzzy logic and neural networks can make a major impact in geophysical applications in the oil industry. These areas include facilitation of automation in data editing and data mining. We also pointed out applications in non-linear signal (geophysical and log data) processing, and discussed better parameterization of the wave-equations using random or fuzzy coefficients, in seismic and other geophysical wave propagation equations, and those used in reservoir simulation. Of significant importance is their use in data integration and reservoir property estimation. Finally, the possibility for quantification and reduction of uncertainty and confidence intervals, using more comprehensive use of fuzzy logic and neural networks, was discussed.

FRED AMINZADEH

President,
DGB-USA & FACT,
14019 Southwest Fwy.,
#301-225,
Sugar Land,
Texas,
TX 77478,
USA
E-Mail: Factn@aol.com

SECTION A

EXPLORATION SEISMOLOGY

A REVIEW OF AUTOMATED FIRST-BREAK PICKING AND SEISMIC TRACE EDITING TECHNIQUES

MICHAEL D. McCORMACK

Optimization Associates, Inc.,
1437 Debon Drive,
Plano, Texas 75075,
USA
E-Mail: m.d.mccormack@worldnet.att.net

Abstract

This chapter describes two software systems, based on artificial neural networks (ANNs), which have largely automated the highly labor-intensive seismic processes of first-break refraction picking and trace editing. The underlying mechanism of these two processes relies heavily on pattern recognition techniques. In contrast, most other seismic processing algorithms depend on more traditional signal processing theory. This explains, at least in part, why first-break picking and trace editing have remained in the domain of human processors, until recently. With the development of ANN theory, powerful pattern recognition algorithms have now become available to address these two problem areas of seismic processing. This chapter describes the various approaches that researchers have employed in automating first-break picking and seismic trace editing, in particular those based on ANNs. These ANN-based systems can achieve accuracies ranging from 94% to 99%, with a fraction of the human effort required for a manual analysis.

1. Introduction

This chapter traces the development of automated first-break refraction picking and seismic trace editing systems that utilize artificial neural networks (ANNs) as the underlying pattern recognition algorithm. These systems represent some of the first successful applications of neural network technology to seismic data processing.

At the time these systems were developed, the prevailing wisdom in the petroleum industry held that these processes could be successfully performed only by humans. Because of the labor-intensive nature of first-break picking and seismic trace editing, substantial efforts to automate these processes have occurred within the industry since computers were first used to process digital seismic data.

W. Sandham and M. Leggett (eds.),
Geophysical Applications of Artificial Neural Networks and Fuzzy Logic, 1-12.

2. Background

Identifying the arrival time of seismic energy at each offset receiver location has been fundamental to the seismic method of oil prospecting since the beginnings of exploration seismology. Prior to digital seismic recording, these "first-break" arrival times were manually picked from analog records, plotted on graph paper and then used to determine the velocity structure of the near surface via various refraction analysis techniques. With this information, the geophysicist could remove variations in the near surface structure that were superimposed on deeper reflection events of interest. The need for this fundamental process remained even after the introduction of digital seismic recording and computers for seismic data processing.

With the introduction of digital computers, new data analysis requirements arose. One of the most time consuming and labor intensive tasks was to remove seismic data traces that were corrupted with excessive levels of noise from the processing stream, in order to optimize the signal-to-noise in the final stacked data sections. Even though computers permitted vastly larger volumes of seismic data to be efficiently processed, first-break picking and seismic trace editing remained two tasks that required extensive human intervention (up to 50% or more of the human effort involved in 2-D seismic processing). This processing bottleneck has been further exacerbated by continuing advances in electronic technology - the number of seismic recording channels has increased exponentially over the last 35 years, consistently doubling approximately every 7 years. It was clear that in order to satisfy the need for high quality and cost effective seismic data that the industry needed for subsurface imaging, automation of first-break arrival picking and seismic trace editing was required.

There have been numerous attempts to develop statistical or parametric algorithms that can automatically perform these two tasks (Ervin et al., 1983; Gelchinsky and Shtivelman, 1983; Neff and Wyatt, 1986; Anderson and McMechan, 1989; Spagnolini, 1991). However, these programs have found limited use since they lack the robustness necessary to accurately and reliably process seismic data when signal and noise conditions change during the course of a survey. Furthermore, because there is always an element of subjectivity in picking first-breaks or editing seismic data records for noise, acceptance of these algorithms varied from one processor to the next depending on whether or not the processor agreed with the answers provided by the program.

Although several independent investigators developed ANN-based automated first-break picking and seismic trace editing programs, it is interesting to note that all systems employ one of two basic pattern recognition approaches. The first, and most common approach, was to pre-process the seismic data to extract various trace amplitude attributes such as energy or the ratio of successive amplitude peaks etc, within a short time window. These attributes, together with human-supplied examples of first-break picks, or noisy seismic data traces, were then used to train a neural network. The neural networks that have been employed for these applications are of supervised training genre, typically a multilayer perceptron (also known as a back-propagation neural network), or a cascade-correlation network. The second approach (applied only to first-break picking) employed a more "visual" methodology, wherein a binary bit string representing a crudely rasterized picture of the seismic data traces in the general neighborhood of the first-break, was trained using a multilayer perceptron.

In Section 3, the formulation and performance of ANN-based first-break picking systems that have been developed over the last decade are described, and their strengths and weaknesses reviewed. This is repeated in Section 4 for ANN-based seismic trace editing systems. The chapter concludes with a discussion of the current status of ANN-based automated first-break picking and seismic trace editing techniques.

3. Neural Networks for Seismic First-break Picking

As mentioned in the previous section, several investigators have developed systems that employ neural networks to automate first-break picking of seismic data records. While significant differences exist in the various types of neural network employed, and the implementation details, all systems developed to date can be classified as one of two categories, from a pattern recognition perspective.

If the automated first-break picking problem is perceived as using a neural network and a set of parameters and features derived from the seismic data to define a region of a multi-dimensional parameter space, where first-break events are separable from reflections, noise and other seismic events, two basic approaches may be recognized:

1. Parametric model employing single and multi-trace attributes extracted from the seismic records. This approach uses many traditional signal processing concepts.
2. A pattern recognition or "visual" model using a compact binary representation of the wiggle trace recordings in the region around the first-break events. This solution attempts to mimic the approach that a human interpreter would use in picking first-break events.

The first approach is fairly straightforward and has been investigated by several researchers. Wagner et al. (1990) implemented an algorithm in which a multilayer perceptron was trained using four seismic trace attributes – peak amplitude, mean power level, power ratio between a forward and reverse sliding window, and the envelope slope for a selected peak and its adjacent peaks. Taner et al. (1979) provides definitions of these attributes. Peak amplitudes A_1, A_2 and A_3 on the seismic trace shown in Figure 1, represent the maximum values on three successive peaks. P_1, P_2 and P_3 are the mean

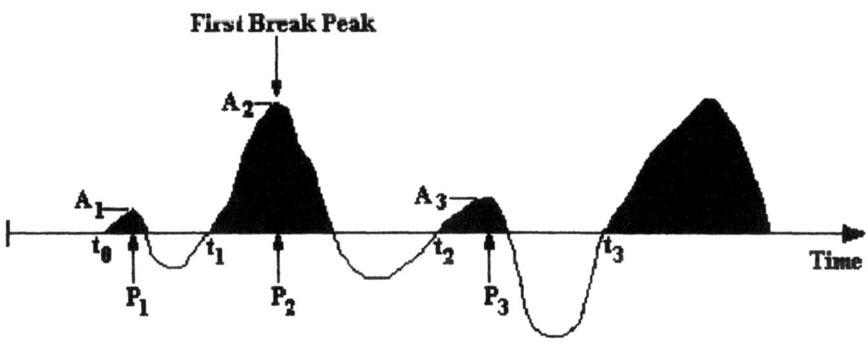

Figure 1. Seismic trace depicting extracted attributes.

power levels in each successive positive black peak. The power ratios R_1, R_2 and R_3 and envelope slope are then computed. A multilayer perceptron with 12 input nodes, 5 hidden-layer nodes and a single output node is then trained using the four seismic attributes from each of three successive peaks, and an output value of 1 or 0 is applied, depending on whether or not the central peak of the three under consideration is the first-break event. Using only these four seismic attributes, the authors achieved 90% or better accuracy in picking the first-break arrivals from surface seismic sources.

Kusuma and Fish (1993) employed an enhanced set of seismic trace attributes together with a cascade-correlation learning architecture to achieve a 10 to 50 fold speed increase in the training of the network. In addition, the cascade-correlation technique permitted a trained network to be incrementally updated so that knowledge from a previous training session could be used to adjust the network for new survey conditions.

The cascade-correlation network was developed by Fahlman and Lebiere (1990) in part to overcome deficiencies in the back-propagation algorithm. The training procedure for a back-propagation algorithm is inefficient because hidden-layer units are trained in parallel and independently of one another i.e. each hidden node adjusts its weight with no knowledge of changes in the weights of adjacent hidden nodes. This makes it difficult for the network to converge efficiently. In contrast, the cascade-correlation network permits only one unit to be modified in any given training iteration. Starting with a minimal network consisting of just an input and output layer, the connections w_{ij} between these two layers, are chosen in order to minimize the mean sum squared error, E:

$$E = \frac{1}{2} \sum_{j=1}^{N} \left[\sum_{i=1}^{M} \left(O_{ij} - d_{ij} \right)^2 \right] \tag{1}$$

where O_{ij} is the actual output of the ith node for the jth training pattern, and d_{ij} is the desired output for the ith node and the jth training pattern. N is the number of training examples and M is the number of output layer nodes. The energy is minimized in the usual manner by taking the derivative with respect to the connection weights w_{ij} (see Rummelhart and McClelland (1986) for details). Hidden units are added one at a time into a new layer between the input and output layers when no significant changes in the residual error are observed. The new hidden unit is chosen from a population of candidate test nodes that are connected to the input layer and existing hidden nodes, but not to the output nodes. The test nodes are trained to maximize the correlation between their outputs and the residual error that remains in the network. The weights are adjusted using the Quickprop algorithm (Fahlman, 1988) until no further reductions in residual energy occur. The candidate node with the highest correlation score is selected as a permanent feature detector in the network with its incoming weights fixed. The connection weights to the output layer including that of the newly added hidden node are retrained. New hidden feature detectors are sequentially added to the network until the residual error is within acceptable limits.

As input to the cascade-correlation network, Kusuma selected three sets of five single-trace attributes: peak amplitude, the ratio of energies above and below the peak, trace offset distance, and the peak amplitude relative to a measure of signal and noise. Signal and noise are determined by finding the mean value on windows above and below a crude estimate of the position of the first-breaks i.e. noise levels are measured on the data before the first-break and signal on the data after the first-break. The three sets of inputs refer to three adjacent peaks on the trace being analyzed. The network is trained to output '1' if the central peak is the first-break event, or '0' otherwise – similar to the method of Wagner described earlier.

In published performance tests of the cascade-correlation first-break picker, training on a 120 trace dynamite data set takes an average of 22 seconds. After training on 240 traces, the network achieved an accuracy of 99% on new records. For seismic data recorded using the Vibroseis method, training times increase to 4 or 5 minutes and accuracy's range from 90% to 99%. Considering the difficulty humans often have in picking first-breaks on Vibroseis data, these results are commendable.

Kusuma and Fish (1993) describe an extension of the above approach that includes multiple traces to improve the robustness of the first-break picker. The addition of multiple traces as neural network inputs, enables the network to learn lateral variations that occur in the arrival time and waveform character of the first-break event. Multi-trace data adds a spatial dimension, which complements the essentially temporal information content of a single trace. The richness of the information content contained in the trace offset dimension will be discussed again later in this section.

Kusuma expanded the single trace network input structure described earlier in this section, to include new temporal features (peak frequency, ratio of average frequency in two adjacent windows) and spatial parameters (phase shift between actual peak times and those predicted by crosscorrelation with adjacent traces, trace offset distance and map coordinates of the sources and receivers). The additional inputs, now totaling nine, often require the addition of a hidden-layer with 6 to 8 nodes in order to train the network successfully. However, the overall performance of the multi-trace network is better than the single trace network. Kusuma notes that the accuracy of the multi-trace network is consistently high - 95% to 99% for impulsive sources and 94% to 98% for Vibroseis sources.

Fuller and Kusuma (1993) applied a modified form of their multi-trace method to three component cross well and seismic data sets with some success. Multicomponent first-break picking is more difficult than single component picking because several traces must be analyzed simultaneously to select the proper first-break event. In this application, extracted features from the P-wave and two S-wave receivers at a specific source-receiver offset are supplied to a cascade-correlation network for training. In addition to the usual suite of amplitude attributes, Fuller included additional extracted features such as eigenvalue ratios computed using the three eigenvalues of the 3x3 covariance matrix formed from windows in the three component traces. The match between neural network and human picks on the same data set was close to 100%, with the neural network being twice as fast.

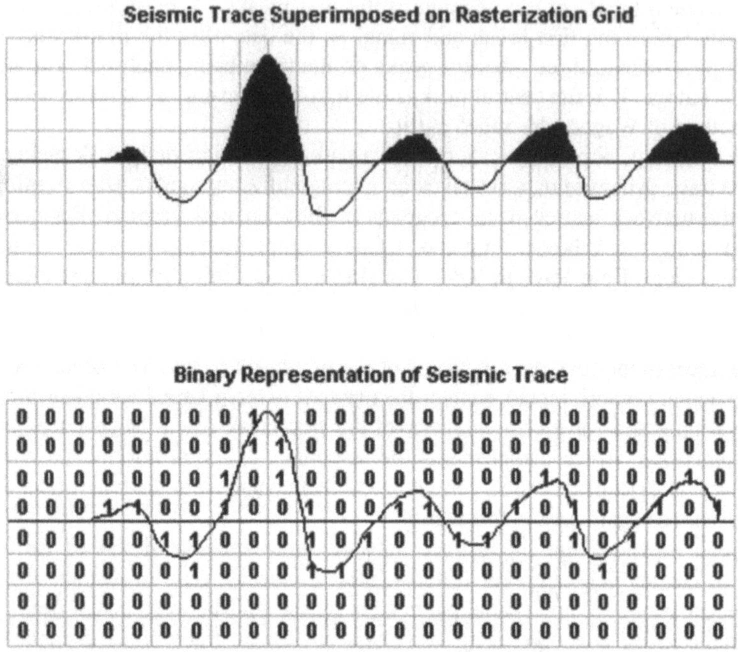

Figure 2. Illustration of the rasterization process for a seismic data trace that is be used as input to the first-break picking MLP network.

The pattern recognition or "visual" model for first-break picking mentioned in the introduction was developed by McCormack et al. (1993). A novel feature of this approach is the manner in which the seismic data is pre-processed prior to being fed into the neural network. The basic thrust is to present the seismic data to the network in a manner which is as close as possible to that employed by a human analyst for picking first-breaks. Since humans pick first-break events using a combination of amplitude, phase, trace-to-trace continuity and wavelet character, a pixel image of the seismic data is created and used as input to the network. The input values consist of a sequence of 1's and 0's representing a two-dimensional binary pixel image of the amplitude trace of the seismic data (Figure 2). A windowed portion of a seismic wiggle trace is re-sampled using a square grid. A grid square is one data sample wide in time. In Figure 2, eight grid divisions are used to represent the total amplitude range of the seismic signal. To construct the pixel image, all grid squares containing a segment of the wiggle trace are assigned a value of '1'; the remaining squares are set to '0'. The amplitude ranges are linearly encoded in this example for clarity; however, other nonlinear representation schemes are possible. The sparse binary representation of the seismic trace illustrated in this figure emphasizes signal polarity transitions while minimizing the effects of noise and spurious signal amplitudes. In normal operation 7 to 11 adjacent traces, each of length 100 samples, and centered in the region of

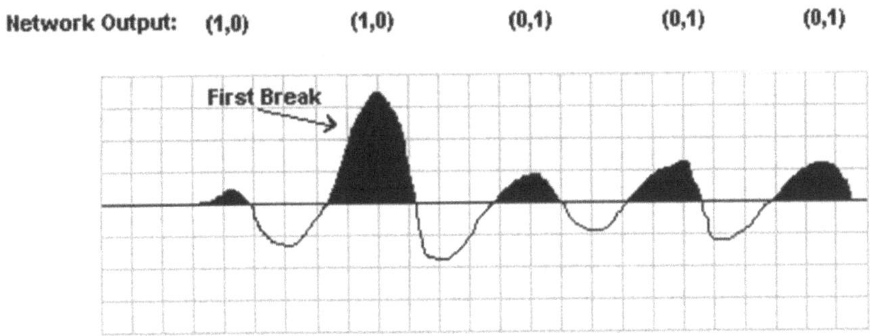

Figure 3. Definition of the output state vector of the first-break picker neural network.

the expected first-break events, are rasterized to create a seismic analysis window that serves as input to the neural network. The window size must be sufficiently large to accommodate any abrupt shifts in the first-break arrival times. Clearly the number of binary bits necessary to create a typical seismic window with reasonable resolution is large.

Figure 3 defines the output state vector of the neural network. To train the neural network, the human processor must first select a set of traces and identify the first-break event on each. The user "shows" the network examples of a first-break on a few selected traces by displaying a shot, receiver or constant offset gather of seismic data on a workstation screen, and then selecting the event representing the first-break for the trace using a mouse. Neural network training is initiated after the user has selected several first-break examples that are representative of variations in amplitude, phase and frequency expected to be encountered in the first-break events along the seismic profile. For each positive amplitude peak within the analysis window, the system first determines whether the first-break that the user identified is above or below this peak in time. If the true first-break position is below the current peak being examined, then the output neurodes are trained to produce (1, 0); if the first-break is above the current peak, then the state of the output vector should be (0, 1). The first-break peak is defined as first of two successive peaks for which the output vector changes from (1, 0) to (0, 1).

The network typically used for first-break picking is a two-layer multilayer perceptron (MLP) consisting of 3000 to 4000 binary inputs derived from the rasterization of the seismic trace discussed previously that are feeding into two output nodes. Normally no hidden-layers are used for the initial network structure. During training, the network iteratively adjusts the connection weights until it selects the same first-break peak as picked by the user on every example trace. It typically takes about 150 iterations and 20 seconds on a Unix workstation to train the network. The system rarely encounters convergence problems during training. In less than one percent of the seismic data processed so far was it necessary for the system to add a hidden-layer to facilitate convergence.

When the MLP has been trained, new traces can be input to the network to determine their first-break times. To do this, the peaks on each trace are successively scanned until the network detects a first-break event using the previously defined convention of a change in the output vector from (1, 0) to (0, 1) on two adjacent peaks. In actual operation however, the network output is rarely a set of binary values – normally, the output will be a floating point pair such as (0.89, 0.11). Furthermore, in scanning a noisy seismic trace, more than one first-break is sometimes detected. To resolve this problem, a reliability factor, *R*, for each first-break indication on a trace is computed and used as a decision and quality control parameter. The function *R* is defined as:

$$R = \tfrac{1}{2}\left(\left|O_1 - O_2\right|_k + \left|O_1 - O_2\right|_{k+1}\right) \tag{2}$$

where O_1 and O_2 are the two network output values, and the subscripts refer to the kth and $(k+1)^{st}$ successive peaks on the trace at which an output state change occurs. *R* ranges between 1 and 0, with 1 representing a high reliability first-break pick, and 0 a low reliability estimate. If there are multiple indications of a first-break on a single trace, the largest value of the reliability factor is used to select the event most likely to represent the first-break. Additionally, *R* is used as a measure of the reliability of each first-break pick for quality control purposes. It should be noted that *R* is a qualitative measure of performance of the neural network, and is not related to statistical probability functions.

From the above discussions automated first-break picking systems with similar performance characteristics can be developed using radically different neural network input parameters. The operating basis for networks that employ trace attribute inputs is fairly obvious, at least in a subjective sense. These systems detect patterns of energy, amplitude and phase distributions in the seismic data that can be correlated with a first-break event. Less intuitive is the underlying pattern recognition dynamics of the "visual" network .model. A sensitivity analysis of the output node values to changes in the inputs of the MLP is described in McCormack and Zaucha (1992). This analysis can reveal the relative importance of each input to the output state, and is useful for understanding the performance mechanisms of the networks that use trace attribute input features. However, in the case of the "visual" model where the MLP inputs are pixels, it is difficult to attach much physical significance to an analysis such as this.

McCormack et al. (1993) conducted a series of experiments using a synthetic shot record to determine the data features that influence the neural network during training. The data set consisted of a 48-trace split spread shot record with two near-surface horizontal refracting layers, together with some deeper random reflection events. The source waveform was a Ricker wavelet with a dominant frequency of about 30 Hz. Figure 4 is a workstation screen print of the synthetic shot record submitted to the first-break picker. The user picked three first-break events on the spread (Traces 1, 12 and 24), all to the left of the source position located between traces 24 and 25, to serve as training examples for the MLP. After the neural network was trained, all traces on the record were automatically picked by the MLP. A reliability factor *R* was computed for each trace and is plotted at the bottom of this figure as a bar ranging between 0 and 1. Neural network picks are indicated in the figure by a black square placed on top of the selected peak.

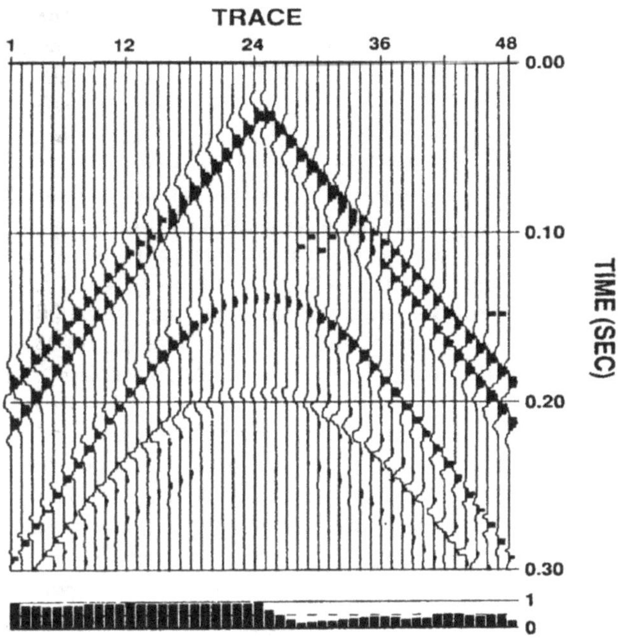

Figure 4. Synthetic shot record with a histogram of neural network first-break reliability factors plotted along the bottom. Note the decrease in first-break pick reliability when the neural network encounters traces with moveout opposite to that used in training.

Note the erroneous neural network picks above and below the true first-break events in the section of the record between traces 25 and 48. From the abrupt change in the reliability factors at the source location it is apparent that a key feature which the network is recognizing is that refraction events have a trace-to-trace time moveout. Since the network was trained with examples from one side of the split spread, it did not learn that first-break moveout could occur with an opposite sign. From this observation we conclude that the neural network is utilizing lateral event continuity and moveout to determine first-breaks, much as human interpreters do. Note that by selecting additional training examples from the right hand side of the spread, the neural network will be able to process all traces with equal reliability.

4. Neural Networks for Seismic Trace Editing

The noise in seismic data sets arises from a number of man-made and natural sources. Noise assumes many stationary and non-stationary forms - short noise spikes, longer noise bursts (e.g., from lightening), high amplitude random noise, and 50 or 60 hertz high line pickup – to name a few. It is common for several noise modes to be present at the same time on a trace. Because of the impact that these extraneous noise sources have on signal-to-noise in the final seismic data section, automated seismic trace editing has been the focus of much research effort since the beginning of seismic data processing in the 1960's. Early efforts employing statistical or parametric techniques, met with limited success due largely to unpredictable variations in noise characteristics from one survey to the next, or even within one seismic line.

Several investigators have applied neural network techniques to the task of removing noisy seismic traces from data sets. An early study by Upham and Cary (1991), compared the trace editing performance of linear and nonlinear neural networks. The authors concluded that trace editing was a nonlinear pattern recognition problem, best solved with a nonlinear MLP. McCormack employed a MLP with a two-layer structure, similar to the first-break picker network, in a seismic trace editing program This system was designed to detect noisy and dead traces in raw field data. Since there are many types of noise, the trace editor system depends upon the user to select representative examples of noisy traces as well as examples of "good" traces. Training and subsequent usage of the trace editor for batch processing is similar to the first-break picker system previously described.

Figure 5 illustrates the data flow through the system. Raw seismic data traces are first analyzed by a feature extractor. A 512-sample Fourier transform amplitude spectrum of the entire trace is computed, along with six intra-trace statistical parameters (average trace frequency and energy, average absolute amplitude, ratio of the first and second largest peak amplitudes, energy decay rate, and normalized trace source-receiver offset distance), and two inter-trace statistical estimates (maximum cross-correlation value of the trace being analyzed with two nearest traces, and average trace energy compared to four adjacent traces). The output from the feature extractor serves as input to a two-layer MLP. The two output nodes of the MLP are trained using user-selected examples of good and bad traces to produce one of two states: (1, 0) if the trace is "good", and (0, 1) if the trace is "bad". Training typically takes 100 to 200 iterations to converge to a solution, or about 15 seconds on an 1990-vintage Unix workstation. Reliability factors for each trace are computed using the absolute difference between the outputs of the two output nodes, and can be used to validate network performance. The MLP-based trace editing system is robust even when there are large variations on noise character. Agreement between the user and the MLP trace editing system typically exceeded 95%.

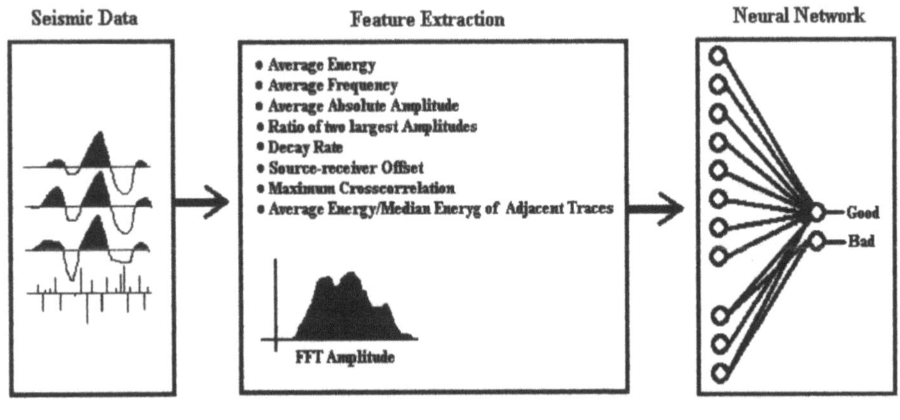

Figure 5. MLP structure for a seismic trace editing system.

Kusuma and Brown (1992) applied a cascade-correlation neural network to the task of trace editing, and achieved results similar to the above. Input data for the neural network included six features extracted from the seismic data – dominant frequency, bandwidth, maximum spike ratio, rms-to-average amplitude ratio, maximum rms-to-average ratio and amplitude decay rate. Training of the cascade-correlation network was fast – an average of 22 seconds to learn a 120 trace dynamite data set up, to 130 seconds on a similar Vibroseis data set. The training time differences between implosive and vibratory sources can be explained in terms of the broader bandwidth of implosive sources. Vibroseis data have significantly more energy before the first-break arrival as an artifact of the crosscorrelation process. Despite these problems the overall accuracy of the trained network is between 95% to 99% for both types of source.

5. Conclusions

Automation of first-break refraction event picking and seismic trace editing has been a major concern of the seismic processing community for three decades due to the labor-intensive nature of these processes. Until the advent of viable neural network algorithms, attempts to automate these procedures using more traditional methods had met with little commercial success. The unpredictable and often large variations in signal and noise that occur in seismic data make it difficult to create robust algorithms based on parametric and statistical models of signal and noise. Neural networks overcome this deficiency by side-stepping the issue of a suitable signal and noise model, and using their ability to learn from examples provided by the human user to create an internal representation of the signal and noise structure.

The success of neural networks in this area is also attributable to the interactions that must occur with the human processor during network training. User acceptance of a new computing system is critical to their commercial success. Model-based algorithms always incorporate the personal biases of the developer, and these biases may or may not be shared by other human users. In contrast, a neural network-based algorithm mimics the unique picking and editing style of the user who trained the system, and thus is more likely to be accepted by the user community.

The sophisticated pattern recognition capabilities of neural networks holds great promise for other applications in seismic data processing. While the two applications discussed in this chapter are available in commercial products, there are a number of other signal processing areas where this technology is ideally suited – moveout and migration velocity analysis, for example. While the use of neural networks to solve these complex pattern recognition problems involves significant technical issues, the previous successes in first-break picking and trace editing will provide the encouragement needed to attack these problems.

Acknowledgments

The author would like to thank the management of the Atlantic Richfield Corporation for permission to publish this chapter. In addition, the author wishes to recognize the significant contributions of his colleagues David E. Zaucha and Dr. Ron Day to the development of key concepts and computer implementation of the algorithms.

References

Anderson, R. G., and McMechan, G. A., 1989, Automatic editing of noisy seismic data: Geophys. Prosp., **37**, 875-892.

Ervin, C. P., McGinnis, L. D., Otis, R. M., and Hall, M. L., 1983, Automated analysis of marine refraction data: a computer algorithm: Geophysics, **48**, 582-589.

Fahlman, S. E., 1988, Faster learning variations on backpropagation: an empirical study: Proceedings of the 1988 Connectionist Models Summer School, Morgan Kaufmann.

Fahlman, S. E. and Lebiere, C., 1990, The cascade-correlation learning architecture: Advances in Neural Information Processing Systems 2, Morgan Kaufmann.

Fuller, B. N. and Kusuma, T., 1993, A method for picking first breaks in 3-component seismic data: 63rd Ann. Internat. Mtg., Soc. Expl. Geophys., Expanded Abstracts, **93**, 325-328.

Gelchinsky, B., and Shtivelman, V., 1983, Automatic picking of first arrivals and parameterization of traveltime curves: Geophys. Prosp., **31**, 915-928.

Kusuma, T., and Brown, M. M., 1992, Cascade-correlation learning architecture for first-break picking and automated trace editing: Expanded Abstracts, 62nd Ann. Internat. Mtg., Soc. Expl. Geophys., Expanded Abstracts, **92**, 1136-1139.

Kusuma, T., and Fish, B. C., 1993, Toward more robust neural-network first break and horizon pickers: 63rd Ann. Internat. Mtg., Soc. Expl. Geophys., Expanded Abstracts, **93**, 238-241.

McCormack, M. D., and Zaucha, D. E., 1992, Automatic trace editing and refraction event first break picking using neural networks: Rev. Inst. Fran. Petr., **47**, 393-405.

McCormack, M. D., Zaucha, D. E., and Dushek, D. W., 1993, First-break refraction event picking and seismic data trace editing using neural networks: Geophysics, **58**, 67-78.

Neff, D. B., and Wyatt, S. B., 1986, Noise suppression by the radial amplitude-slope rejection method: Geophysics, **51**, 844-850.

Rummelhart, D. E., and McClelland, J. L., 1986, Parallel distributed processing - explorations in the microstructure of cognition Volume 1: Foundations, The MIT Press, Cambridge, Mass.

Spagnolini, U., 1991, Adaptive picking of refracted first arrivals: Geophys. Prosp., **39**, 293-312.

Taner, M. T., Koehler, F., and Sheriff, R. E., 1979, Complex seismic trace analysis: Geophysics, **44**, 1041-1063.

Upham, W. and Cary, P., 1991, Comparison of linear and nonlinear neural networks for seismic trace editing: 61st Ann. Internat. Mtg., Soc. Expl. Geophys., Expanded Abstracts, **91**, 293-294.

Wagner, D. E., Ehlers, J. W., Veezhinathan, J., 1990, First break picking using neural networks: 60th Ann. Internat. Mtg., Soc. Expl. Geophys., Expanded Abstracts, **90**, 370-373.

AUTOMATED PICKING OF SEISMIC FIRST-ARRIVALS WITH NEURAL NETWORKS

DOUGLAS I. HART

Western Geophysical,
1625 Broadway, Suite 1300,
Denver, Colorado 80202,
USA
E-Mail: doug.hart@westgeo.com

Abstract

Neural network implementations of first-arrival picking use a representative set of data and associated first-arrival pick times to create a multilayer perceptron neural network. The success of this technique depends on the statistical properties of the features input to the neural network, and the ability of the neural network to approximate a wide class of functions. Using methods from statistical pattern recognition, it is possible to determine the properties of the features input to the neural network that impact upon the reliability of the picking process. Interpreting the output of the neural network as the probability that the associated feature is the first-arrival, requires an appropriate probability model. This model is based on the multinomial distribution, since it can reflect the fact that there is only one first-arrival event on each data trace. Optimization of the neural network weights with an error function based on this probability distribution, produces a neural network that properly estimates the probability that the associated feature is a first-arrival event.

1. Introduction

The time of the first-arrival in seismic data, is essential for performing refraction statics computations or diving-wave tomography calculations. Picking first-arrival times is a pattern recognition problem and, as such, often involves a substantial amount of human effort. Modern 3-D seismic surveys, with their increased quantities of recorded data, have only exacerbated this bottleneck in the productivity of processing seismic data. Previous efforts to automate the picking of first-arrival times, rely on various models of the refraction process (Hatherly, 1982; Gelchinsky and Shtivelman, 1983; Coppens, 1985; Spagnolini, 1991). Recently, neural networks have been brought to bear on many pattern recognition problems. Quite naturally then, they have been applied to the first-

W. Sandham and M. Leggett (eds.),
Geophysical Applications of Artificial Neural Networks and Fuzzy Logic, 13-30.

arrival picking problem (Veezhinathan et al., 1991; Murat and Rudman, 1992; McCormack et al., 1993).

The neural network approach to first-arrival picking is a classic example of a supervised learning problem. Using example data, a multilayer perceptron (MLP) neural network is trained to pick the first-arrival events for previously unseen data. Statisticians will recognize the first-arrival picking problem as a binary statistical pattern recognition problem. There are two classes of events, events that are first-arrivals and events that are not first-arrivals. By exploiting the statistical differences it is possible to formulate a rule that differentiates the two types of events.

This paper discusses the classification of first-arrival events using neural networks and statistical pattern classification. In Section 2, the generic problem of applying neural networks to first-arrival picking, is presented. The main ideas of statistical pattern classification are then discussed in Section 3. In Section 4, these ideas are used to determine the type of feature attributes that will be effective in picking first-arrival events. A number of feature attributes are then presented in Section 5 which have been found useful for picking first-arrival events. Finally, error functions are discussed that enable probabilistic interpretations to be made from the output of a neural network.

2. Neural Network Approach for Picking First-arrivals

Supervised learning, in the neural network context, implies training a neural network with the use of example data. The example data, also called a training set, contain instances of input data along with the desired response from the neural network. After training the neural network to produce the desired outputs for the example data, the network can be used to compute outputs from previously unseen data.

2.1. TRAINING

To train a neural network to pick the first-arrival event, the analyst must first select a representative set of seismic data, and decide what feature in the data corresponds to the first-arrival event. Typically, an easily recognizable characteristic of the seismic trace is selected, such as a local peak amplitude. Once the analyst decides on the desired feature, the first-arrival feature for the complete training set is picked. The time corresponding to the first-arrival feature is the first-arrival pick time. Figure 1 shows the first-arrivals from three seismic sources.

A number of exemplars are now available for training a neural network: the seismic data itself, a decision regarding what type of feature is the first-arrival, and the first-arrival pick times for the data. Collectively these data are the training set for the neural network. Suppose that the feature on the seismic trace that the analyst has chosen to pick is a positive peak amplitude. Within the training set, all positive peak amplitudes must be found and labeled as either first-arrival features (or not), using the provided first-arrival pick times.

To pick the actual seismic data with a neural network would require a network with a large number of input nodes. Although in principle this does not present a problem, it does represent significant computation.

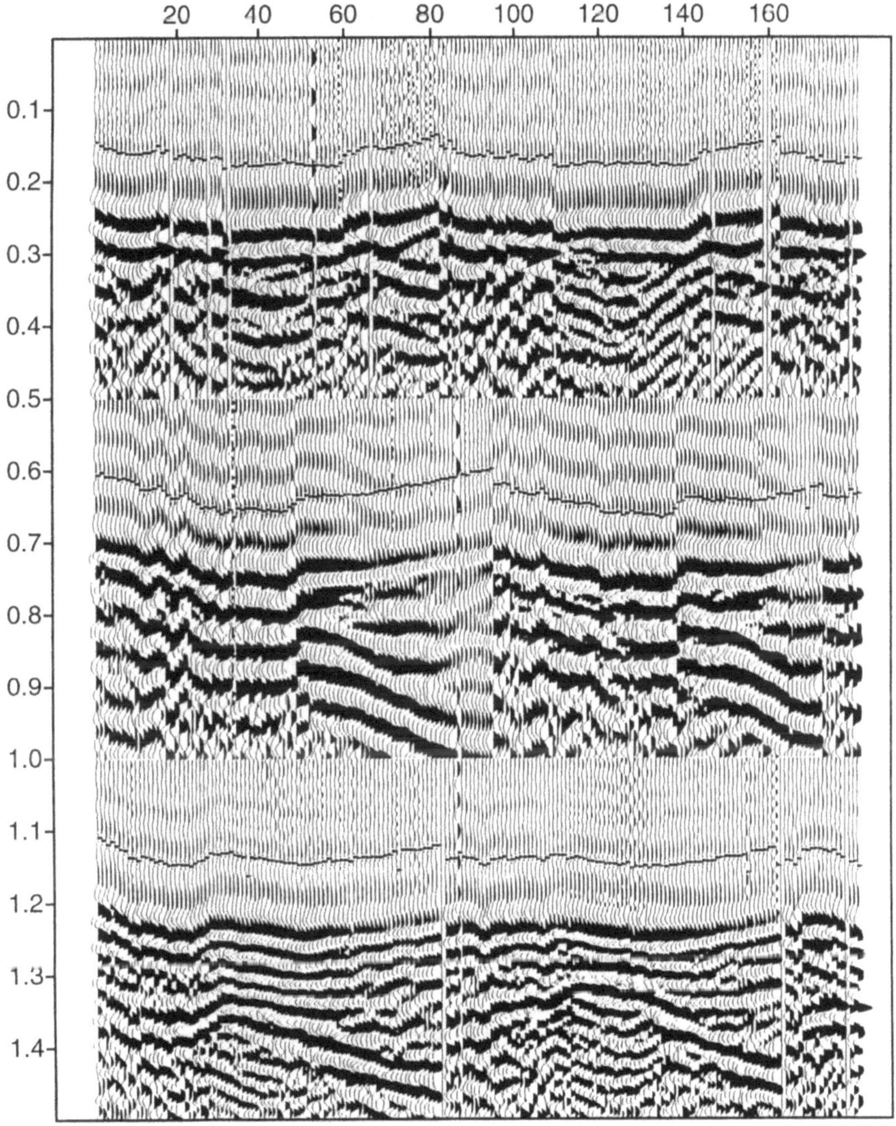

Figure 1. Display of seismic first-arrival events from three seismic sources. The seismic traces are numbered consecutively along the horizontal axis. The vertical axis marks time in seconds. The recordings from each source occupy a 0.5-second window of the display. A positive spike amplitude has been place on each trace 0.1 second above the first-arrival event as chosen using a neural network.

To circumvent this quandary, what is most often done is to compute data attributes for each feature. The exact nature of these attributes is discussed later. However, once computed, these values may be aggregated into a vector. Suppose that n different attributes for each feature in the data, are computed. Denote this vector of feature attributes as $\mathbf{x} = (x_1,\ldots,x_n)^T$. Furthermore, suppose that a '0' is used to label those features that are *not* first-arrivals, and a '1' used to label those features that *are* first-arrivals. Then, if the i-th feature in the data is a first-arrival and the (i - 1)-st and (i + 1)-st features are not first-arrivals, the list of data in the training set would appear as shown in Table 1.

If **w** is a set of neural network weights, the output of the neural network for the i-th feature-attribute vector $\mathbf{x}_i$, is $y_i = y(\mathbf{x}_i, \mathbf{w})$ where the function y represents the computation done by the neural network. The task of training the neural network is to determine a set of weights **w**, that minimizes the difference between the *actual* outputs of the neural network y_i, and the *desired* outputs of the neural network i.e. the correct class labels.

2.2. PICKING FIRST-ARRIVALS

After training a neural network i.e. deriving the weights, it can now be used to automate the first-arrival picking process. To pick the first-arrival feature on a trace, all potential first-arrival features on the data trace are computed, for instance all positive peak amplitudes in a time window surrounding the first-arrival event. For each feature, the vector of feature-attribute values $\mathbf{x}_i$ are computed and input to the neural network, to produce an output value y_i. The first-arrival feature is then the one whose feature-attribute values generate an output from the neural network that is closest to '1'.

3. Statistical Pattern Classification with Neural Networks

One of the earliest uses of neural networks was to solve classification problems. In this case, the neural network learns to produce a desired output, a class label, when presented with a given input from a set of training data.

TABLE 1. The data for training the neural network contain a class label and a vector of feature-attribute values for every feature in the training set. Part of the list of training data is shown here. The ith feature is a first-arrival. The (i – 1)-st and (i + 1)-st features are not first-arrivals.

Feature Index	Class Label	Associated Vector of Feature Attribute Values
⋮	⋮	⋮
$i-1$	0	$\mathbf{x}_{i-1}$
i	1	$\mathbf{x}_i$
$i+1$	0	$\mathbf{x}_{i+1}$
⋮	⋮	⋮

Once the neural network has learned the information in the training set, it may be used to classify previously unseen input data. Bishop (1995) and Ripley (1996) both describe the general use of neural networks for solving statistical pattern classification problems. Only an overview is given here of statistical pattern classification as it relates to the problem of picking first-arrival features.

3.1. CONCEPTUAL STATISTICAL PATTERN CLASSIFICATION

The basic problem faced in automating first-arrival picking is determining whether a feature is a first-arrival or not. Suppose that $\mathbf{C}_0$ denotes the class of not being a first-arrival and $\mathbf{C}_1$ denotes the class associated with those features that are first-arrivals. To aid in determining whether a feature is a first-arrival or not, there is a set of feature-attribute values, $\mathbf{x} = (x_1, \ldots, x_n)^T$, computed for each feature. Advantage is taken of the differences in the statistical properties of the feature-attribute values for the two classes of events.

Suppose that the two conditional probability densities of the feature-attribute values $p(\mathbf{x} \mid \mathbf{C}_0)$ and $p(\mathbf{x} \mid \mathbf{C}_1)$ are known, or can be estimated from the data. Bayes' rule gives the probability of class membership when given the vector of feature-attribute values,

$$\Pr(\mathbf{C}_0 \mid \mathbf{x}) = \frac{p(\mathbf{x} \mid \mathbf{C}_0)\Pr(\mathbf{C}_0)}{p(\mathbf{x})} \tag{1}$$

and

$$\Pr(\mathbf{C}_1 \mid \mathbf{x}) = \frac{p(\mathbf{x} \mid \mathbf{C}_1)\Pr(\mathbf{C}_1)}{p(\mathbf{x})} \tag{2}$$

The difference in these two conditional probabilities enables a classification rule to be defined that assigns a class label to each vector of feature-attributes. The classification procedure may be expressed as follows:

$$\text{If} \quad \Pr(\mathbf{C}_0 \mid \mathbf{x}) > \Pr(\mathbf{C}_1 \mid \mathbf{x}) \tag{3}$$

then the associated feature is deemed to belong to class $\mathbf{C}_0$,

$$\text{and conversely, if} \quad \Pr(\mathbf{C}_0 \mid \mathbf{x}) < \Pr(\mathbf{C}_1 \mid \mathbf{x}) \tag{4}$$

the associated feature is assigned to class $\mathbf{C}_1$.

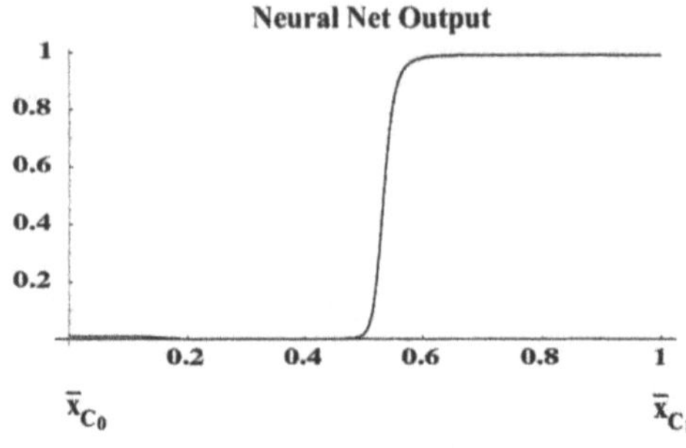

Figure 2. The output of a neural network trained to pick first-arrival features. The input varies from regions of the feature-attribute space that are classified as non first-arrivals to regions of the feature-attribute space that are classified as first-arrivals.

The neural network solution to the classification problem involves regression. The network uses the example data to create a functional approximation to the classification rule that continuously maps vectors of feature-attribute values to class labels. Consequently, the conditional probability densities for the data do not need to be explicitly determined.

As an example, consider the output from a neural network that has been trained to pick first-arrivals (Figure 2). The left end-point of the graph starts at the computed mean value of the feature-attribute values for those features that are not first-arrivals, or $\bar{\mathbf{x}}_{C_0}$. The right end-point of the graph terminates at the computed mean value of the feature-attribute values for those features that are the first-arrivals, denoted $\bar{\mathbf{x}}_{C_1}$. An output value of zero indicates that the associated feature is not a first-arrival and an output value equal to one indicates that the associated feature is a first-arrival. In Figure 2, the neural network outputs values close to zero and almost one, near input values $\bar{\mathbf{x}}_{C_0}$ and $\bar{\mathbf{x}}_{C_1}$, respectively. The output of the neural network quickly changes between zero and one. This step-function character is desirable at the output of the neural network since the role of the network is to unambiguously assign either of the two class labels when presented with a vector of feature-attribute values.

3.2. ASSESSING THE RELIABILITY OF CLASSIFICATION

Since a given feature is either a first-arrival or not, and probabilities must sum to one,

$$\Pr(\mathbf{C}_0 \mid \mathbf{x}) + \Pr(\mathbf{C}_1 \mid \mathbf{x}) = 1 \tag{5}$$

The classification procedure says that as long as $\Pr(\mathbf{C}_0 \mid \mathbf{x}) > 1/2$, the associated feature is classified as a non first-arrival. If $\Pr(\mathbf{C}_0 \mid \mathbf{x})$ is just slightly greater than 1/2, consequently there is a probability slightly less than 1/2 that the event is the first-arrival feature. This is the probability of mis-classifying the feature. Minimizing the probability of mis-classifying a feature requires the conditional probability to be as close to one as possible, so that the associated probability of mis-classifying the feature is as small as possible. Recall that Bayes' rule, as defined in eq.(1) and eq.(2), states that the conditional probabilities are proportional to the conditional densities of the feature-attribute values. As a result, to ensure reliable first-arrival picking, feature-attribute values are required that accurately distinguish the first-arrival feature from other features.

Consider the distribution of the feature attribute shown in Figure 3. Note that for

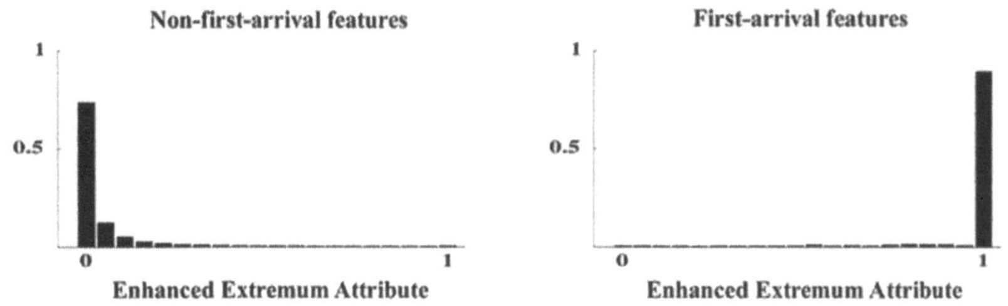

Figure 3. The left histogram shows the distribution for one of the enhanced extremum values for features that are not first-arrivals. On the right is the distribution of values for the first-arrival features. The two distributions do not overlap significantly, indicating that this attribute, by itself, can distinguish the first arrival.

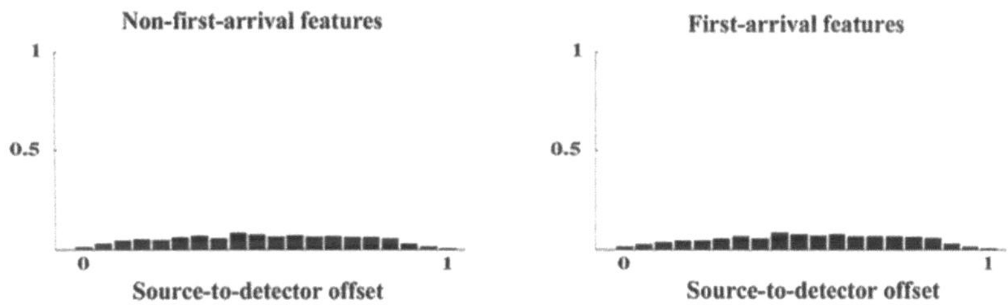

Figure 4. The left histogram shows the distribution of the normalized source-to-detector offset distance for non-first-arrival features. On the right is the distribution for the first-arrival features. These two distributions overlap substantially, indicating that this feature, by itself, does not distinguish the first arrival

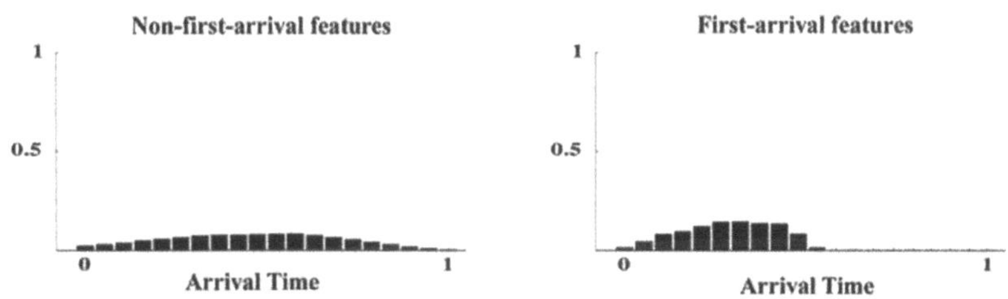

Figure 5. The left histogram shows the distribution of the normalized arrival-time for non-first-arrival features. On the right is the distribution for the first-arrival features. Like the distribution of the source-to-detector offset in Figure 4, this attribute does not distinguish the first arrival by itself.

those features that are not first-arrivals, the value of the attribute is close to zero. For those features that are first-arrivals the value of the attribute is close to one. Since these two histograms do not overlap significantly, the value of this attribute can reliably determine whether the associated feature is the first-arrival or not.

Now consider the distribution of the source-to-detector offset, the distance traveled by the first-arrival event, shown in Figure 4. The distribution of values is almost identical for both types of feature. This means that, by itself, the source-to-detector offset distance does not reliably classify features. The arrival time of the feature, shown in Figure 5, has properties that are somewhat similar to the offset attribute. Again, its value cannot reliably classify the feature. Taken together, however, the source-to-detector offset distance and the arrival time form a two-dimensional attribute space that defines the velocity of the seismic energy. Figure 6 shows that the first-arrival feature falls along a narrow velocity corridor. If the values of these two attributes fall outside the velocity corridor then the feature is almost certainly not a first-arrival.

4. The Feature Attributes

In order for a neural network, or any classification process, to be reliable, the feature-attribute values must differentiate the first-arrival feature from the other features. Arguably, the central characteristic of the first-arrival in exploration seismic data is that it marks the onset of energy from the seismic source. Consequently, one of the main properties that the feature-attribute values should exhibit is detecting the onset of energy. Another property of the first-arrival is that it occurs at the transition between the reception of random noise and the genuine seismic signal. Accordingly, feature-attribute values that discriminate between the properties of signal and noise will also aid in determining what feature is the first-arrival.

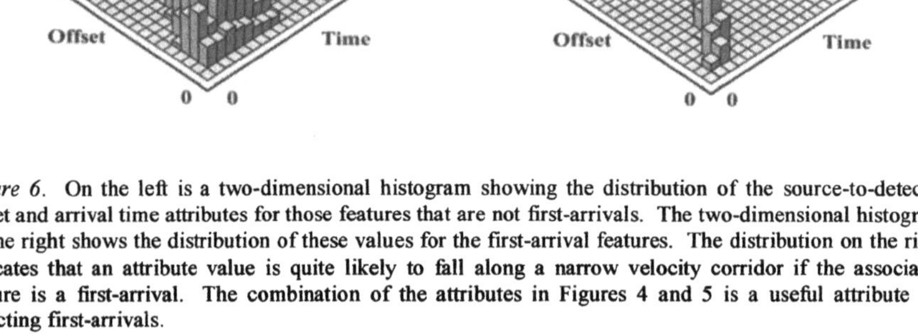

Figure 6. On the left is a two-dimensional histogram showing the distribution of the source-to-detector offset and arrival time attributes for those features that are not first-arrivals. The two-dimensional histogram on the right shows the distribution of these values for the first-arrival features. The distribution on the right indicates that an attribute value is quite likely to fall along a narrow velocity corridor if the associated feature is a first-arrival. The combination of the attributes in Figures 4 and 5 is a useful attribute for detecting first-arrivals.

4.1. BASIC FEATURE ATTRIBUTES

A seismic trace is a sequence of values, say $s(t)$, where t is a regularly sampled time interval. Given $s(t)$, there is a sub-sequence of values $\tilde{s}(k)$ at which the seismic trace takes on *interesting values*: local peak amplitudes, zero crossings etc. To aid in this development is a function that maps index values in the sub-sequence to index values (time values) in the original sequence i.e. there is a function $g(\cdot)$, such that $g(k) = t$.

Consider the portion of the seismic trace shown in Figure 7, and furthermore consider the computation of feature-attribute values for the k-th peak amplitude. The time of the peak, $g(k)$, and the amplitude of the peak, $\tilde{s}(k)$, are two simple feature-attribute values which may be used as input to the neural network. As noted previously, the attribute values by themselves may not differentiate the first-arrival feature. Combinations of attribute values may be useful for distinguishing the first-arrival feature. Hence, along with the time of the feature, the source-to-detector offset distance of the seismic trace is input to the neural network.

4.2. RATIO ATTRIBUTES

An increase in energy is expected near a first-arrival event. The ratio of the energy in two temporally adjacent windows on the seismic trace will be large near a transition from low to high energy. If the energy level in the two windows is the same then the ratio is one. Specifically, consider a window that starts at the midpoint between the k-th and $(k - 1)$-st feature and stops at the midpoint between the $(k + 1)$-st and $(k + 2)$-nd feature. The power-ratio attribute for the k-th feature is defined as the ratio of the energy in this window against the energy in a window of the same length immediately preceding the window.

For instance, for the trace shown in Figure 7, $g(k - 1) = 354$, $g(k) = 381$, $g(k + 1) = 409$, and $g(k + 2) = 456$. The midpoint between the k-th and $(k + 1)$-st feature is

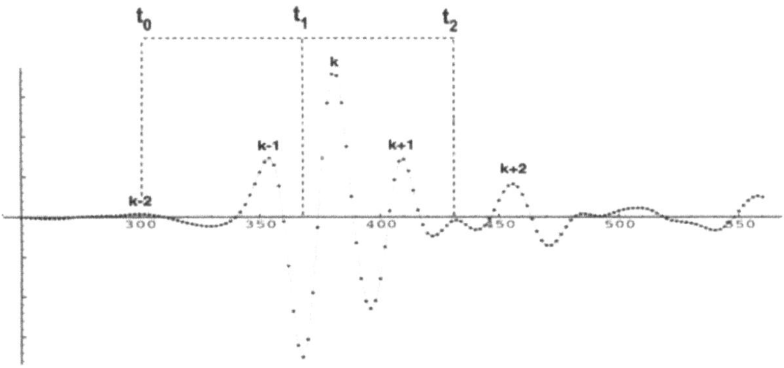

Figure 7. A portion of a seismic data trace around the first-arrival feature. The sampled trace values are denoted by dots that are connected with lines. The k-th feature on the trace is labeled, along with its neighboring features. Temporal windows pertinent for computing feature-attribute values for the k-th feature are shown with dashed lines.

$t_1 = [g(k) + g(k-1)] / 2 = (381 + 354) / 2 = 376.5$. The midpoint between the $(k+1)$st and $(k+2)$nd feature is $t_2 = [(g(k+1) + g(k+2)] / 2 = (409 + 456) / 2 = 432.5$. The length of the window is $l = t_2 - t_1 = 65$. The start time for the preceding window is $t_0 = t_1 - l = 367.5 - 65 = 302.5$. The power-ratio feature for the k-th feature is then

$$\frac{\sum_{t=t_1}^{t_2} s^2(t)}{\sum_{t=t_0}^{t_1} s^2(t)} \tag{6}$$

Dashed lines in Figure 7 mark the current and preceding windows for the k-th feature.

The arithmetic-ratio attribute is similar to the power-ratio attribute in that it has a large value at a transition between low and high amplitudes. It is the ratio of the average amplitude of features before and after the selected feature.

Specifically, let n be the number of features to consider. Then the arithmetic ratio is,

$$\frac{\sum_{i=1}^{n} \tilde{s}(k+i-1)}{\sum_{i=1}^{n} \tilde{s}(k-i)} \tag{7}$$

If $n = 2$, the arithmetic-ratio attribute is the ratio of the average of the current and subsequent peak amplitudes, against the average of the two previous peak amplitudes. For the k-th feature in Figure 7, when $n = 2$, the arithmetic-ratio attribute is,

$$\frac{\tilde{s}(k) + \tilde{s}(k+1)}{\tilde{s}(k-1) + \tilde{s}(k-2)} \tag{8}$$

The geometric-ratio attribute is similar in concept to the arithmetic ratio. Again, let n be the number of features to consider. The geometric ratio for the k-th feature is,

$$\frac{\prod_{i=1}^{n} \tilde{s}(k+i-1)}{\prod_{i=1}^{n} \tilde{s}(k-i)} \tag{9}$$

If $n = 2$, the geometric ratio for the k-th feature is,

$$\frac{\tilde{s}(k) \cdot \tilde{s}(k+1)}{\tilde{s}(k-1) \cdot \tilde{s}(k-2)} \tag{10}$$

It may be anticipated that the level of noise in the data is less than the level of the signal. If this is the case then the variability of the data recorded after the onset of the signal should be greater than the variability of the data when only recording noise. The standard-deviation-ratio attribute quantifies this idea. If $\sigma(\cdot)$ represents the operator that computes the standard deviation of the values in a sequence, then the standard-deviation attribute for feature *k* in Figure 7 is,

$$\frac{\sigma[s(t_1),\ldots,s(t_2)]}{\sigma[s(t_0),\ldots,s(t_1)]} \tag{11}$$

4.3. HILBERT TRANSFORM-LIKE ATTRIBUTES

The instantaneous-frequency attribute measures the rate of change in the instantaneous period associated with each feature. A change in the period is expected at the transition between receiving ambient noise and the seismic signal on the seismic data trace. For the *k*-th feature, the time of the previous feature $g(k-1)$, subtracted from its time $g(k)$, is an estimate of the instantaneous period. This estimate is subtracted from the previous estimate of the instantaneous period, which is $g(k-1)-g(k-2)$ to give the instantaneous frequency. The change in the instantaneous period,

$$[g(k)-g(k-1)]-[g(k-1)-g(k-2)] \tag{12}$$

is an estimate of the instantaneous frequency for the *k*-th feature.

The instantaneous-amplitude attribute is the ratio of the average of the instantaneous amplitude after a feature, to the average of the instantaneous amplitude before the feature. Again, this feature acts as an energy onset detector. It will have a large value for a feature that is near the first-arrival and a smaller value for other features.

4.4. ENHANCED EXTREMUM VALUES

The enhanced-extremum-value attribute is also an energy onset detector. It is defined as the ratio the amplitude of the feature under consideration against the amplitude of previous features. As before, at the onset of energy in the seismic trace, these features will have large values.

The enhanced-extremum value, using all previous features of the same polarity (APSP), for the *k*-th feature in Figure 7 is,

$$\frac{\tilde{s}(k)}{\frac{1}{(k-1)}\sum_{i=1}^{k-1}\tilde{s}(i)} \tag{13}$$

This value is the ratio of the current peak amplitude to the average of all previous positive peak amplitudes.

The enhanced extremum value, using all previous features of all polarities (APAP), differs in the fact that it is the ratio of the amplitude of the current feature against the average absolute value of all previous positive and negative valued peaks.

Finally, the enhanced extremum value attribute, uses one previous feature of the same polarity (OPSP) and is defined as the ratio of the current peak amplitude against the previous peak amplitude. Referring to Figure 7, this is $\tilde{s}(k)/\tilde{s}(k-1)$ for the k-th feature.

4.5. BOUNDED VARIATION QUANTITY ATTRIBUTE

The bounded variation quantity attribute (Hirota and Iijima, 1978) sums the absolute values of the slope between samples in a window around the selected feature. For feature k in Figure 7 it is defined as,

$$BVQ(k) = \sum_{t=g(k)-j}^{g(k)-j+n} \frac{|s(t+1)-s(t)|}{n} \tag{14}$$

where j is the relative start for the summation and n is the length of the window.

4.6. DISCUSSION OF FEATURE ATTRIBUTES

Table 2 shows the mean values of those feature-attributes for an example training set (Hart, 1996). These feature-attribute values have been normalized onto the interval $[0,1]$. Because many of these attributes are designed to be energy onset detectors, the feature-attribute values for those features that are first-arrivals are, on average, larger than the feature-attribute values for the features that are not first-arrivals. However, this is not universal. More complicated statistical variations between the attributes, such as the time and source-to-detector offset distance, may be found implicitly and utilized by the neural network.

TABLE 2. Mean values of the feature-attributes for an example training set.

Attribute	Class '0' Mean	Class '1' Mean
Amplitude	0.033176	0.125977
Time	0.378068	0.412743
Offset Distance	0.420561	0.428682
Power Ratio	0.025519	0.225338
Standard Deviation Ratio	0.025050	0.325757
Arithmetic Ratio	0.013993	0.976740
Geometric Ratio	0.000594	0.996740
Instantaneous Frequency	0.192433	0.710983
Instantaneous Amplitude	0.202073	0.042026
Enhanced Extremum Value APSP	0.005107	0.997609
Enhanced Extremum Value APAP	0.003880	0.997431
Enhanced Extremum Value OPSP	0.156836	0.986686
BVQ	0.420561	0.114145

5. Optimizing Neural Network Weights

In terms of feature classification, a neural network may be regarded as a function that maps a vector of feature-attribute values to an output class label. It assumes the role of the classification procedures defined by eq.(3) and eq.(4), by outputting the value '1' in those regions of the feature-attribute space that correspond to first-arrival features, and '0' elsewhere. The ability of neural networks to approximate a large class of functions (White, 1992) makes them ideally suited for approximating the step function nature of the classification procedure. The problem that remains is to determine a set of neural network weights **w**, so that the neural network outputs the correct class label for an input vector of feature-attribute values. This usually involves solving a nonlinear optimization problem. Sarle (1999) has a monthly Usenet posting that contains up-to-date references for many of these algorithms. Additionally, Bishop (1995) describes many of the essentially nonlinear optimization algorithms that are used for finding neural network weights.

In this section, optimization functions are discussed so that appropriate probabilistic interpretations of the output from the neural network can be made.

5.1. LEAST-SQUARES ERROR

Suppose there are n distinct features in the training set and let d_i denote the desired output class label for the i-th element in the training set. Then, the least-squares error estimate of the neural network weights is found by solving the following minimization problem,

$$\min_{\mathbf{w}} \sum_{i=1}^{n} y(\mathbf{x}_i;\mathbf{w}) - d_i{}^2 \tag{15}$$

where $y(.)$ is the output of the neural network when the feature-attribute vector $\mathbf{x}_i$ and the neural network weights $\mathbf{w}$, are given.

This procedure generally works well. However, it is important to consider the following two situations: (1) the maximum value output from the neural network is near zero for all features on a seismic trace, or (2) the neural network yields a value close to one for two or more features on a seismic trace. The appropriate interpretation is that the neural network has determined that there is no first-arrival feature or that there is more than one first-arrival feature for the seismic data trace, respectively. It is desirable to interpret the output of the neural network as a probability that the feature is a first-arrival.

5.2. CROSS-ENTROPY ERROR

Consider a neural network with a single output y where,

$$\Pr(\mathbf{C}_0 \mid \mathbf{x}) = y \tag{16}$$

and consequently,

$$\Pr(\mathbf{C}_1 \mid \mathbf{x}) = 1 - y \tag{17}$$

A target-value coding scheme for the neural network output is,

$$d = 1 \text{ if } \mathbf{x} \text{ belongs to } \mathbf{C}_0 \tag{18}$$

and

$$d = 0 \text{ if } \mathbf{x} \text{ belongs to } \mathbf{C}_1 \tag{19}$$

The probability of observing either target value, when given the feature attribute vector $\mathbf{x}$, is,

$$\Pr(d \mid \mathbf{x}) = y^d (1-y)^{1-d} \tag{20}$$

This is the Bernoulli distribution.

With repeated independent sampling from this distribution, indexed by i, the likelihood function is,

$$\prod_i (y_i)^{d_i}(1-y_i)^{1-d_i} \tag{21}$$

The negative logarithm of the likelihood function yields the cross-entropy error function,

$$E=-\sum_{i=1}^{n} d_i \log y(\mathbf{x}_i;\mathbf{w})+(1-d_i)\log 1-y(\mathbf{x}_i;\mathbf{w}) \tag{22}$$

Optimizing the neural network weights with the error function enables the output of the neural network to be interpreted as a probability.

5.3. THE MULTINOMIAL LIKELIHOOD FUNCTION

The probabilistic interpretation of the neural network output, presented in the previous section, neglects an important element of the first-arrival-picking problem; there is only one first-arrival feature for each seismic data trace. The Bernoulli distribution is appropriate for modeling two random alternatives. The first-arrival-picking problem actually requires choosing one of n alternatives. Consequently, a multinomial distribution is the appropriate probability model.

Suppose that there are k+1 potential first-arrival features on a seismic data trace. Let $s_1, s_2, \ldots, s_{k+1}$ denote the outcomes and let $y_j = \Pr(s_j), j=1,\ldots,k+1$, with $\sum_{j=1}^{k+1} y_j = 1$. If d_j is a random variable associated with s_j such that $d_j = 1$ if the outcome of the trial is s_j, and zero otherwise, then the probability mass function is,

$$f(d_1,\ldots,d_k;y_1,\ldots,y_k)=\prod_{j=1}^{k+1} {y_j}^{d_j} \tag{23}$$

where

$$d_j = 0 \text{ or } 1, \text{ for } j=1,\ldots,k+1$$
$$\sum_{j=1}^{k+1} d_j = 1 \text{ (one } d_j \text{ is one, all others are zero),}$$
$$0 \le y_j \le 1, \text{ for } j=1,\ldots,k+1, \text{ and}$$
$$\sum_{j=1}^{k+1} y_j = 1$$

The first of these conditions results from the fact that there is only one first-arrival for each data trace. The second condition follows directly from the first. The inequality, expressed in the third condition, is a consequence of the interpretations of these values as probabilities. Finally, the last condition says that the probabilities must sum to one. Equivalently, this means there is a first-arrival on each seismic data trace.

If an independent sample of size n is drawn from this distribution, then the likelihood function is,

$$\prod_{i=1}^{n} f(d_1,\ldots,d_k;y_1,\ldots,y_k) = \prod_{i=1}^{n}\prod_{j=1}^{k+1} y_{ij}^{d_{ij}} \tag{24}$$

Taking the negative of the logarithm of the likelihood function yields the error function to use for optimizing the neural network weights,

$$E = -\log\left(\prod_{i=1}^{n}\prod_{j=1}^{k+1} y_{ij}^{d_{ij}}\right) = -\sum_{i=1}^{n}\sum_{j=1}^{k+1} d_{ij}\log y(\mathbf{x}_{ij};\mathbf{w}) \tag{25}$$

This function is to be minimized subject to the following constraint,

$$\sum_{j=1}^{k+1} y(\mathbf{x}_{ij};\mathbf{w}) = 1 \text{ for each } i \tag{26}$$

Optimizing the neural network weights with the error function in eq.(25) subject to the constraint in eq.(26) enables the interpretation of the output of the neural network as the probability that the associated feature is a first-arrival. The constraints on the optimization problem require the output of the neural network to sum to one for all the features on a given seismic data trace.

The traditional neural network solution to solving a 1-of-n type classification problem is to use a neural network with n output nodes and an error function based on the softmax activation function (Bishop, 1995). In addition to the computational complexity of adding additional output nodes to the neural network, there is one central problem when applying this strategy to the first-arrival picking problem. This model does not take into account that the number of features that are considered as candidate first-arrival events might be random, and consequently the number of features cannot be fixed at some predefined value. The multinomial likelihood function only alters the optimization of the neural network weights; it does not change the way that first-arrival features are subsequently picked. The simplicity of sequentially inputting a feature-attribute vector for one potential feature and computing an output from the neural network is maintained. All feature-attribute vectors for the seismic trace need not be input to a neural network immediately. Conditioning the multinomial probability for the number of features on the seismic data trace enables the number of features on the seismic trace also to be a random variable. A neural network, whose weights have been computed using a likelihood function based on this probability model, can assess the probability associated with the observed number of features on a given seismic data trace.

6. Conclusions

Connecting the technique of picking first-arrivals using a neural network, with the underlying statistical pattern classification process, enables the determination of characteristics of the feature attributes input to the neural network, which will be useful in discriminating the first-arrival feature from other features. Attributes whose values distinguish the first-arrival feature from the other features will be useful. Additionally, combinations of features, that by themselves may not distinguish the first-arrival feature, may also be useful. A benefit of the neural network approach to first-arrival picking is that these statistical relationships do not need to be explicitly specified.

Each seismic data trace has only one first-arrival. Optimizing the neural network weights using a likelihood function based on the multinomial probability distribution, allows the first-arrival property to be captured. A neural network created using an error function based on the multinomial distribution, produces estimates of the probability that the associated feature is the first-arrival.

Acknowledgments

The development of the feature attributes described in Section 4, should be credited to my colleagues, who are now unfortunately unidentified to me. These feature attributes were used to pick the data in Stresau *et al.* (1992). However, to the best of my knowledge, there are no references that describe these computational techniques.

References

Bishop, C. M., 1995, Neural networks for pattern recognition: Oxford University Press.

Coppens, F., 1985, First arrival picking on common-offset trace collections for automatic estimation of static corrections: Geophys. Prosp., **33**, 1212-1231.

Gelchinsky, B., and Shtivelman, V., 1983, Automatic picking for first arrivals and parameterization of traveltime curves: Geophys. Prosp., **31**, 915-928.

Hart, D. I., 1996, Enhancing the reliability of first-break picking using neural networks: 66th Ann. Internat. Mtg., Soc. Expl. Geophys., Expanded Abstracts, 1662-1665.

Hatherly, P. J., 1982, A computer method for determining seismic first arrival times: Geophysics, **47**, 1431-1436.

Hirota, K., and Iijima, T., 1978, The bounded variation quantity (B.V.Q.) and its application to feature extraction: Proceedings of the Fourth International Conference on Pattern Recognition, November 7-10, Kyoto, Japan, 457-461. (Also available through the Colorado Technical Reference Center, University of Colorado, Boulder (303) 492-8774 Engineering TK 7882.P3.)

McCormack, M. D., Zaucha, D. E., and Dushek, D. W., 1993, First-break refraction event picking and seismic data trace editing using neural networks: Geophysics, **58**, 67-78.

Murat, M. E., and Rudman A. J., 1992, Automated first arrival picking: a neural network approach: Geophys. Prosp., **40**, 587-604.

Ripley, B. D., 1996, Pattern recognition and neural networks: Cambridge University Press.

Sarle, W. S., ed., 1999, Neural Network FAQ, part 2 of 7: Learning: periodic posting to the Usenet newsgroup comp.ai.neural-nets, URL: ftp://ftp.sas.com/pub/neural/FAQ.html

Spagnolini, U., 1991, Adaptive picking of refracted first arrivals: Geophys. Prosp., **40**, 587-604.

Stresau, W. R., Farrell, R. C., and Ford, K. P., 1992, 3-D refraction analysis and modeling using surface-consistent decomposition on the workstation: 62nd Ann. Internat. Mtg., Soc. Expl. Geophys., Expanded Abstracts, 1116-1119.

Veezhinathan, J., Wagner, D., and Ehlers, J., 1991, First break picking using a neural network, *in* Aminzadeh, F., and Simaan, M., Eds., Expert systems in exploration: Soc. Expl. Geophys., 179-202.

White, H., 1992, Artificial neural networks: approximation and learning theory: Blackwell Publishers.

AUTOMATED 3-D HORIZON TRACKING AND SEISMIC CLASSIFICATION USING ARTIFICIAL NEURAL NETWORKS

MILES LEGGETT[†], WILLIAM A. SANDHAM[‡] and TARIQ S. DURRANI[‡]

† *Jason Geosystems bv, Plaza Building, Weena 598, P.O. Box 1573, 3000BN, Rotterdam, THE NETHERLANDS.*
E-Mail: miles@jason.nl

‡ *Signal Processing Group, Institute of Communications and Signal Processing, Dept. of Electronic and Electrical Engineering, University of Strathclyde, Glasgow G1 1XW, SCOTLAND.*
E-Mail: w.sandham@eee.strath.ac.uk
t.durrani@eee.strath.ac.uk

Abstract

Seismic surveys are routinely carried out in three dimensions, resulting in large volumes of high-resolution seismic data and a corresponding increase in the workload of an interpreter. An automatic tracker is described in this chapter, based on artificial neural networks (ANNs), which enables horizons to be tracked in three dimensions with less input from an interpreter compared to most commercial automatic trackers. More time can therefore be spent by the interpreter investigating geologically complex areas. A hybrid ANN is employed which combines both unsupervised (self-organising feature map) and supervised (multilayer perceptron) network paradigms. The tracker is demonstrated on a real three-dimensional (3-D) seismic data set, and is shown to be a viable technique for use as a standard tool and for enhancing efficiency in 3-D seismic interpretation. 1.5-D and 2-D methods have also been demonstrated successfully, which account for the seismic character above, below, behind and ahead of the current tracking position.

1. Introduction

One of the principal aims of interpretation in the structural analysis of a three-dimensional (3-D) seismic data set, is to identify and track the most distinct and relevant geological boundaries in the data i.e. horizon tracking[1]. These also form important velocity boundaries, which are essential for time-to-depth conversion of the seismic

[1] Also known as horizon picking or event tracking/picking. Event and horizon tend to be used interchangeably in a tracking context. Automated horizon trackers are also called autotrackers.

W. Sandham and M. Leggett (eds.),
Geophysical Applications of Artificial Neural Networks and Fuzzy Logic, 31-44.

data, and therefore must be tracked with a relatively high degree of accuracy. Before interpretation can begin, well-log data is used to generate a synthetic seismogram along the well track, and the seismic data is tied in to this information. This synthetic seismogram is not error-free and is itself open to interpretation. Well-log data allows a horizon to be specified at a few points in the 3-D volume, however the interpretation can be complicated by horizons showing changes in reflection character, amplitude and polarity - most often seen when the horizon is an unconformity.

The wavelets which constitute horizons on a seismic section are usually distinguished by comparable amplitude, morphology, and two-way traveltime, appearing consistently from trace-to-trace and across lines. However, variations typically occur and horizons are often affected by faulting, folding, intrusions and other discontinuities e.g. salt domes and channels. Wavelets forming such horizons exhibit great variation in reflection character, and may even change polarity. Normally, an interpreter will investigate the morphology or character of the reflection i.e. wavelet shape, and not simply base decisions on reflection amplitude. An overall view of the regional geology, and knowledge of the stratigraphic and structural relationships of the various geological formations, is also essential in the interpretation of horizons, especially in areas that are geologically complex. Interpretation can therefore be a tedious process involving many steps and iterations, before a consistent understanding of the geology emerges (Brown, 1991,1998).

Horizon tracking forms the basis of many automated seismic interpretation techniques. These techniques generally concentrate on classical pattern recognition methods (Bowie and Durrani, 1988) and are usually more robust than techniques simply employing reflection amplitude or crosscorrelation.

Current pattern recognition methods rely on *a-priori* knowledge of the seismic events and are therefore fairly limited. Since horizon tracking is essentially a pattern recognition process, it is an ideal candidate for the application of artificial neural networks (ANNs) (Haykin, 1999). ANNs differ from conventional pattern recognition techniques in their ability to discriminate or “learn” through repeated exposure to examples. They also possess a high tolerance to component failure and are fairly robust in the presence of high noise levels. In the horizon tracking context, ANNs only require knowledge regarding which horizons the interpreter wishes to track.

This chapter discusses a hybrid network of an unsupervised ANN, the Kohonen self-organising feature map (SOFM), and a supervised ANN, the multilayer perceptron (MLP). In Section 2, a review of automated horizon trackers is presented. The SOFM-MLP hybrid autotracker and its implementation details are discussed in Sections 3 and 4, respectively. In Section 5, the tracker is demonstrated on a number of horizons for a 3-D seismic data set from the UK sector of the North Sea. The issue of seismic character recognition using ANNs is presented in Section 6, and the chapter concludes with a general discussion and conclusions in Section 7.

2. Review of Automated Horizon Tracking

As discussed above, horizon tracking is essential for the interpretation of a 3-D seismic data set, though it is often a difficult procedure to achieve automatically with minimum user interaction (Harrigan, 1992). A large variety of algorithms have therefore evolved over the years which attempt to track seismic horizons automatically.

The simpler algorithms are not based on ANNs, and are useful for tracking horizon wavelets which can be easily distinguished from the surrounding data on an amplitude basis. They are also the foundation of most commercial trackers. However, these trackers often fail when (a) the signal-to-noise ratio becomes poor (Dyk and Eisler, 1951), (b) the wavelets show variations in amplitude, or (c) when neighbouring horizon wavelets possess similar amplitudes. Other amplitude-based methods use the dip of the horizon being tracked as a further guide to which event is the true one. These methods work reasonably well for shallow horizons, in which the wavelets exhibit a consistent amplitude across lines and traces, and which have little geological structure associated with them. However, they tend to fail very quickly in deeper areas of high geological complexity, where the horizon wavelet exhibits a profound lateral variation in reflection amplitude. A number of commercial trackers employ crosscorrelation (used extensively in seismic data processing), to detect a known wavelet which is corrupted by noise. The predicted wavelet i.e. the anticipated wavelet expected from that horizon, is crosscorrelated with the seismic section and produces a maximum amplitude when the desired signal is encountered. However, this method does not perform well when the seismic signal is temporally stretched or compressed. Unfortunately, this is often the case with seismic data, and therefore crosscorrelation trackers fail in many cases. Most commercial trackers also require a dense seed grid of points, which have to be input manually, in order to track horizons accurately.

Paulson and Merdler (1968) detected the amplitude maxima of the horizon wavelets using the zero-crossings of the first-order derivative with respect to time. Garotta (1971) used a transform similar to the straight-line Hough transform, to detect horizons on synthetic seismograms based on dip and traveltime. This method assumes that horizons may be approximated by straight lines, which is generally untrue for most seismic data. This problem has been partially addressed by Huang *et al.* (1987), who modelled horizons using a third-order polynomial, thus allowing one anticline and one syncline within a single horizon.

The use of complex seismic trace attributes (Taner *et al.*, 1979), and pattern recognition techniques based on fuzzy set theory (Zadeh, 1965), have been used to automatically track horizons and identify termination points (Lashgari, 1991; Lashgari and Estill, 1991)

Kalman filtering, target tracking and probabilistic data association have been developed extensively over the last number of decades for radar and sonar applications, enabling real-time tracking of multiple targets in a cluttered environment. This theory has been adapted and applied successfully to the horizon tracking problem (Harrigan, 1992; Harrigan and Durrani, 1991; Harrigan *et al.*, 1992a; Tu *et al.*, 1993; Leggett *et al.*, 1994a; Woodham *et al.*, 1995a, 1995b).

Autotrackers based on ANNs were introduced in the early 1990s. Wavelet extraction for horizon tracking was first investigated by Harrigan *et al.* (1991a, b, 1992b) and Harrigan (1992), using an MLP consisting of two hidden layers. The tracker performed very well in geologically complex scenarios exhibiting considerable event discontinuities such as salt domes. Prochnow (1999) also demonstrated an MLP-based tracker capable of tracking over faults and transgressions. This method uses a technique known as automatic attribute analysis to determine the optimum geometry and combination of seismic attributes that characterise a target horizon. The method also employed postprocessing to remove mis-picks.

Multiple horizon tracking using a hybrid methodology has been developed which incorporates (1) cross-correlation to determine small segments of horizons in the data analysis window, (2) training of an ANN to rate peaks in the longest segments, (3) application of the trained ANN to rate peaks in other segments, and (4) identification of segments for the horizon corresponding to the longest segment using the ANN ratings and a branch-and-bound search technique (Kemp *et al.*, 1992; Veezhinathan and Threet, 1993). The ANN employed in Step (2) was a multilayer perceptron with a single hidden layer of two nodes and an output layer consisting of a single node. The technique requires less human intervention than commercial autotrackers, and is able to track horizons in the presence of significant discontinuities. The basic approach was later modified by Alberts and Moorfeld (1999), and Alberts *et al.* (2000).

Kusuma and Fish (1993), applied a cascade-correlation neural network (Fahlman, 1988; Fahlman and Lebiere, 1990) to the horizon tracking problem. The cascade-correlation network has advantages of efficient training and incremental learning over the conventional multilayer perceptron/back-propagation training paradigm. Using nine local attributes, the cascade-correlation network successfully tracked a number of horizons in a land seismic survey which exhibited considerable geological structure and was contaminated by background noise. First-break picking using the network was also demonstrated successfully.

Horizon tracking using self-organizing feature maps (SOFMs) has been investigated by a number of researchers (Huang *et al.*, 1990; Leggett *et al.*, 1993, 1994b, 1995). A hybrid SOFM-MLP network has also been developed for tracking horizons in 3-D data sets (Leggett *et al.*, 1996), and is described further in the following sections.

Clark *et al.* (1996) employed a probabilistic neural network (Specht, 1990) for picking events automatically in pre-stack migrated common reflection point (CRP) gathers. The authors report a probability of correct classification of 93%, with a 95% confidence interval (lower bound 89% and upper bound 96%).

Huang (1997) investigated automated horizon tracking using a Hopfield neural network (Hopfield, 1982, 1984; Hopfield and Tank, 1985, 1986; Tank and Hopfield, 1987). A seismic wavelet clustering approach has also been developed (Huang and Wang, 2000) for autotracking, which involves pre-processing the data using an MLP, followed by an unsupervised self-organising clustering network based on adaptive resonance theory (Carpenter and Grossberg, 1987; Carpenter *et al.*, 1987).

3. SOFM-MLP Hybrid Neural Network Horizon Tracking

The hybrid neural network mentioned in the last section, is shown schematically in Figure 1. It consists of a SOFM layer (after unsupervised organisation) in cascade with an MLP network (Leggett *et al.*, 1996).

The active nodes in the Kohonen Layer (Kohonen, 1984) i.e. those nodes which correspond most strongly to the training data and have been classed as such, are used as the input layer to the MLP. The Kohonen layer thus serves to form an alternative representation of the input pattern vectors and reduces dimensionality. This method alleviates the pattern recognition problem typical of seismic data exhibiting such great variation in reflection character.

Training the hybrid SOFM-MLP ANN is carried out as follows:

1. A window of each seismic trace, with the required horizon at the centre, is input to the SOFM layer. No scaling is necessary, since the data have already been scaled after input from SEG-Y tape. This scaling is usually performed such that 99.9% of the data remain unclipped.
2. After organisation of the SOFM layer, the active nodes are used as input to the MLP. The active nodes are defined as those that have the smallest Euclidean distance from the input vector.
3. A new set of training data is also supplied, together with the original data, for MLP training. Each input vector is presented to the SOFM layer, the Euclidean distances calculated for the active nodes and these values scaled to lie between -1.0 and +1.0 before being input to the MLP.
4. The MLP is trained using the back-propagation algorithm. The desired output is [+1.0,-1.0] for a true horizon, and [-1.0,+1.0] for a false horizon. Shifting the true horizon data up and down relative to the required horizon provides the training data for a false horizon.

An error plot from a training sequence is shown in Figure 2. This network has 48 inputs to a 200 node SOFM layer, and 22 active nodes feeding into an MLP consisting of two hidden layers with 96 nodes in the first and second hidden layers. The output layer consists of two nodes.

After training, tracking is initiated at a user-defined position in the 3-D data set. A window of a trace containing the horizon of interest is presented to the network, and the tracking effectively propagates laterally to produce an appropriate output from the MLP. The output is examined to see if the horizon has been detected. These data are presented to the MLP part of the hybrid network for a number of iterations, until the change in error stabilises.

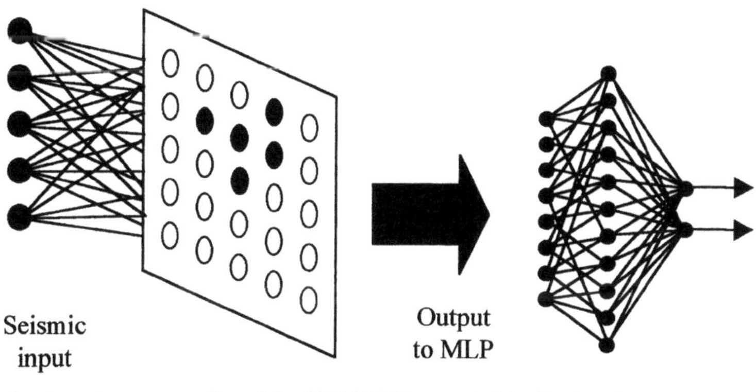

Figure 1. A schematic representation of the SOFM-MLP network. The black nodes in the Kohonen layer represent those that correspond most strongly to the input horizon.

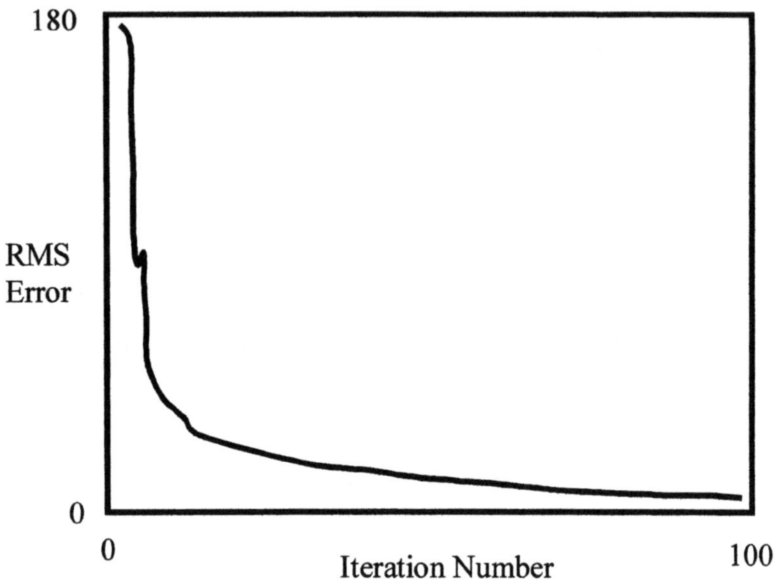

Figure 2. Training error (RMS) for the SOFM-MLP network.

4. Implementation

The tracking technique described above has been partially integrated within the Landmark™ 3-D Seismic Interpretation Package (SeisWorks 3-D) using the SeisWorks Development Toolkit. This enables any volume of seismic data and any areal extent of horizon data to be read from, or written to, disk. The software can easily be ported to link in with other interpretation packages, if the correct interface programs are available. The autotracker can also be incorporated as an interactive tool for use during interpretation of in-lines and cross-lines, similar to the current tools available.

5. Tracking Results

This tracking algorithm has been applied to a number of horizons in a 3-D seismic data set. Figure 3 shows one line from this data set, with the required horizons (A and B) delineated by arrows. These two horizons have been chosen because they are among those that are usually interpreted as important velocity boundaries in this particular area of the North Sea. A number of events from various lines and traces were selected as training data and input to the SOFM. For Horizon A, 90 training traces were chosen, using a window of 34 samples, or 132 ms, around the horizon. The SOFM has 200 nodes in the Kohonen layer and it feeds into an MLP consisting of 46 input nodes, 96 nodes in the first hidden layer and 48 nodes in the second layer, with 2 output nodes. In

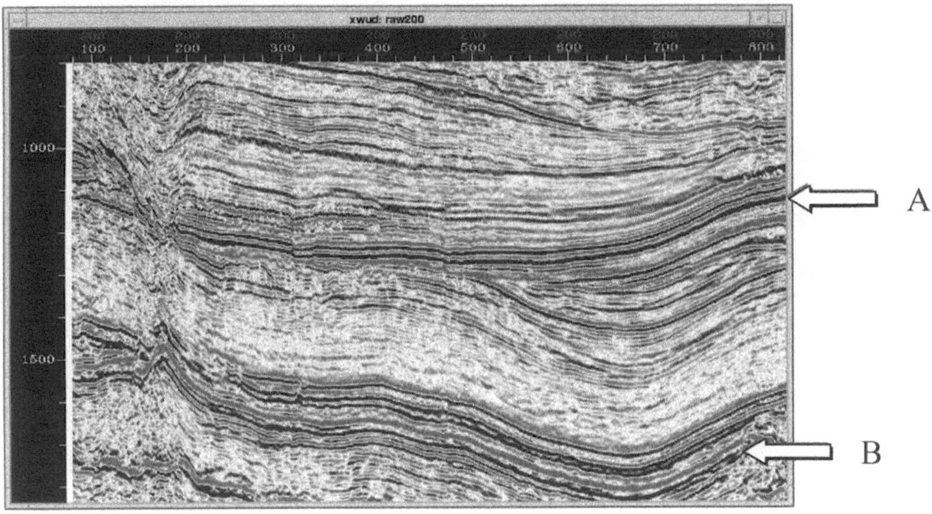

Figure 3. Seismic line from a 3-D data set. The two horizons to be tracked (marked A and B) are indicated with arrows.

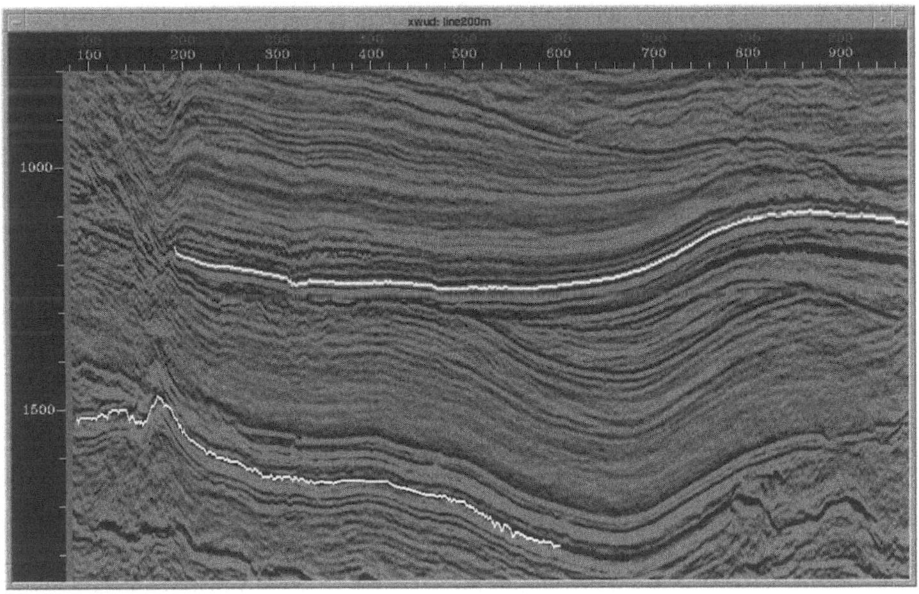

Figure 4. Tracked Horizons A and B from Figure 3, delineated in white and superimposed on the seismic data.

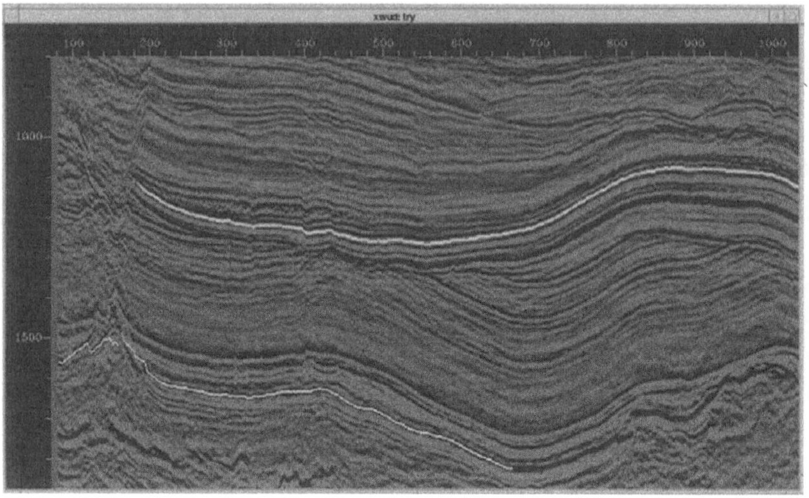

Figure 5. Tracked Horizons A and B from Figure 3 on another seismic line, delineated in white and superimposed on the seismic data.

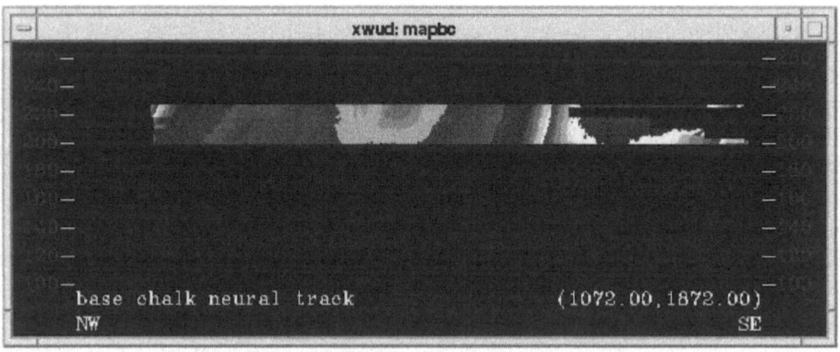

Figure 6. Map view of Horizon A tracked by the neural network.

the case of Horizon A, 80 training traces were used, the SOFM had 200 nodes and the active nodes fed into an MLP with 62 input nodes and 24 nodes in each of two hidden layers, with 2 output nodes.

A seismic line from the tracked data volume is shown in Figure 4. Horizons A and B from the survey have been successfully tracked in 3-D and these are displayed as white lines overlaying the seismic data (Horizon A is the shallower). Another example is shown in Figure 5. A map view of Horizon A is displayed in Figure 6 and covers approximately 27 lines and 800 traces.

6. Seismic Character Recognition

This chapter has concentrated upon the use of seismic character in a 1-D sense (i.e. simply taking a vector of values from the seismic data, centred on the horizon(s) of interest. However, extensions to this methodology are possible, for example to 1.5D, 2-D and 3-D. The scope of the present work is limited to a 2-D extension, but further ideas are presented for application to 3-D seismic data.

Seismic character, as recognised by an interpreter, is not simply a 1-D process but a 2-D process, taking into account the relationship of the information above, below, behind and ahead of the current point being tracked. Methods to take these into account, in order to recognize the seismic character in a section using ANNs, are described below.

6.1. 1.5-D Method

The 1.5-D Method is an extension of the 1-D vector application described above; by simply using more than a single 1-D vector, the algorithm can look above and below, behind and ahead of the current position, also training the ANN in the current "context" of the horizon. The training and recognition inputs are designed in connected 1-D vectors, which overlap with the next position (see Figure 7 for details). This method enables the tracker to work with greater efficiency as it now has more information about the seismic character to work on.

This method, of course, tries to mimic the approach used by an interpreter, who will track horizons not as distinct lines but as boundaries between layers showing different seismic character. The interpreter will always keep in mind the character around the horizons of interest and also the way that this character changes across the survey. In reality, a 1.5D approach that constantly updates the "memory" of the character may be better.

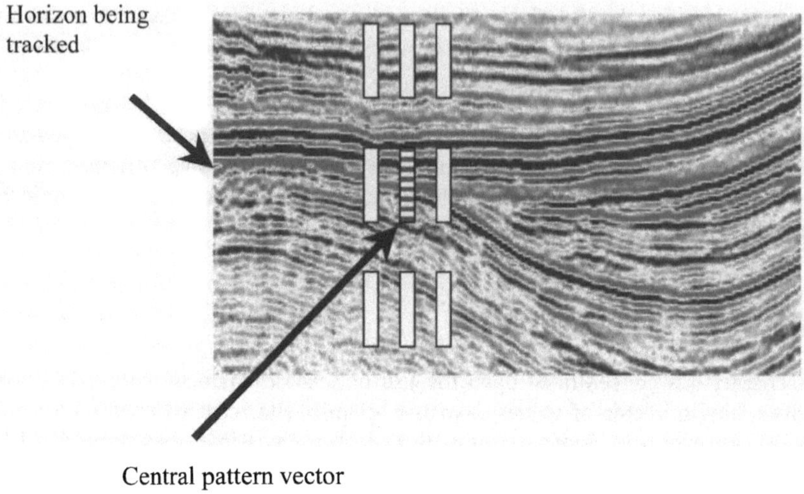

Figure 7. **Schematic representation of the 1.5-D tracking method, using a context window that looks ahead, behind, above and below the point being tracked.**

6.2. 2-D Method

In the 2-D Method, the network is reconfigured to take a small window of seismic data as input. This is dealt with, mathematically, in the same way as for a 1-D vector. The network simply has more weights per node. However, the input vector will have a different form from that of a 1-D input. The 2-D input vector, given a window of seismic data that has relatively continuous horizons, will repeat for every portion of a trace. Portions of seismic data that have continuous horizons but are different in amplitude will also give different signatures in the SOFM. However, if discontinuities are present or the character changes, the vectors will be different for every portion of a trace. As an illustration of this, Figure 8 shows 3 examples of seismic data, each with different characteristics, as indicated.

For training the network on 2-D inputs, the procedure is the same as for the 1-D or 1.5D method, except that the input vector is now an "unfolded" version of the 2-D window. Figure 9 shows a section of seismic data with areas of different character. By training the network on windows of data, the seismic section can be segmented into different layers (in the section below the seismic data). Although simplistic, this method has been fairly successful. By including extra information or by decomposing the seismic data into some significant components, a better result may also be obtained.

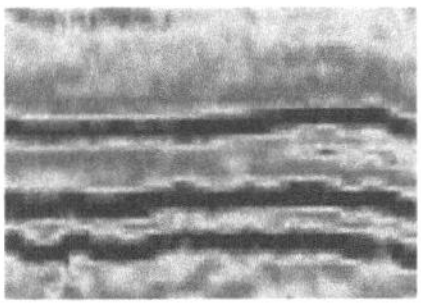

Flat, coherent events

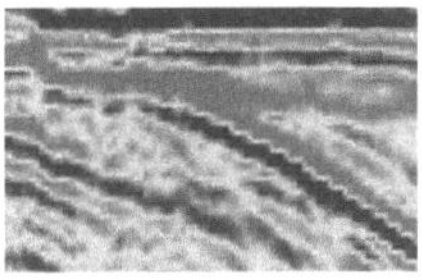

Unconformity, truncation

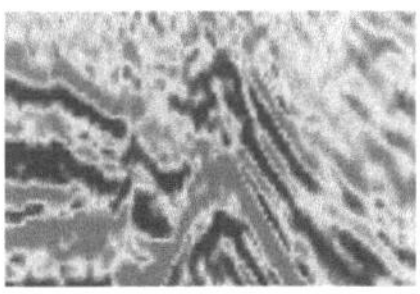

Faulted, folded

Figure 8. Examples of common seismic character from a 3-D survey.

7. Discussion and Conclusions

The results show that the SOFM-MLP network architecture is a successful technique for automatic tracking of horizons through a 3-D seismic data volume. A major advantage of this method over existing technologies is that the automatic tracker requires very little input (i.e. no dense seed grid) from an interpreter, therefore reducing the amount of time required to carry out a 3-D interpretation. This technique enables an interpreter to spend a longer period of time on areas that are geologically complex while other horizons are tracked in the background. This should enable faster turnaround of 3-D seismic interpretation and appraisal of potential prospects. The ANN tracker could also be used for pre-processing (i.e. skeletonisation) prior to full tracking using some alternative technique. Other challenges for an automatic tracker include tracking across larger faults (e.g. with a throw of 200ms) and discontinuities, both of which are feasible using this methodology.

1.5-D and 2-D methods, which account for the nature of the seismic character above, below, behind and ahead of the current tracking position, have also been demonstrated successfully. Utilising the 3-D nature of the data, i.e. its true 3-D attributes, as well as other more standard approaches, enhancements could be made to the tracking algorithm. These could include such attributes as instantaneous dip and azimuth, based purely on the data and not on horizons, 3-D seismic character and the wavelet transform.

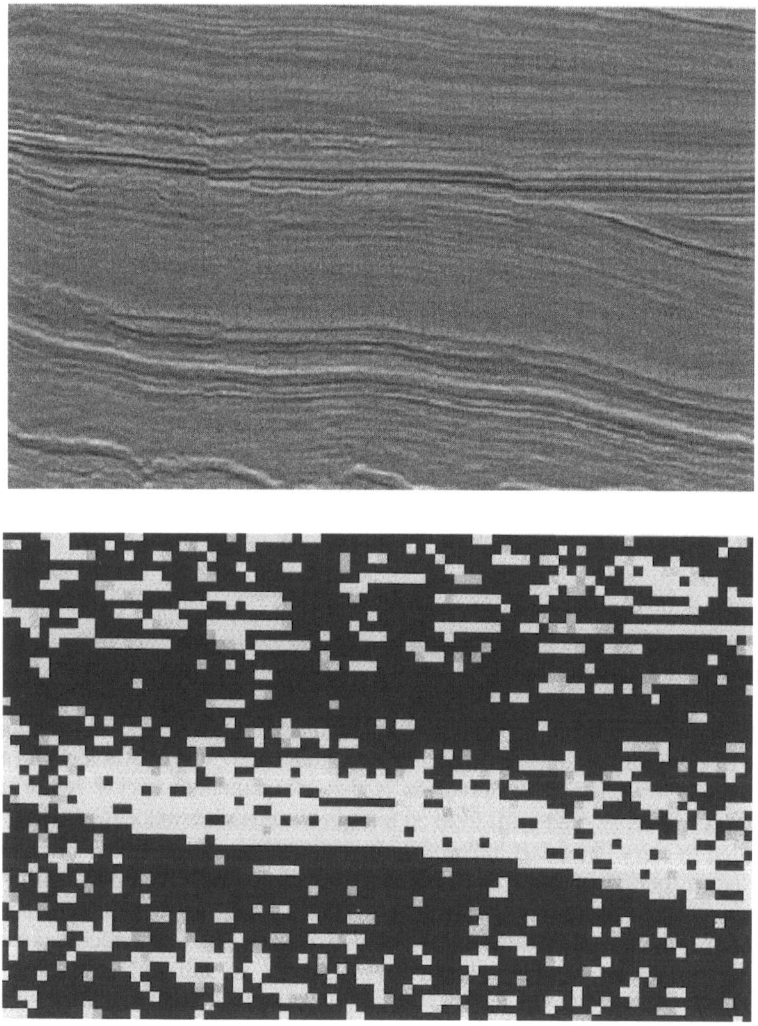

Figure 9. The top image shows a seismic section exhibiting areas of different seismic character. The lower image is the segmented output from the neural network.

Acknowledgements

The authors wish to thank Shell UK Exploration and Production for funding this research, providing the 3-D seismic data, and for granting permission to publish this material. The authors also wish to thank the EAGE for permission to publish this material.

References

Alberts, P., and Moorfeld, K., 1999, Artificial neural networks for seismic horizon tracking across discontinuities: 61st Ann. Internat. Mtg., Eur. Assoc. Geosci. Eng., Extended Abstracts, Paper 6-31.

Alberts, P., Warner, M., and Lister, D., 2000, Artificial neural networks for simultaneous multi-horizon tracking across discontinuities: 70th Ann. Internat. Mtg., Soc. Expl. Geophys., Expanded Abstracts, INT 5.4.

Bowie, J. L., and Durrani, T. S., 1988, Event tracking in a seismic wavefield: 50th Ann. Internat. Mtg., Eur. Assoc. Expl. Geophys. Extended Abstracts.

Brown, A. R., 1991, Interpretation of three-dimensional seismic data: AAPG Memoir 42, Tulsa, Oklahoma.

Brown, A. R., 1998, Interpreter's corner-picking philosophy for 3-D stratigraphic interpretation: The Leading Edge, **17**, 1198-1200.

Carpenter, G. A., and Grossberg, S., 1987, A massively parallel architecture for a self-organizing neural pattern recognition machine: Computer Vision, Graphics, and Image Processing, **37**, 54-115.

Carpenter, G. A., Cohen, M. A., and Grossberg, S., 1987, Technical comments on "Computing with neural networks.": Science, **235**, 1226-1227.

Clark, G. A., Glinsky, M. E., Sandhya Devi, K. R., Robinson, J. H., Cheng, P. K. Z., and Ford, G. E., 1996, Automatic event picking in pre-stack migrated gathers using a probabilistic neural network: 66th Ann. Internat. Mtg., Soc. Expl. Geophys., Expanded Abstracts, 735-738.

Dyk, K., and Eisler, J. D., 1951, A study of the influence of background noise on reflection picking: Geophysics, **16**, 450-455.

Fahlman, S. E., 1988, Faster-learning variations on back-propagation: an empirical study, *in* Touretzky, D. S., Hinton, G., and Sejnowski, T., Eds., Proceedings of the 1988 connectionist models summer school, San Mateo, CA: Morgan Kaufmann, 38-51.

Fahlman, S. E., and Lebiere, C., 1990, The cascade-correlation learning architecture, *in* Touretzky, D. S., Ed., Advances in neural information processing systems 2, San Mateo, CA: Morgan Kaufmann, 524-532.

Garotta, R., 1971, Selection of seismic picking based upon the dip, move-out and amplitude of each event: Geophys. Prosp., **19**, 357-370.

Harrigan, E., 1992, Seismic event tracking: Ph.D. thesis, University of Strathclyde, Glasgow, Scotland.

Harrigan, E., and Durrani, T. S., 1991, Automated horizon picking by multiple target tracking: 53rd Ann. Internat. Mtg., Eur. Assoc. Expl. Geophys., Extended Abstracts, 440-441.

Harrigan, E., Kroh, J. R, Sandham, W. A., and Durrani, T. S., 1991a, Seismic horizon picking using an artificial neural network: Proc. IEEE Internat. Workshop on Parallel Architectures for Seismic Data Processing, 2-3 September 1991, Glasgow, Scotland.

Harrigan, E., Kroh, J. R, Sandham, W. A., and Durrani, T. S., 1991b, Seismic wavelet extraction using artificial neural networks: Proc. 2nd IEE Internat. Conf. on Artificial Neural Networks, 18-20 November 1991, Bournemouth, UK.

Harrigan, E., Sandham, W. A., and Durrani, T. S., 1992a, 3-D event tracking using cross-probabilistic data association (XPDA): 54th Ann. Internat. Mtg., Eur. Assoc. Expl. Geophys., Extended Abstracts.

Harrigan, E., Kroh, J. R, Sandham, W. A., and Durrani, T. S., 1992b, Seismic horizon picking using an artificial neural network: Proc. IEEE Internat. Conf. on Acoustics, Speech and Signal Processing, **3**, 105-108.

Haykin, S., 1999, Neural networks – a comprehensive foundation (2nd ed): Prentice-Hall, Inc.

Hopfield, J. J., 1982, Neural networks and physical systems with emergent collective computational abilities: Proc. National Academy of Sciences, **79**, 2554-2558.

Hopfield, J. J., 1984, Neurons with graded response have collective computational properties like those of two-state neurons: Proc. National Academy of Sciences, **81**, 3088-3092.

Hopfield, J. J., and Tank, D. W., 1985, Neural computation of decisions in optimization problems: Biological Cybernetics, **52**, 141-152.

Hopfield, J. J., and Tank, D. W., 1986, Computing with neural circuits: Science, **233**, 625-633.

Huang, K. Y., 1997, Hopfield neural network for seismic horizon picking: 67th Ann. Internat. Mtg., Soc. Expl. Geophys., Expanded Abstracts, INT 1.02.

Huang, K. Y., Sheen, T. H., Cheng, S. W., Lin, Z. S., and Fu, K. S., 1987, Seismic image processing, *in* Aminzadeh, F., Ed., Pattern recognition and image processing: handbook of geophysical exploration: Geophysical Press, **20**, 79-109.

Huang, K. Y., Chang, W. R. I., and Yen, H. T., 1990, Self-organising neural network for picking seismic horizons: 60th Ann. Internat. Mtg., Soc. Expl. Geophys., Expanded Abstracts, 313-316.

Huang, K. Y., and Wang, S. P., 2000, Neural networks for seismic wavelet extraction and clustering: 70th Ann. Internat. Mtg., Soc. Expl. Geophys., Expanded Abstracts, INT P1.11.

Kemp, F., Threet, J. R., and Veezhinathan, J., 1992, A neural net branch and bound seismic horizon tracker: 62nd Ann. Internat. Mtg., Soc. Expl. Geophys., Expanded Abstracts, 10-13.

Kohonen, T., 1984, Self-organization and associative memory: Springer-Verlag, Berlin.

Kusuma, T., and Fish, B. C., 1993, Toward more robust neural network first break and horizon pickers: 63rd Ann. Internat. Mtg., Soc. Expl. Geophys., Expanded Abstracts, 238-241.

Lashgari, B., 1991, Fuzzy classification with applications to geophysical data, *in* Aminzadeh, F., and Simaan, M., Eds., Expert systems in exploration: Soc. Expl. Geophys., 161-178.

Lashgari, B., and Estill, R. E., 1991, Application of fuzzy pattern recognition to seismic stratigraphy: 53rd Ann. Internat. Mtg., Eur. Assoc. Expl. Geophys., Extended Abstracts, A040.

Leggett, M., Sandham, W. A., and Durrani, T. S., 1993, Seismic event classification using a self-organising Kohonen network: Proc. IEE/IEEE Workshop on Natural Algorithms in Signal Processing, 14-16th November 1993, Chelmsford, U.K.

Leggett, M., Woodham, C. A., Sandham, W.A., and Durrani, T. S., 1994a, Seismic event tracking with PDA in an interpretation environment: Proc. 7th Internat. Conf., Eur. Assoc. Sig. Proc., 225-228.

Leggett, M., Sandham, W. A., and Durrani, T. S., 1994b, Pre-processing strategies for 3-D seismic event tracking using artificial neural networks: 56th Ann. Internat. Mtg., Eur. Assoc. Expl. Geophys, Extended Abstracts, BO49.

Leggett, M., Smyth, M., Manning, A., Prescott, C. N., and Edwards, H., 1995, Neural networks and paper seismic interpretation: 65th Ann. Internat. Mtg., Soc. Expl. Geophys., Expanded Abstracts, 142-144.

Leggett, M., Sandham, W. A., and Durrani, T. S., 1996, 3-D Seismic horizon tracking using an artificial neural network: First Break, **14**, 413 – 418.

Paulson, K. V., and Merdler, S. C., 1968, Automatic seismic reflection picking: Geophysics, **33**, 431-440.

Prochnow, U., 1999, Horizon tracking in 3D seismic datasets using neural networks: 61st Ann. Internat. Mtg., Eur. Assoc. Geosci. Eng., Extended Abstracts, 6-32.

Specht, D. E., 1990, Probabilistic neural networks: Neural Networks, **3**, 109-118.

Taner, M. T., Koehler, F., and Sheriff, R. E., 1979, Complex seismic trace analysis: Geophysics, **44**, 1041-1063.

Tank, D. W., and Hopfield, J. J., 1987, Collective computation in neuronlike circuits: Scientific American, 257, 104-114.

Tu, P., Mason, I., and Zisserman, A., 1993, An automated system for picking seismic events: 63rd Ann. Internat. Mtg., Soc. Expl. Geophys., Expanded Abstracts, 234-237.

Veezhinathan, J., and Threet, J., 1993, Multiple horizons tracking: Proc. World Congress on Neural Networks 1993, I(379) – I(382).

Woodham, C. A., Sandham, W. A., and Durrani T. S., 1995a, 3-D seismic tracking with probabilistic data association: Geophysics, **60**, 1088-94.

Woodham, C. A., Sandham, W. A., and Durrani, T. S., 1995b, Error analysis for the seismic PDA tracker: Geophysics, **60**, 1451-1456.

Zadeh, L., 1965, Fuzzy sets: Information and Control, **8**, 338-353.

SEISMIC HORIZON PICKING USING A HOPFIELD NETWORK

KOU-YUAN HUANG

Department of Computer and Information Science,
National Chiao Tung University,
Hsinchu,
TAIWAN
E-Mail: kyhuang@cis.nctu.edu.tw

Abstract

A Hopfield neural network is used to solve the problem of seismic horizon picking. The input seismogram is pre-processed to produce seismic peak data. Pre-processing steps include envelope processing, thresholding, peak detection, and compression in time. Each peak represents a single seismic wavelet, and each pre-processed data item corresponds to one neuron in the Hopfield network. The constraint conditions for detecting seismic horizons are used to construct the Liapunov energy function, which is then used to extract connection weights between neurons. From the equation of motion, the next state value of each neuron can be calculated. Changing the value of a neuron decreases the energy. The system becomes stable when the value of each neuron no longer changes. A single horizon is extracted using the algorithm, and the extracted horizon is removed from the original seismic data before the next horizon is extracted. This process is repeated until no more horizons are extracted. Applying the technique to bright spot data, horizons extracted using the neural network were found to match those obtained by visual inspection.

1. Introduction

In the 1980s, Hopfield proposed his model of neural networks (Hopfield, 1982, 1984, 1987). Several interesting applications adopted the Hopfield model for associative memory, pattern recognition, and optimization problems (Hopfield and Tank, 1985, 1986; Tank and Hopfield, 1986, 1987). Lippman (1987), Pao (1989), Zurada (1992), and Haykin (1999) analyzed analog and discrete types of the Hopfield model. In seismic oil exploration, Huang et al. (1989) applied the Hopfield model to the detection of bright spot patterns. In this chapter, the Hopfield model is used to solve the problem of seismic horizon picking.

W. Sandham and M. Leggett (eds.),
Geophysical Applications of Artificial Neural Networks and Fuzzy Logic, 45-56.

2. Discrete Hopfield Neural Network

Figure 1 shows the discrete version of the Hopfield model. It is a recurrent neural network, the response of which is dynamic. Each node (neuron) has a number of inputs and outputs. The input to each node is derived from the outputs of other nodes. Inputs are modified by the activation function and produce an output, which is then fed back to the other nodes of the system. Initially, external inputs are assigned to the initial state value of each node, the next state value of which can be calculated by the motion equation. Changing the state value of each node can decrease the Liapunov energy of the network. The updating process terminates when there is no further change to the state value of each node i.e. the system reaches a steady state.

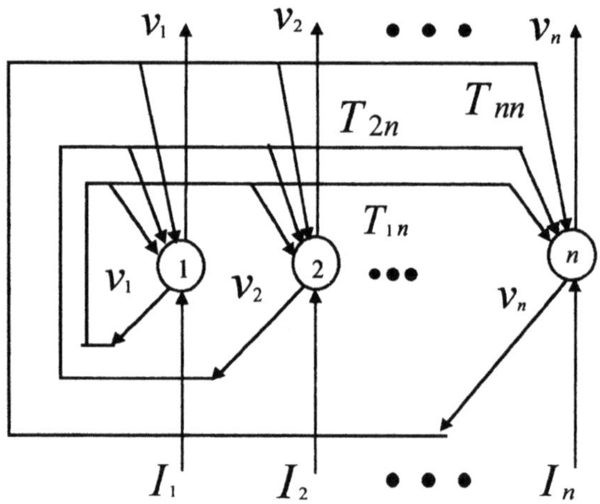

Figure 1. Discrete Hopfield neural network.

In Figure 1, there are n neurons in the discrete Hopfield model.

$V_i(t)$ state value '1' or '0' of the i-th neuron at time step t

I_i external input to the i-th neuron

θ_i threshold value at the i-th neuron

Each neuron has a hardlimiter activation function between its inputs and output,

$$f_h(s)=1 \text{ if } s > 0$$
$$f_h(s)=0 \text{ if } s < 0$$

T the matrix of the connection weights between the neurons (nodes)

T_{ij} of **T** connection weight between neuron i and neuron j.

Hopfield and Tank established the energy function of the network which they called the Liapunov function (Hopfield and Tank, 1985), and defined it as follows (Pao, 1989; Zurada, 1992):

$$E=-\frac{1}{2}\sum_i \sum_j T_{ij} V_i(t) V_j(t) - \sum_i I_i \ V_i(t) + \sum_i \theta_i \ V_i(t) \tag{1}$$

In the application, Hopfield and Tank used the constraints to construct the Liapunov function E of the Hopfield network to solve the travelling salesman problem, in order to find the minimum distance for visiting n cities (Hopfield and Tank, 1985). Using eq.(1), T_{ij} was extracted. Then, using the equation of motion, the problem was solved.

With the conditions $T_{ij} = T_{ji}$ and $T_{ii}=0$, the energy change of eq.(1) is,

$$E(t+1)-E(t)=\Delta E=-[\sum_j T_{ij} V_j(t)+ I_i - \theta_i]\Delta V_i = -s\Delta V_i \tag{2}$$

where $\Delta V_i = V_i(t+1) - V_i(t)$, $[\sum_j T_{ij} V_j(t)+ I_i - \theta_i]=s$, $V_i(t+1)=f_h(s)$.

In order to decrease the energy, ΔE must be non-positive. Hence, there are two results:

(1) If $s > 0$ ($f_h(s)=1=V_i(t+1)$), then $\Delta V_i = V_i(t+1)-V_i(t) \geq 0$.
Thus, if $V_i(t)=1$, then $V_i(t+1)=1$,
if $V_i(t)=0$, then $V_i(t+1)=1$.

(2) If $s < 0$ ($f_h(s)=0=V_i(t+1)$), then $\Delta V_i = V_i(t+1)-V_i(t) \leq 0$.
Thus, if $V_i(t)=1$, then $V_i(t+1)=0$,
if $V_i(t)=0$, then $V_i(t+1)=0$.

From the above, the output $V_i(t+1)$ is determined by the equation of motion (Hopfield and Tank, 1985; Pao, 1989):

$$\begin{aligned} V_i(t+1)&=1 \quad \text{if } [\sum_j T_{ij} V_j(t)+ I_i - \theta_i] > 0, \text{ and} \\ V_i(t+1)&=0 \quad \text{if } [\sum_j T_{ij} V_j(t)+ I_i - \theta_i] < 0 \end{aligned} \tag{3}$$

Using vector and matrix expressions, **V**(t) is the present state neuron vector, **V**(t+1) is the next state neuron vector, **T** is the connection weight matrix, **I** is the external input vector,and Θ is the threshold vector of the nodes. **TV**(t)+ **I** – Θ passes through the hardlimiter activation function and produces the output **V**(t+1).

If the external inputs **I** are **0** and the thresholds of the neurons Θ are **0**, the algorithm of the Hopfield neural network in updating neurons is as follows (Hopfield and Tank, 1985; Lippman, 1987; Pao, 1989) (the updating can be synchronous or asynchronous):

Algorithm 1: Hopfield network for synchronous updating

Step 1. Design the Liapunov function E of the Hopfield system according to the application problem.

Step 2. Obtain the connection weights T_{ij} between the neurons from E.

Step 3. Initialize with an unknown input vector **Y,** and let the initial state neuron vector **V**(0)=**Y**

Step 4. Calculate the next state neuron vector

$$\mathbf{V}(t+1)= f_h[\mathbf{TV}(t)],$$

where $f_h[\,]$ is the hardlimiter activation function.

Step 5. Repeat by going to Step 4 until the values of the neurons do not change.

For asynchronous updating, the neurons are updated one-by-one with the newest neuron values; the formula in Step 4 is

$$V_i(t+1)= f_h[\sum_j T_{ij} V_j(t)] \qquad i=1, 2, \ldots, n. \; n \text{ is the number of neurons.}$$

3. A Hopfield Network for Seismic Horizon Picking

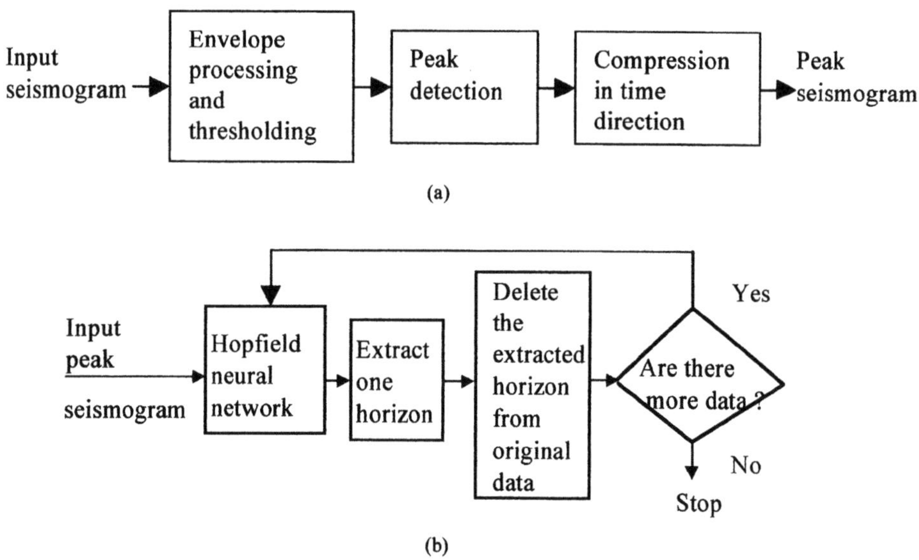

Figure 2. (a) Pre-processing of input seismogram. (b) Seismic horizon picking using the Hopfield network.

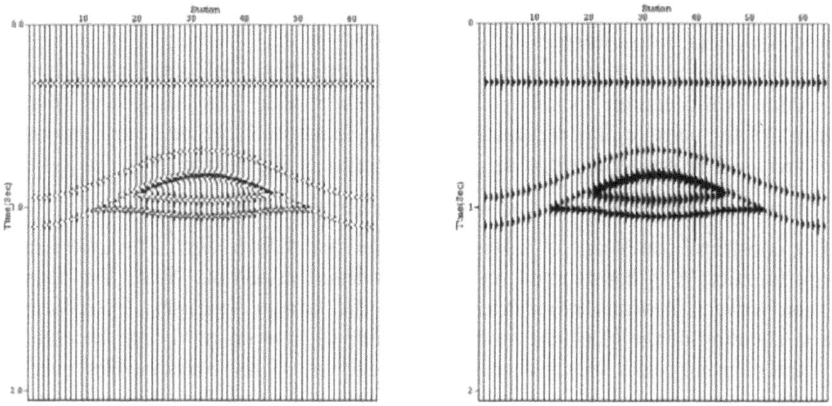

(a) (b)

```
000000000000000000000000000000000000000000000000000000000000
000000000000000000000000000000000000000000000000000000000000
111111111111111111111111111111111111111111111111111111111111
000000000000000000000000000000000000000000000000000000000000
000000000000000000000000000000000000000000000000000000000000
000000000000000000000000000000000000000000000000000000000000
000000000000000000000000000000000000000000000000000000000000
000000000000000000000000000000000000000000000000000000000000
000000000000000000000000000000000000000000000000000000000000
000000000000000000000000000000000000000000000000000000000000
000000000000000000000000000000000000000000000000000000000000
000000000000000000000000000000000000000000000000000000000000
000000000000000000000000000000000000000000000000000000000000
000000000000000000000000111111111111000000000000000000000000
000000000000000000001110000000000000011100000000000000000000
000000000000000001110000000000000000000011100000000000000000
000000000000000110000000000000000000000000110000000000000000
000000000000111000000000000011100000000000000111000000000000
000000000011000000000011111100011111100000000000110000000000
000000011100000000001100000000000000001100000000001110000000
000111100000000011111000000000000000010001110000000001111100
001000000000001000001111111111111111100000001110000000000011
000000000001110000000000000000000000000000000000011000000000
000000001110001111110000000000000000000111111110011100000000
000000110000000000000111111111111111110000000000000011100000
001111000000000000000000000000000000000000000000000001110000
110000000000000000000000000000000000000000000000000000000011
000000000000000000000000000000000000000000000000000000000000
```

(c)

Figure 3. **(a) Bright spot seismogram. (b) Envelope. (c) Thresholding, peaking, and vertical compression.**

Figure 2(a) illustrates the pre-processing steps applied to the input seismogram. The seismogram passes through envelope processing, thresholding, and peak detection (Huang, 1990). In order to reduce discontinuities in the seismic horizon, the seismic data are compressed in the vertical time direction. The output of the pre-processing step are the detected peaks from the wavelets. The output value '1' represents a peak, and the output value '0' represents a non-peak. Figure 2(b) illustrates the steps in using a Hopfield neural network to pick seismic horizons one-by-one. The peaks can be linked as horizons by the Hopfield neural network. One example from Figures 3(a), (b), and (c) displays the simulated seismogram of a bright spot, its envelope, and the pre-processed seismic peak data, respectively.

The input to the Hopfield model is the pre-processed seismic section. Each pixel in the section is the neuron location of the Hopfield model. The position at row x and column i of the seismic data represents the exact location of the neuron. Two indices x and i are used in the representation of the location of the neuron $V_{x,i}$ in 2-D space. The connection weight between two neurons is represented by $T_{x,i;y,j}$.

The constraint conditions of a seismic horizon are defined first. The total Liapunov energy function E is constructed for each constraint condition. Next, the connection weights between neurons are extracted from the energy function. Then, using the motion equation of the Hopfield model, the value of neurons at '1' can be changed to '0' if the neuron (peak) does not satisfy the conditions of the horizon. However, if the value of neurons (non-peak) is '0', then the process is skipped. Finally, peaks satisfying the conditions of the horizon are linked as a single horizon.

From eq.(1), the standard form of the system energy function is defined as follows.

$$E=-\frac{1}{2}\sum_x \sum_i \sum_y \sum_j T_{x,i;y,j}\, V_{x,i}(t) V_{y,j}(t)-\sum_x \sum_i I_{x,i}\, V_{x,i}(t)+\sum_x \sum_i \theta_{x,i}\, V_{x,i}(t) \quad (4)$$

The five constraint conditions for the seismic horizon, described in the following figures, are used to construct the energy function. Compare the energy function construction with eq.(4). The weighting coefficients $T_{x,i;y,j}$ can be extracted from the comparison. Here, the threshold for each neuron is chosen as $\theta_{x,i}=0$ and $I_{x,i}=0$. Initially the pre-processed seismic peak data are assigned to the value of each neuron $V_{x,i}$.

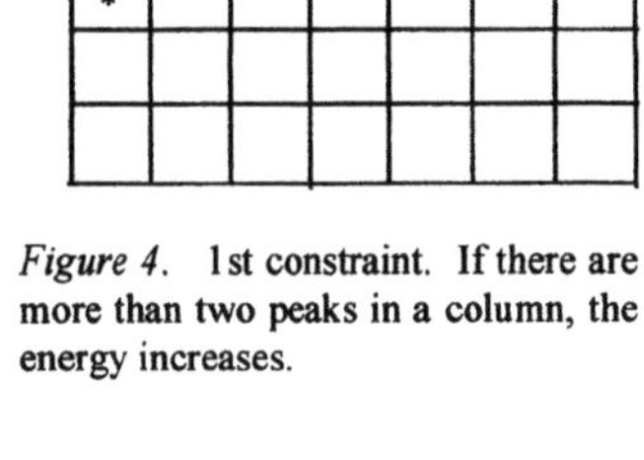

Figure 4. 1st constraint. If there are more than two peaks in a column, the energy increases.

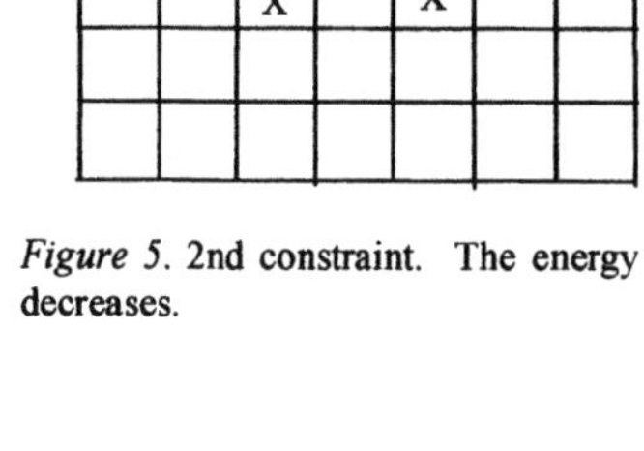

Figure 5. 2nd constraint. The energy decreases.

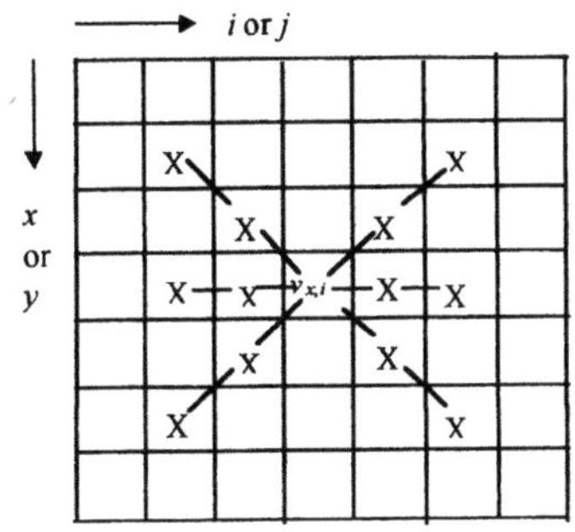

Figure 6. 3rd constraint. The energy decreases.

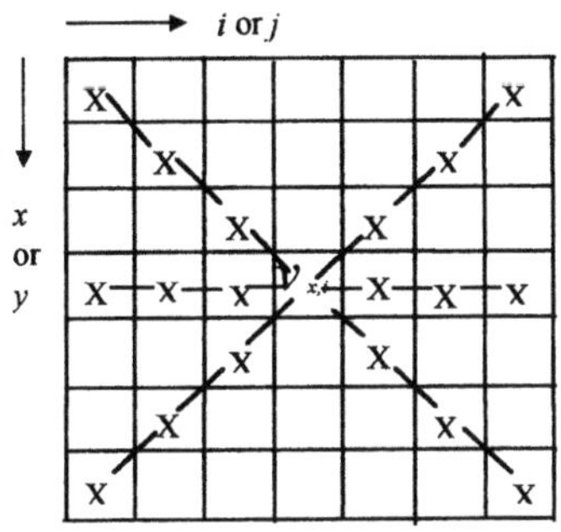

Figure 7. 4th constraint. The energy decreases.

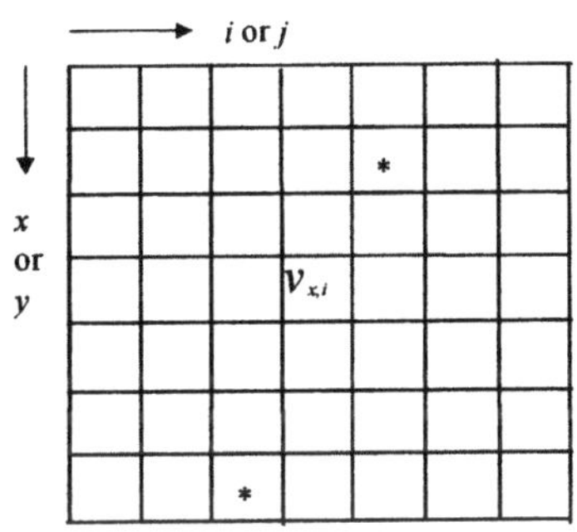

Figure 8.. 5th constraint. A point not in the horizon increases the energy.

In Figure 4, the horizon extracted step-by-step from Algorithm (2) results in a peak in one column using the first constraint condition. If there are more than two peaks in one column, then the energy increases for the peaks below the first peak in the same column.

$$E_1 = \frac{a}{2} \sum_x \sum_i \sum_{y>x} V_{x,i} V_{y,i}$$

where a is a positive constant.

The reason for the constraint conditions in Figures 5, 6, and 7 is the same. A peak can be an element of a horizon depending on the shape of its neighboring peaks. In Figure 5, the second constraint is that if $V_{x,i}$ is a peak and one of its six neighbors is a peak, then $V_{x,i}$ can be an element of a horizon, and the energy decreases.

$$E_2 = -\sum_x \sum_i \Big[\frac{b_1}{2}(V_{x,i-1} V_{x,i} + V_{x,i} V_{x,i+1})$$
$$+\frac{b_2}{2}(V_{x-1,i-1} V_{x,i} + V_{x,i} V_{x+1,i+1} + V_{x+1,i-1} V_{x,i} + V_{x,i} V_{x-1,i+1})\Big]$$

where b_1 and b_2 are positive constants.

In Figure 6, if $V_{x,i}$ is a peak and two other neighboring peaks are in a straight line, then the energy decreases.

$$E_3 = -\sum_x \sum_i \Big[\frac{c_1}{2}(V_{x,i} V_{x,i+1} V_{x,i+2} + V_{x,i-2} V_{x,i-1} V_{x,i})$$

$$+ \frac{c_2}{2}(V_{x,i} V_{x+1,i+1} V_{x+2,i+2} + V_{x,i} V_{x-1,i+1} V_{x-2,i+2}$$

$$+ V_{x-2,i-2} V_{x-1,i-1} V_{x,i} + V_{x+2,i-2} V_{x+1,i-1} V_{x,i})\Big]$$

where c_1 and c_2 are positive constants.

In Figure 7, if $V_{x,i}$ is a peak and three other neighboring peaks are in a straight line, then the energy decreases.

$$E_4 = -\sum_x \sum_i \Big[\frac{d_1}{2}(V_{x,i} V_{x,i-1} V_{x,i-2} V_{x,i-3} + V_{x,i} V_{x,i+1} V_{x,i+2} V_{x,i+3})$$

$$+ \frac{d_2}{2}(V_{x,i} V_{x+1,i-1} V_{x+2,i-2} V_{x+3,i-3} + V_{x,i} V_{x-1,i+1} V_{x-2,i+2} V_{x-3,i+3}$$

$$+ V_{x,i} V_{x-1,i-1} V_{x-2,i-2} V_{x-3,i-3} + V_{x,i} V_{x+1,i+1} V_{x+2,i+2} V_{x+3,i+3})\Big]$$

where d_1 and d_2 are positive constants.

In Figure 8, if the distance between $V_{x,i}$ (peak) and the peaks of its neighboring two columns are far apart, then $V_{x,i}$ cannot be on a horizon and the system energy increases.

$$E_5 = \frac{e}{2} \sum_x \sum_i \sum_y (h_{x,i;y,i+1} V_{x,i} V_{y,i+1} + h_{x,i;y,i-1} V_{x,i} V_{y,i-1})$$

where e is a positive constant.

The constant h is defined as follows. For peaks in the 8 neighbors of $V_{x,i}$, $h_{x,i;y,i+1}=0$ and $h_{x,i;y,i-1}=0$, where $|x-y| \leq 1$. Otherwise, for discontinuous peaks at the two-side neighboring columns, $h_{x,i;y,i+1}=|x-y|$ and $h_{x,i;y,i-1}=|x-y|$.

The total energy function is given by:

$$E = E_1 + E_2 + E_3 + E_4 + E_5 \tag{5}$$

Substituting each energy function into eq.(5), and comparing it with eq.(4), $T_{x,i;y,j}$ can be extracted, as given by eq.(6) below.

$$
\begin{aligned}
T_{x,i;y,j} =& -a\,\delta_{i,j}\,\Delta_{x,y} \\
&+b_1(\delta_{x,y}\,\delta_{i-1,j}+\delta_{x,y}\,\delta_{i+1,j}) \\
&+b_2(\delta_{x+1,y}\,\delta_{i+1,j}+\delta_{x-1,y}\,\delta_{i+1,j}+\delta_{x-1,y}\,\delta_{i-1,j}+\delta_{x+1,y}\,\delta_{i-1,j}) \\
&+c_1(\delta_{x,y}\,\delta_{i-2,j}\,V_{x,i-1}+\delta_{x,y}\,\delta_{i+2,j}\,V_{x,i+1}) \\
&+c_2(\delta_{x+2,y}\,\delta_{i+2,j}\,V_{x+1,i+1}+\delta_{x-2,y}\,\delta_{i+2,j}\,V_{x-1,i+1}+\delta_{x-2,y}\,\delta_{i-2,j}\,V_{x-1,i-1} \\
&\quad+\delta_{x+2,y}\,\delta_{i-2,j}\,V_{x+1,i-1}) \\
&+d_1(\delta_{x,y}\,\delta_{i-3,j}\,V_{x,i-2}\,V_{x,i-1}+\delta_{x,y}\,\delta_{i+3,j}\,V_{x,i+1}\,V_{x,i+2}) \\
&+d_2(\delta_{x-3,y}\,\delta_{i+3,j}\,V_{x-1,i+1}\,V_{x-2,i+2}+\delta_{x+3,y}\,\delta_{i+3,j}\,V_{x+1,i+1}\,V_{x+2,i+2} \\
&\quad+\delta_{x+3,y}\,\delta_{i-3,j}\,V_{x+2,i-2}\,V_{x+1,i-1}+\delta_{x-3,y}\,\delta_{i-3,j}\,V_{x-2,i-2}\,V_{x-1,i-1}) \\
&-e(\delta_{i+1,j}\,h_{x,i;y,j}+\delta_{i-1,j}\,h_{x,i;y,j})
\end{aligned}
\tag{6}
$$

where $\delta_{p,q}=1$ if $p=q$, $\delta_{p,q}=0$ if $p\neq q$; $\Delta_{x,y}=1$ if $y>x$, $\Delta_{x,y}=0$ if $y\leq x$; h are in E_5.

In eq.(6), the weighting coefficients $T_{x,i;y,j}$ are not fixed and can be changed by the values of neurons in every step. Here, synchronous updating finds $T_{x,i;y,j}$, and the neurons are updated after each iteration (each survey). An algorithm using a Hopfield network for seismic horizon picking is proposed as follows.

Algorithm 2: Hopfield network for seismic horizon picking.
Input: Seismic peak data.
Output: Extracted seismic horizons.
Method:
Step 1. Set up the starting state of neurons $V_{x,i}$. Let the input seismic peak data be the initial value of neurons.
Step 2. Calculate $T_{x,i;y,j}$ according to eq.(6).
Step 3. Input seismic peak data from top to bottom, and left to right. If a peak is present at location (x, i), then calculate the sum, $\sum_y \sum_j T_{x,i;y,j} V_{y,j}$, as the input for the neuron at location (x, i).
Step 4. Change the value of the neuron $V_{x,i}$.

If $\sum_y \sum_j T_{x,i;y,j} V_{y,j}>0$, $V_{x,i}=1$,

if $\sum_y \sum_j T_{x,i;y,j} V_{y,j}<0$, $V_{x,i}=0$

Step 5. Repeat steps (2), (3), and (4) until $V_{x,i}$ does not show any detectable change.

4. Experiments and Discussion

The Hopfield neural network was applied to the 10x10 pre-processed peak data of the bright spot pattern given in Figure 9(a). There are 100 neurons and 10,000 connection weights $T_{x,i;y,j}$ in the Hopfield model. The constants $a = 5.0$, $b_1 = b_2 = c_1 = c_2 = d_1 = d_2 = 1.0$, and $e = 0.0$ were chosen. Figure 9(b) is the first extracted horizon after 10 complete surveys. Figure 9(c) is the second extracted horizon after 20 complete surveys. Figure 9(d) is the third extracted horizon after 28 complete surveys. Figure 9(e) is the fourth extracted horizon after 32 complete surveys. The extracted horizons match the results obtained by visual inspection. For another set of constants, namely $a = 5.0$, $b_1 = b_2 = 1.0$, $c_1 = c_2 = d_1 = d_2 = e = 0.0$, the extracted horizons are the same.

Bright spot	Iteration 10	Iteration 20
0000000000	0000000000	0000000000
0000xx0000	0000xx0000	0000000000
000x00x000	000x00x000	0000000000
00x0000x00	00x0000x00	0000000000
0x00xx00x0	0x000000x0	0000xx0000
x00x00x00x	x00000000x	000x00x000
00xxxxxx00	0000000000	00x0000x00
0x000000x0	0000000000	0x000000x0
xxxxxxxxxx	0000000000	x00000000x
0000000000	0000000000	0000000000
	# 1 horizon	# 2 horizon
(a)	(b)	(c)

Iteration 28	Iteration 32
0000000000	0000000000
0000000000	0000000000
0000000000	0000000000
0000000000	0000000000
0000000000	0000000000
0000000000	0000000000
0000000000	000xxxx000
0000000000	0000000000
0xxxxxxxx0	0000000000
0000000000	0000000000
# 3 horizon	# 4 horizon
(d)	(e)

Figure 9. Extracted seismic horizons. "x" represents peak.

The constraints on E_2, E_3, and E_4 are the conditions for the detection of a horizon. The constraints on E_1 and E_5 achieve one horizon at one survey and exclude the other candidates. Other heuristic constraints may be considered to enhance the robustness of detection.

5. Conclusions

A Hopfield neural network has been used to solve the problem of seismic horizon picking. The input seismogram is pre-processed to produce seismic peak data. Each pre-processed data element corresponds to one neuron in the Hopfield network. The constraint conditions for detecting seismic horizons are used to construct the Liapunov energy function. The connection weights between neurons are extracted from this energy function. From the motion equation, the next state value of each neuron can be calculated. Changing the value of a neuron decreases the energy. The system becomes stable when the value of each neuron no longer changes. One horizon is extracted by using the algorithm in exactly one survey. Experimental results on seismic peak data derived from a bright spot, show that the horizons extracted using the neural network, match those obtained by visual inspection.

The weighting coefficients $T_{x,i;y,j}$ are not fixed; rather, they are changed whenever the neurons are changed. This is quite different from Algorithm 1 of the conventional Hopfield network. The values '1' or '0' of the constants *a*, b_1, b_2, c_1, c_2, d_1, d_2, and *e*, indicate the inclusion or exclusion of the constraint conditions.

Acknowledgements

The author would like to thank Mr. I. H. Huang for the preliminary programming work. This investigation was supported by the National Science Council, Taiwan, under NSC-89-2218-E-009-021.

References

Haykin, S., 1999, Neural networks – a comprehensive foundation (2nd ed): Prentice-Hall Inc.

Hopfield, J. J., 1982, Neural networks and physical systems with emergent collective computational abilities: Proceedings of the National Academy of Sciences USA, **79**, 2554-2558.

—— 1984, Neurons with graded response have collective computational properties like those of two-state neurons: Proceedings of the National Academy of Sciences USA, **81**, 3088-3092.

—— 1987, Learning algorithms and probability distributions in feed-forward and feed-back networks: Proceedings of the National Academy of Sciences USA, **84**, 8429-8433.

Hopfield, J. J., and Tank, T. W., 1985, 'Neural' computation of decisions in optimization problems: Biological Cybernetics, **52**, 141-152.

—— 1986, Computing with neural circuits: A model: Science, **233**, 625-633.

Huang, K. Y., 1990, Branch and bound search for automatic linking process of seismic horizons: Pattern Recognition, **23**, 657-667.

Huang, K. Y., Liu, W. H., and Chang, I. C., 1989, Hopfield model of neural networks for detection of bright-spot: 59th Ann. Internat. Mtg., Soc. Expl. Geophys., Expanded Abstracts, 444-446.
Lippman, R. P., 1987, An introduction to computing with neural nets: IEEE ASSP Magazine, **4**, 4-22.
Pao, Y. H., 1989, Adaptive Pattern Recognition and Neural Networks: Addison-Wesley.
Tank, D. W., and Hopfield, J. J., 1986, Simple neural optimization networks: An A/D converter, signal decision circuit and a linear programming circuit: IEEE Trans. Circuits and Systems, **33**, 533-541.
—— 1987, Neural computation by concentrating information in time: Proceedings of the National Academy of Sciences USA, **84**, 1896-1900.
Zurada, J. M., 1992, Introduction to artificial neural systems: West Publishing Co.

REFINEMENT OF DECONVOLUTION BY NEURAL NETWORKS

ENDERS A. ROBINSON

Columbia University,
Krumb School of Mines,
New York,
NY 10027,
USA
E-Mail: earl1@columbia.edu

Abstract

This chapter presents two seismic deconvolution applications of artificial neural networks. The first application involves a network based upon an adaptive linear combiner (ALC), in order to refine the results of deconvolution. For example, the conventional method of designing a shaping filter for a minimum-delay wavelet, requires the computation of a fixed filter by the method of least-squares. An alternative approach, considered in this chapter, involves an ALC incorporating a learning rule. A fixed filter vector, computed using least-squares, is used as the initial value of the weight vector for the ALC. The ALC then upgrades or refines this weight vector, in order to obtain the final refined filter coefficients. The second application uses a Hopfield neural network to correct for phase rotation, in order to produce an improved zero-phase seismic section, based upon well data.

1. Introduction

The seismic method may be considered as an instrument for remote detection, and represents a non-invasive technique to determine the structure of an inaccessible body. The work of the geophysicist is to "look" beneath the surface of the earth in the search for new deposits of oil and natural gas. The subsurface geologic structures of interest can be as deep as four or five miles (Robinson and Durrani, 1986).

The exploration geophysicist illuminates the subsurface by means of an energy source which generates seismic waves. These waves propagate downward from the source and are reflected at geological interfaces separating the subsurface rock layers. The sequence of reflection coefficients of the interfaces that occur directly below a given surface point make up the so-called reflectivity series ε, or reflectivity, at that point. The reflected waves then propagate upward from the interfaces. A primary

W. Sandham and M. Leggett (eds.),
Geophysical Applications of Artificial Neural Networks and Fuzzy Logic, 57-70.

(reflection) is defined as a reflection caused by a wave that travels directly down to the interface, and then directly back up to the surface. Similarly, a multiple (reflection) is defined as a reflection caused by a wave that bounces back and forth among various interfaces, including the earth's surface, as it proceeds on its journey.

All primary and multiple reflected waves are detected at the surface by receivers; geophones for land exploration and hydrophones for marine exploration. The digitized signal recorded at each receiver point for a given source point is called a field trace, and collectively these traces constitute the raw data. The purpose of digital seismic processing is to transform the raw data into computer-generated images of the subsurface geological structures. The images, in the form of maps and cross-sections are then interpreted by exploration geologists and geophysicists, in order to determine the most favourable drilling sites for oil or gas wells.

Deconvolution is a signal processing operation performed on each and every trace recorded in the seismic exploration for oil and gas (Robinson, 1954; Robinson and Treitel, 1980). What is required is a fast and inexpensive method to improve these deconvolution results. Such a method could drastically improve the efficacy of oil exploration if used consistently over the entire prospect. This need is more important now since it is extremely important to tie the deconvolved traces to the actual lithology. Any method which improves deconvolution, is an important step in bringing together seismic methods with petrophysical analysis in both new field exploration and reservoir characterization.

The use of artificial neural networks (ANNs) is an effective way of performing deconvolution refinement. The first part of this chapter describes the use of an ANN to refine the deconvolution result, which is based upon an adaptive linear combiner (ALC). This system can improve the quality of the final seismic record section in routine processing, and indicates more involved networks that can be developed to handle more complex scenarios. The second part of this chapter illustrates a simple example of the use of a Hopfield neural network to correct for a constant phase rotation, in order that the seismic section will better tie with a synthetic seismogram computed from downhole information.

2. The Convolutional Model and Deconvolution

It is assumed that the seismic trace has been corrected for amplitude decay, due to spherical spreading over the seismic time-scale of interest (say from 0 to 6 seconds). In an ideal seismic experiment, an impulsive source (i.e. a spike source) produces the reflection impulse response h. In other words, the reflection impulse response represents the response of the layered system (defined by the reflectivity ε) to an ideal spike source. However, in reality there are other effects that must be considered. One of these is the source wavelet, also referred to as the source signature s.

When seismic waves propagate through the earth, some of the elastic energy of the waves is transformed into heat energy. The absorption response a represents the loss of elastic energy to heat. This earth filter is necessarily minimum-phase (also known as minimum-delay). Modifications also occur due to the receiver and associated instrumentation r. These receiver effects are also considered minimum-phase.

Under various approximations, serious or minor, the successive effects that produce the field seismic trace can be represented by a linear time-invariant convolutional model. This model states that the field trace x may be expressed as the convolution of the source signal, absorption response, receiver response, and reflection impulse response, as follows,

$$x = s * a * r * h \tag{1}$$

The dynamic convolutional model (Robinson, 1954) displays the reflection impulse response h as a time-varying convolution of the reflectivity with causal dynamic wavelets. Because of the time-varying nature of the field trace, it is necessary to segment the trace into shorter sections, called design gates. Ideally, a design gate is chosen such that within the gate the reflectivity series ε is small and white. That is, the reflection coefficients corresponding to primary reflections within the gate should be uncorrelated, and their magnitudes should be much less than one. In such cases, within the time gate, the reflection impulse response can be approximated by the following linear time-invariant convolutional model,

$$h = m * \varepsilon \tag{2}$$

The wavelet m in this equation represents ghosts, reverberations, short-period multiples, and long-period multiples. The wavelet m is necessarily minimum-phase. Thus, the field trace x within the specified time gate, can be represented by the time-invariant convolutional model given by,

$$x = s * a * r * m * \varepsilon \tag{3}$$

In eq.(3), the component,

$$w = s * a * r * m \tag{4}$$

is called the seismic field wavelet. The seismic convolutional model, eq.(3), may now be written simply as,

$$x = w * \varepsilon \tag{5}$$

The convolutional model is important for the understanding of the seismic method. There has recently been a major resurgence of interest in the model, because of its applicability in direct hydrocarbon detection, porosity mapping, and the dynamic removal of multiple reflections.

In an ideal situation, the source signature s would be chosen to be an impulsive signal, ideally a Dirac delta function. In seismic exploration carried out on land, an explosion of dynamite can be used as the source signal. In such a case, the signature begins as a spike, but soon loses high frequencies and broadens into a minimum-phase wavelet of some width. However, in many situations non-minimum phase sources are used, such as vibroseis arrays for land exploration and air gun arrays for marine exploration. In such cases, the shape of the signature is either directly measured or estimated (Ziolkowski, 1984). The process of removing a known non-minimum phase signature from the trace is called signature deconvolution, and the resulting trace is called the signature deconvolved trace. Signature deconvolution enhances the seismic resolution, and requires knowledge of the shape of the source signature. The removal of all the non-minimum phase components transforms the field trace x into an intermediate trace z which has the following convolutional model,

$$z = u * \varepsilon \tag{6}$$

In eq.(6), u is a minimum-phase wavelet and ε represents the same reflectivity series as before.

The synthetic seismogram is an example of the direct use of the convolutional model. The reflectivity series is computed from downhole information obtained from an oil well. This reflectivity is convolved with a seismic wavelet to produce the synthetic seismogram. For a synthetic seismogram, the wavelet in the convolutional model is usually zero-phase.

However, the operation can also be reversed i.e. the trace can be spike deconvolved to produce the reflectivity. For the application of spike deconvolution it is assumed that a minimum-phase convolutional model holds. As such, spike deconvolution represents a type of seismic inversion. The deconvolution operation actually is achieved through convolution i.e. by the convolution of the deconvolution filter with the trace. The deconvolution filter is required to be causal, and the deconvolved trace provides an estimate of the reflectivity series.

The purpose of spike deconvolution is to remove the wavelet u from the trace y while leaving the reflectivity ε intact. The spike deconvolution filter f is the least-squares inverse of the minimum-delay wavelet u i.e. $f \approx u^{-1}$. When the spike deconvolution filter is convolved with the trace, the spike deconvolution filter is, in fact, convolved in turn with all of the wavelets that make up the trace. This operation converts each of these wavelets, as far as possible, into a spike, greatly improving the seismic resolution. The amplitude of each spike is proportional to the corresponding reflection coefficient, and the sharpness is a function of the frequency bandwidth. The spike can be no sharper than the best approximation possible with the available bandwidth.

Deconvolution is one of the most effective tools for the digital inversion of seismic traces. It constitutes the keystone of many current seismic processing methods, and is the most effective way to remove short-period reverberations on marine seismic records. More generally, it is possible to arrive at deconvolution filters which remove repetitive events having specified periodicities. A multiple reflection represents seismic energy that has been reflected more than once. If we strictly adhere to this definition, then virtually all seismic energy involves multiple reflections. A deconvolution filter can be viewed as an operator that removes multiple energy by its predictability on the seismic trace.

Deconvolution, in general, consists of two steps; the design phase and the application phase. In the design phase, the prediction error operator (PEO) is computed from the autocorrelation of the trace. This operator is the required deconvolution filter, and the computation is, in essence, the classical method of least-squares. In the application phase, the PEO is convolved with the trace to obtain the prediction errors which make up the deconvolved trace i.e. the convolution of the PEO and the trace yields the deconvolved trace as output.

A given convolutional model is assumed not to hold for the entire trace, but only to hold within a specific time gate. In practice, several different gates are specified on a trace. If two adjacent gates do not overlap, the first deconvolution filter is applied down to the end of the first gate, then with decreasing linear taper down to the beginning of the second gate. If two adjacent gates overlap, the operator designed for the first gate is applied down to the beginning of the zone of overlap. The operator designed for the second gate is applied from beyond the zone of overlap. Within the zone of overlap, the filter is obtained as a time-varying weighted average of the filter designed in the first gate and the filter designed in the second gate.

In the application of deconvolution, it is assumed that within the time gate (a) the wavelet is unknown, but is minimum delay, and (b) the frequency content of the reflectivity is completely white. The frequency content of the trace wavelet can be obtained by computing the autocorrelation of the trace. The reason for this is that the power spectrum of a white signal (i.e. the reflection coefficient series) is completely flat, so the colored frequency content of the wavelet is the same as that of the trace. In customary practice, the deconvolution filter is computed by the least-squares method, which leads to a set of Toeplitz normal equations. The known quantities in the Toeplitz normal equations are the autocorrelation coefficients of the trace. The deconvolution filter is found by solving the Toeplitz normal equations. Computer codes for deconvolution and spectral analysis are given in Robinson and Osman (1996).

3. The Adaptive Linear Combiner (ALC)

The adaptive linear combiner (ALC) is not only the simplest, but also one of the most useful types of neural networks. The ALC has many input channels, but only one output channel. Each input channel has a weight associated with it, and the net input is the sum of the products of the inputs and their respective weights. In the case of the ALC the output function is the identity function, and hence the output is the same as the net input.

Consider a set of input vectors, each having its own correct, or desired, output. The problem is to find a single weight vector that associates each input vector with its desired output vector. The process of finding the weight vector is called training the ALC. This is a supervised learning technique in the sense that there is an "external teacher" that knows the correct response for each given input vector. The learning rule can be embedded in the device itself. The device can self-adapt as inputs and desired outputs are presented to it. Adjustments are made to the weight values as each input-output combination is processed, until the ALC produces the correct outputs (Cioffi and Byun, 1993).

We will now consider the case in which the ALC is an adaptive digital filter. An adaptive system has the means of monitoring its own performance (Bozic, 1979). It can vary and improve its performance using a closed-loop action. Adaptive signal processing is used in areas as diverse as antenna arrays, radar, sonar, satellite transmission, telephony, and speech (Haykin, 1996).

An adaptive filter differs from fixed coefficient filters in that the filter coefficients vary as a function of time. There are two input time series: the filter input $x_0, x_1, \ldots, x_n$ and the desired output $d_0, d_1, \ldots, d_n$. The adaptive filter at time k forms an m-term time-varying linear combination of the input vector $x_k, x_{k-1}, \ldots, x_{k-m}$ in order to estimate the value of the desired output d_k. The elements of the ALC weight vector are the filter coefficients. This weight vector, which depends upon time k, is denoted by,

$$h = h_0(k), h_1(k), \ldots, h_m(k) \tag{7}$$

The output of the filter at time k is,

$$y_k = h_0(k)x_k + h_1(k)x_{k-1} + \ldots + h_m(k)x_{k-m} \tag{8}$$

The error is defined as the difference between the desired output and the filter output i.e. the error sequence is given by $\varepsilon_k = d_k - y_k$.

The gradient descent method, which employs a local approximation to the error surface, can be used to find the filter coefficients. The approximation is used for each input vector. The input vector $x_k, x_{k-1}, \ldots, x_{k-m}$ at time k is used to compute the filter output y_k. Then the error ε_k is computed, and this error value is used to upgrade the filter coefficients (Freeman, 1994).

For example, suppose we want to design a shaping filter for a minimum-delay wavelet. The conventional design method is to compute a fixed filter by the method of least-squares. An alternative approach is to employ an ALC incorporating a learning rule. The fixed filter vector is used as the initial value of the weight vector for the ALC, which then upgrades, or refines, this weight vector in order to obtain the final refined filter coefficients.

Let the weight vector at time k be $h = h_0(k), h_1(k), \ldots, h_m(k)$. The initial weight vector i.e. the weight vector for k=0, is the fixed filter that was computed by batch least-squares. For each step in the refining process, from the initial value at k=0 to the final value at k=m, the weight vector is upgraded. Suppose at time k, the error is,

$$\varepsilon_k = d_k - y_k = d_k - h_0(k)x_k - h_1(k)x_{k-1} - \ldots - h_m(k)x_{k-m} \tag{9}$$

As an approximation to the mean squared error, the local value of the squared error at time k is used. This local value is

$$I_k = \varepsilon_k^2 = d_k - h_0(k)x_k - h_1(k)x_{k-1} - \ldots - h_m(k)x_{k-m}{}^2 \tag{10}$$

Since I_k is a function of the filter coefficients, the gradient is computed as the following derivatives,

$$\begin{aligned}
\frac{\partial I_k}{\partial h_0(k)} &= -2\ \ d_k - h_0(k)x_k - h_1(k)x_{k-1} - \ldots - h_m(k)x_{k-m}\ \ x_k = -2\varepsilon_k x_k \\
\frac{\partial I_k}{\partial h_1(k)} &= -2\ \ d_k - h_0(k)x_k - h_1(k)x_{k-1} - \ldots - h_m(k)x_{k-m}\ \ x_{k-1} = -2\varepsilon_k x_{k-1} \\
&\ \ \vdots \\
\frac{\partial I_k}{\partial h_m(k)} &= -2\ \ d_k - h_0(k)x_k - h_1(k)x_{k-1} - \ldots - h_m(k)x_{k-m}\ \ x_{k-m} = -2\varepsilon_k x_{k-m}
\end{aligned} \tag{11}$$

Each filter coefficient is then adjusted by a small amount in the direction opposite to the gradient. In other words, the filter coefficients are updated by means of the equations,

$$\begin{aligned}
h_0(k+1) &= h_0(k) + \eta\ \varepsilon_k\ x_k \\
h_1(k+1) &= h_1(k) + \eta\ \varepsilon_k\ x_{k-1} \\
&\ \ \vdots \\
h_m(k+1) &= w_m(k) + \eta\ \varepsilon_k\ x_{k-m}
\end{aligned} \tag{12}$$

The learning rate parameter η is a positive number, usually chosen to be a value much less than one. This learning law is called the LMS rule, or delta rule. By repeated application of this rule from k=0 to the final value k=m, the point on the error surface moves down the slope toward the minimum point, though it does not necessarily follow the exact gradient of the surface. As the weight vector moves toward the minimum point, the error value decreases.

The above method can be used to refine the prediction filter used in deconvolution. In such a case the desired output is taken to be the future value (namely, the value one step ahead) of the trace, that is,

$$d_k = x_{k+1} \tag{13}$$

Once the prediction filter is upgraded, the prediction error operator is formed to give the refined deconvolution filter. The deconvolution design gates must be specified, since the convolutional model is only approximately valid for certain sections of the trace. Also, the deconvolution operator length must be specified. After deconvolution is performed, the question arises as to how a better result may be achieved through refining the process. We will compare two methods. The first is called *redeconvolution*, in which the deconvolved trace is deconvolved again. The other is the use of an ALC neural network to obtain what we call *deconvolution refined.*

Figure 1(a) shows a section of a synthetic seismic trace within a specified time gate, and Figure 1(b) shows the reflectivity series that gave rise to this trace. Figure 2(a) shows the deconvolved trace. Comparison of this deconvolved trace with the reflectivity sequence shows that, although the agreement is good, there is a considerable amount of noise present which could adversely affect the interpretation in some cases. Next, the deconvolved trace in Figure 2(a) is subjected to the redeconvolution process and the result is shown in Figure 2(b). Comparison of this redeconvolved trace with the reflectivity sequence in Figure 1(b) shows that the agreement is approximately the same as before. In this case, redeconvolution has not rendered any significant improvement.

The next step is to refine the results of Figure 2 by the ALC method. The deconvolved trace in Figure 2(a) is refined by the ALC process, and the deconvolved refined result, is shown in Figure 3(a). Similarly, the redeconvolved trace in Figure 2(b) is refined, and the redeconvolved refined result, is shown in Figure 3(b). Comparison of each of these refined traces with the reflectivity in Figure 1(b), shows a much better agreement than before. This example shows that the ALC method can provide a means to improve the results of conventional deconvolution.

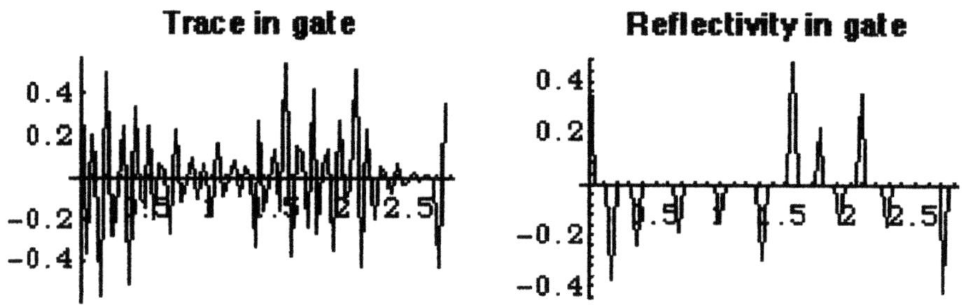

(a) (b)

Figure 1. (a) Section of trace within the given time gate. (b) Corresponding reflectivity series.

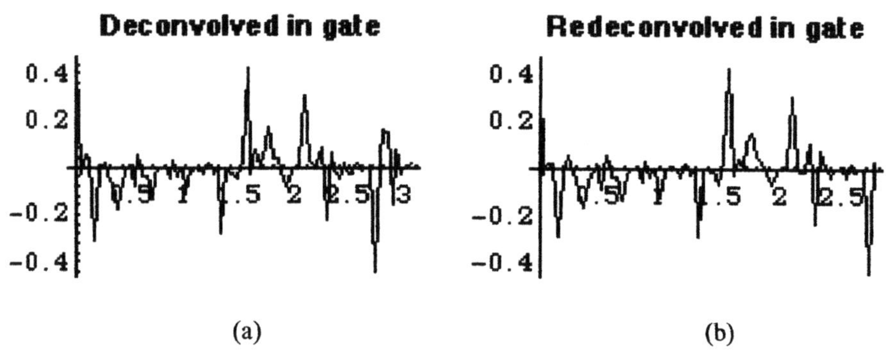

(a) (b)

Figure 2. (a) Result of spike deconvolution on the trace shown in Figure 1(a). (b) Result of redeconvolution on the deconvolved trace shown in Figure 2(a).

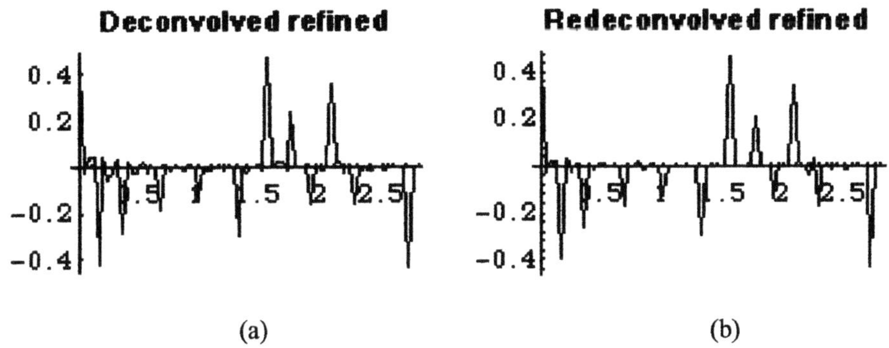

Figure 3. (a) Result of ALC refinement on the deconvolved trace shown in Figure 2(a). (b) Result of ALC refinement on the redeconvolved trace shown in Figure 2(b).

4. Phase Corrections

Conventional seismic data can be improved by adequate phase corrections. For example, the phase distortion imposed on the received seismic signal by the data acquisition system must be accounted for and corrected. Geophysicists tend to think in terms of positive and negative polarity on their seismic data when two surveys do not match. However, the traditional concepts of normal and reverse polarity are insufficient to compensate for the wide variation of phase shifts which occur in recording and processing. A simple reverse of polarity on one section may improve the tie at some levels, but usually it will make other ties worse. Accurate phase corrections are vital when dealing with synthetic logs. Seismic records show phase shifts of various numbers of degrees other than 180° for different configurations of recording instruments. As a result, seismic sections from different surveys will not adequately tie. Proper phase matching is vital.

Seismic data processing can be classified into two parts. One is geophysics, in which the emphasis is on the physics of wave motion. The other is petrophysics, where the concern is with the physics of the sedimentary rock layers for more accurate mapping of petroleum reservoirs. When dealing with the geophysics of wave propagation, it is necessary to carry out the processing under the minimum-phase assumption. Minimum-phase is a characteristic of many physical systems involved in the wave propagation process. The minimum-phase assumption is in accordance with the physical action of the earth as the seismic wave travels through the various strata. Also, the recording instruments are minimum-phase, with a linear phase-shift added for the case of a constant time delay. Deconvolution depends upon the minimum-phase assumption in order to increase seismic resolution. However, at some point in the data processing scheme, the physics of wave propagation has to be abandoned and the attention turned to petrophysics. The final images must reveal the rock layers in the greatest detail. In this phase of the processing, the minimum-phase component should be removed to yield a section made up of zero-phase wavelets. The zero-phase wavelet is better suited for stratigraphic resolution and yields better synthetic logs.

If two surveys do not tie, it is sometimes advantageous to reverse the polarity on one of the ties. There may be an improvement of fit at one point, but deterioration at another. Such a procedure is usually not warranted because phase differences between instruments tend to be roughly 40° or 60°, depending on their configuration. Different field equipment usually have smaller intermediate phase differences than 180°. Signature deconvolution seeks to reduce each event to zero-phase. Its success depends upon how well the necessary conditions can be met in practice. Saw-tooth effects on synthetic logs may be due to phase shifts. Two sets of seismic data at a common tie point, but acquired with different instruments, can produce synthetic logs that have improper phase characteristics. A constant phase shift applied to one of the synthetic logs can result in a modified log that has a better phase appearance.

The process of rotating the phase of an input trace can be described as follows. First, the Hilbert transform is used to compute the conjugate trace. Then the output trace is formed by combining the input trace and the conjugate trace, each modified by the specified phase angle. A 90° rotation will produce the conjugate trace as the output. A 180° rotation will reverse the polarity of the input trace i.e. a 180° rotation is exactly equivalent to a polarity reversal. For example, depending upon the switch settings used in the recording, vibroseis data can be 90° out of phase with other data recorded in the same area. A 90° counter-rotation can correct the problem.

The convolutional model consists of two components: the wavelet and reflectivity series. Wavelets are usually determined by recording them directly in the field, or by estimating them in data processing by making some assumption regarding their phase properties. The autocorrelation function of the wavelet provides the amplitude characteristic. Because the autocorrelation does not contain phase information, the shape of the wavelet cannot be determined from the autocorrelation alone. This problem can be solved by invoking the minimum-phase assumption. With this assumption, the deconvolution operator can be determined.

In the construction of a synthetic seismogram, a zero-phase assumption is made regarding the wavelet. In Figure 4 the center trace represents a synthetic seismogram computed from downhole data provide by an oil well at that location. The reflectivity series is computed from the well logs. The wavelet is a zero-phase wavelet that preserves the frequency content of the seismic data. The upper bank of traces represent a seismic section acquired by a surface seismic survey taken on one side of the well. The lower bank of traces represent the corresponding seismic section on the other side of the well. The seismic sections have undergone standard wavelet processing. The poor match of the synthetic seismogram with the seismic section on each side should be observed. The problem is to improve this match so that the seismic data will tie with the downhole data taken in the oil well.

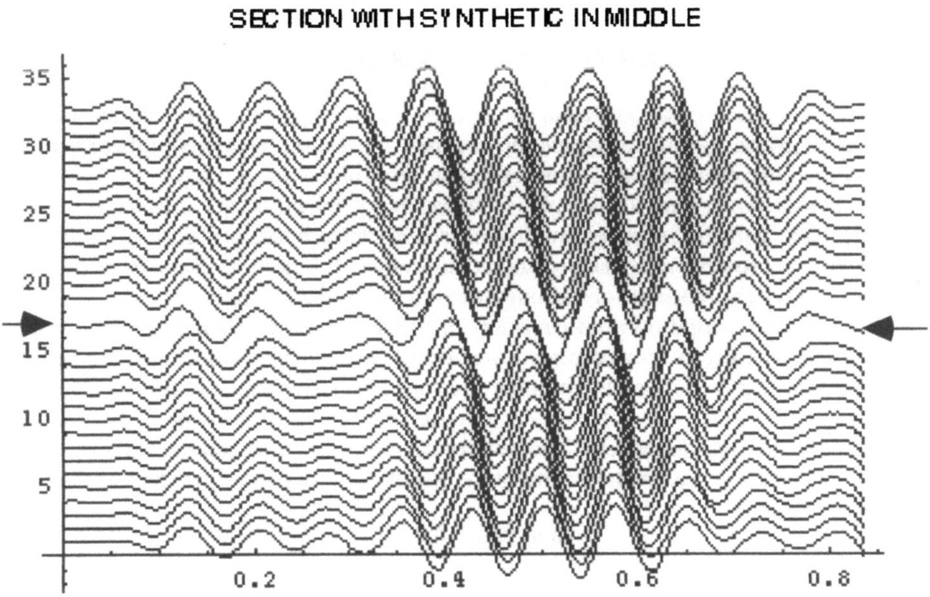

Figure 4. The center trace (indicated by the arrows) is a synthetic seismogram computed from downhole data taken in an oil well. The traces above and below the center trace represent a seismic survey taken each side of the well. It is seen that the seismic data do not tie with the synthetic seismogram.

5. The Hopfield Network

Consider a list of vector pairs $(x_1, y_1), (x_2, y_2), \ldots, (x_n, y_n)$. An input vector x_i should yield the corresponding vector y_i as output. Furthermore, a noisy or corrupted version of x_i should yield the same output y_i. A network with this characteristic is referred to as associative memory. The network stores a list of vector pairs. For any actual input vector, the network searches for the stored input vector x_i closest to the input vector. The system then gives the corresponding vector y_i as output. In the case of a Hopfield network, the output vectors are the same as the corresponding input vectors. Thus, a Hopfield network is a system with auto-associative memory.

Wang and Mendel (1992) use Hopfield neural networks for adaptive minimum prediction-error deconvolution, and source wavelet estimation. In their work, the Hopfield neural network is used to implement a new adaptive minimum prediction-error deconvolution (AMPED) procedure. Unlike commonly-used deconvolution methods, a random reflectivity model is not required for their technique. Their approach relates the cost functions of the deconvolution and wavelet estimation problem with the energy functions of these Hopfield neural networks. The stable states are taken as points for which the energy functions are locally minimized. When these neural networks reach these stable states, the outputs of the networks give the solution to the deconvolution and wavelet estimation problem. Their method decomposes deconvolution and wavelet estimation into three sub-processes: reflectivity location detection, reflectivity magnitude estimation, and source wavelet extraction. Three Hopfield neural networks are constructed to implement the three sub-processes. The three networks are then connected in an iterative manner to implement the entire deconvolution and wavelet estimation procedure. The criteria functions of these neural networks are adaptive versions of the prediction-error of the seismic trace.

In what follows, a simple example of the use of a Hopfield neural network is given. Consider the synthetic seismogram shown as the center trace in Figure 4, which is subjected to a sequence of phase rotations, as shown in Figure 5. This set of rotated traces is used as the sequence of training traces in the Hopfield neural network.

The Hopfield neural network considers the seismic trace on each side of the well, and decides whether it is a corrupted version of one of the members of the set of training traces. In each case, the training trace with a phase rotation of 50° was chosen by the neural network. The value of 50° indicates that a counter-rotation of 50° should be applied to the seismic section. This counter-rotation correction converts the wavelets on the section to approximately zero-phase wavelets. The phase-corrected section is shown in Figure 6.

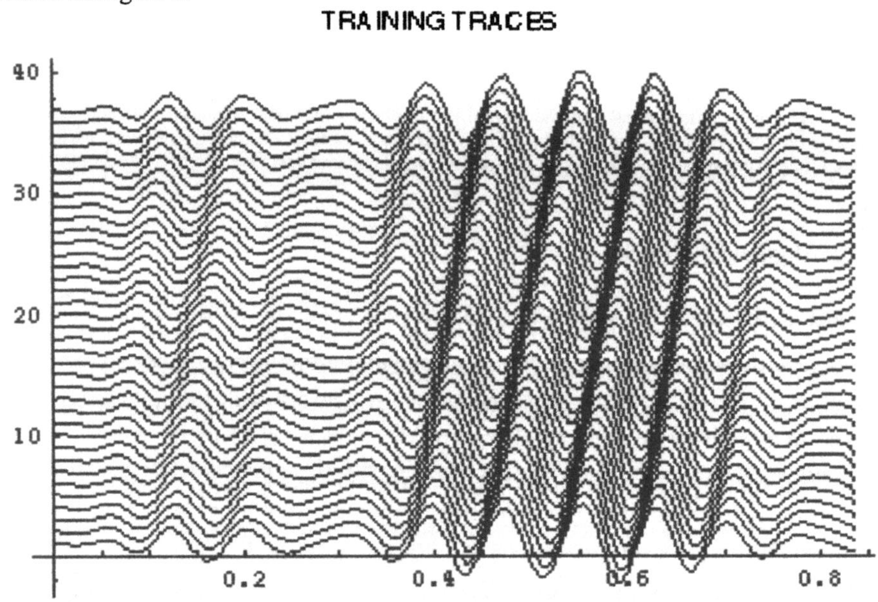

Figure 5. The synthetic seismogram (the center trace indicated by the arrows in Figure 4) is subjected to a sequence of phase rotations. This figure shows the synthetic seismogram subjected to constant phase rotations in 10° increments for the range 0-360°. The bottom trace is the synthetic seismogram. The trace above is the synthetic seismogram subjected to a phase rotation of 10°, the one above that to 20°, and so on, until the top trace is reached with a phase rotation of 360°. Since a 360° phase rotation is one complete circle, the top trace is identical to the bottom trace. This set of traces is used as the set of training traces in the Hopfield neural network.

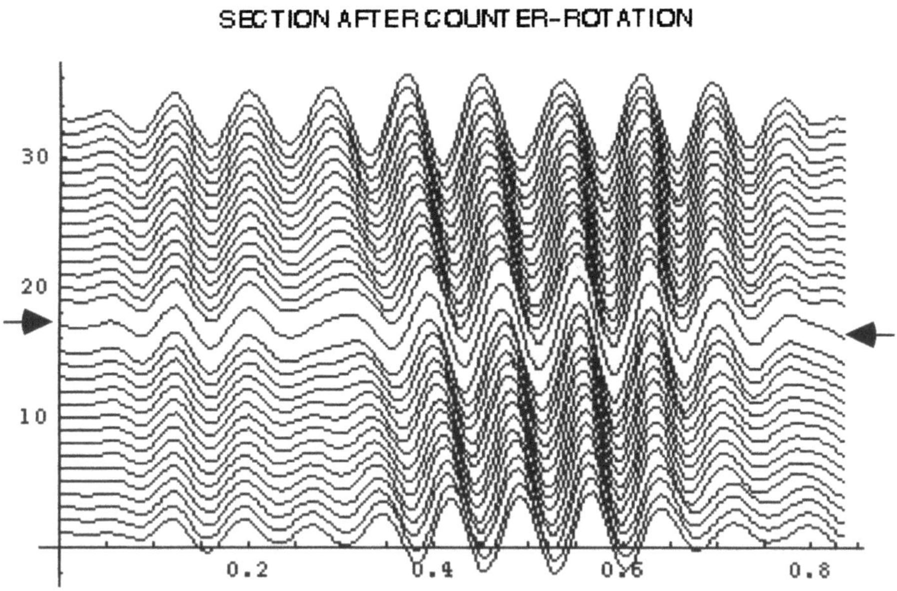

Figure 6. The center trace (indicated by the arrows) is the synthetic seismogram computed from the downhole data. The seismic traces above and below the center trace have been counter-rotated by the amount determined by the Hopfield neural network. It is seen that the seismic traces now tie with the synthetic seismogram.

The application of counter phase rotation to produce processed waveforms which are approximately zero-phase is commonly used. The amount of counter-rotation is often determined by eye, and is subjective. However, as seen here, a Hopfield neural network can remove the guess-work and produce superior results.

6. Conclusions

This chapter has discussed two seismic deconvolution applications of artificial neural networks. The first application involves a network based upon an adaptive linear combiner (ALC), in order to refine the results of deconvolution. A fixed filter vector, computed using least-squares, is used as the initial value of the weight vector for the ALC, which then upgrades or refines this weight vector, in order to obtain the final refined filter coefficients. Simulation results show that the ALC method can provide a means to improve the results of conventional deconvolution.

The second application uses a Hopfield neural network to correct for phase rotation, in order to produce an improved zero-phase seismic section, based upon well data.

References

Bozic, S. M., 1979, Digital and Kalman Filtering, Edward Arnold.

Cioffi, J. M., and Byun, Y., 1993, Adaptive filtering, *in* Mitra, S. K., and Kaiser, J. F., Eds., Handbook for digital signal processing, John Wiley and Sons, New York, 1085-1142.

Freeman, J. A., 1994, Simulating neural networks, Addison-Wesley, Reading, MA.

Haykin, S. S., 1996, Adaptive filter theory, 3rd ed., Prentice-Hall, Upper Saddle River, NJ.

Robinson, E. A., 1954, Predictive decomposition of time series with application to seismic exploration, Ph.D. thesis, M.I.T., Cambridge, MA, (reprinted in Geophysics, **32**, 1967, 418-494).

Robinson, E. A., and Treitel, S., 1980, Geophysical signal analysis, Prentice-Hall, Englewood Cliffs.

Robinson, E. A., and Durrani, T. S., 1986, Geophysical signal processing, Prentice-Hall International, UK Ltd, London.

Robinson, E. A., and Osman, O. M., 1996, Deconvolution, Society of Exploration Geophysicists, Tulsa, Oklahoma.

Wang, L., and Mendel, J. M., 1992, Adaptive minimum prediction-error deconvolution and source wavelet estimation using Hopfield neural networks, Geophysics, **57**, 670-679.

Ziolkowski, A., 1984, Deconvolution, International Human Resources Dev. Corp., Boston, MA.

IDENTIFICATION AND SUPPRESSION OF MULTIPLE REFLECTIONS IN MARINE SEISMIC DATA WITH NEURAL NETWORKS

ROBERT ESSENREITER[†], MARTIN KARRENBACH[‡]
and SVEN TREITEL[*]

† *PGS Seres AS,*
Strandveien 4,
P.O, Box 354,
N-1326 Lysaker,
NORWAY

‡ *Geophysical Institute,*
University of Karlsruhe,
Hertzstr. 16,
76185 Karlsruhe,
GERMANY

E-Mail: Robert.Essenreiter@oslo.pgs.com

* *TriDekon, Inc.*
Suite 308, Sinclair Building 6 E. 5. St.,
Tulsa,
Oklahoma 74103,
USA

Abstract

The process of generating multiple reflections in the earth is generally non-linear. We present a technique for identifying and attenuating multiple reflections in seismic data using non-linear filters i.e. multilayer perceptron neural networks. The aim is to separate the multiple wavefield from the primary wavefield and suppress it in order to create a multiple-free seismic section. A neural network is trained with modeled data that are generated from well-log information and then applied to the areas between the boreholes. Input to the neural network is either the seismic trace itself, or a number of selected, representative attributes computed from the seismic trace. The method is demonstrated on synthetic data and is validated by comparing it to the linear Wiener-filter technique.

W. Sandham and M. Leggett (eds.),
Geophysical Applications of Artificial Neural Networks and Fuzzy Logic, 71-88.

1. Introduction

Multiple reflections in seismic data are generally considered unwanted noise that often seriously impede correct mapping of the subsurface geology in the search of oil and gas reservoirs. We train a multilayer perceptron (MLP) in order to recognize and remove these multiple reflections and thereby bring out the primary reflections underneath. In our first approach the training data consist of both modeled data containing free-surface and internal multiples and the corresponding seismic sections containing only the primary arrivals. We compare the results to one of the many existing conventional deconvolution techniques, the method of predictive deconvolution based on Wiener filters. First we suppress multiples in zero-offset seismograms and then show the application to whole seismic sections. In the latter case we employ an ensemble of neural networks, where every component is responsible for one offset in the section. The basis for the training is subsurface information in form of well-log data. The neural network represents one of the few approaches to multiple attenuation that allows for making use of existing a-priori knowledge about the geology. A second approach applies a combined domain multiple suppression method to synthetic model data. Instead of separating multiples from primaries in *one* parameter domain - as is commonly done, e.g. in $f - k$ filtering - we train a neural network with attributes selected from *various* domains.

2. The Problem of Multiple Reflections

In seismic exploration the problem of multiple reflections contaminating seismograms and thus disguising important information about subsurface reflectors is well-known but

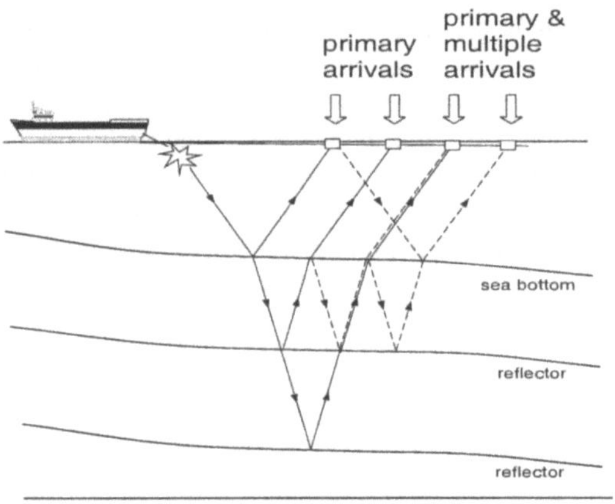

Figure 1. Marine seismic data acquisition: travel paths of the primary reflections (solid lines) and several multiple reflections (dashed lines).

not yet totally solved. Today, the majority of all oil and gas resources are discovered offshore in continental shelf areas in shallow water. Before oil-producing wells can be drilled, geophysicists have to provide an image of the subsurface geology that shows where reservoirs can be expected. In marine exploration we encounter the problem that the water layer often behaves as a wave trap (Backus, 1959), where seismic waves are multiply reflected between sea surface and sea bottom. Waves that are transmitted through the seabottom can also reverberate between deeper reflectors. The energy of these interbed multiples and water layer reverberations can become so strong that the primary reflection arrivals of deeper target reflectors become completely invisible. As a result, marine seismograms often show a ringy character with strong multiples masking most of the primary arrivals from deeper reflectors.

For correctly locating a target reflector that might indicate an oil reservoir, these disturbing multiple reflections have to be eliminated. Figure 1 shows a typical marine data acquisition method. Travel paths of primary reflections (solid lines) and several multiple reflections (dashed lines) illustrate the problem of interference of these two types of arrivals. Especially where primary and multiple reflections arrive at the same zero-offset travel time and show little move-out difference, it is very hard to separate the two signals or even to tell if the event is a primary or a multiple. Today, seismic signal processing is often based on very simple linear models, whose theory rests on assumptions that are often not met in practice. An example is seismic deconvolution using Wiener filters, a method that tries to predict and subtract multiples from the seismic trace with the aid of auto- and cross-correlations of the input trace signal. This method suffers from limitations: for example, it tends to remove all periodic signals even if they are not multiples but originate from cyclic stratigraphy. Nevertheless, these traditional methods are implemented very successfully in many cases, but there is a variety of cases where they fail.

The neural network does not depend on such restricting assumptions about the underlying model, and has the potential to work in areas where conventional methods have difficulties. It is adaptive and able to learn highly non-linear interrelations in the data, should they exist. A further advantage of neural nets is that they are able to make extensive use of a-priori knowledge about the geologic subsurface structure that exists in the form of velocity and density logs from a borehole. This advantage, however, can turn into a disadvantage if there is not sufficient subsurface information available to train the neural network.

So far there is no multiple attenuation technique that works universally. Common methods used in the industry today can be divided into six different major categories:

1. Methods based on the periodicity of multiples in contrast to the primary arrivals: Predictive deconvolution using Wiener filters (Robinson and Treitel, 1980; Peacock and Treitel, 1969) is one successful approach that designs a filter operator which predicts the multiples and subtracts them from the seismic trace. However, at far offsets this method fails, since multiples are no longer periodic. Another drawback of this method is that it requires assumptions such as that the seismic

trace be stationary and the reflector sequence be a random series of spikes. Taner (1980) showed a way to overcome the non-periodicity problem at far offsets by applying predictive deconvolution in the radial trace space. Similarly, in the $\tau - p$ domain multiples are periodic for all p-values (Carrion, 1986), but a horizontally layered medium must be presumed.

2. Stacking methods that exploit the move-out (travel time) difference between the primary and the multiple reflection hyperbolae: In principle, these methods sum over all possible hyperbolae by scanning through all zero-offset times and curvatures. If the summation goes right along a hyperbola, a high amplitude signal is generated, while in all other cases the signal will be weaker. Since the curvature of the hyperbola relates to the velocity of the corresponding arriving wave, this method is called velocity analysis (Yilmaz, 1987, e.g.). In the velocity spectrum an experienced interpreter can distinguish primaries (generally with higher velocities) from multiples (generally with lower velocities). Normal move-out (NMO) correction with the picked primary velocity model and subsequent common-midpoint (CMP) stacking already reduces the amount of multiple energy (Schneider et al., 1965). If the NMO correction is applied in a way that primaries are overcorrected and multiples are undercorrected, the two signals map onto different half-spaces in the $f - k$ domain (Yilmaz, 1987). However, these correlation methods often fail because arrivals from deeper reflectors or complex geologic structures increasingly deviate from the hyperbolic relationship. Thus primaries often cannot be separated from multiples in the velocity or $f - k$ spectrum, due to interference effects. These methods also profit from existing subsurface information.

3. Methods that predict multiples by extrapolation of the wave field and subsequently subtract them from the data (Berryhill and Kim, 1986; Wiggins, 1988): Here, the wavefield is propagated down and up through the water layer so that the primaries become first-order multiples which are then subtracted adaptively from the original data. An estimate of the water-bottom topography and reflectivity is needed, then only water-bottom multiples can be removed.

4. The method of predicting free-surface related multiples by iterative autoconvolutions in time and space (Berkhout and Verschuur, 1997; Verschuur and Berkhout, 1997), assuming knowledge about the source signature and the existence of zero- and near-offset data.

5. Methods based on computing coherency measures (using Singular Value Decomposition) after an NMO correction that flattens multiple reflections (Kneib and Bardan, 1997). The eigenimages with the highest eigenvalues (pertaining to the multiples) are removed, leaving the primary information in the ideal case.

6. Inverse scattering methods that express the total wave field as the wavefield from a homogenous background plus the wave field generated by scatterers (Weglein et al., 1992). Multiples are represented by higher order terms in a non-linear inverse

scattering series and can thus be removed. However, here both source signature and a background model have to be known.

Coming from another point of view, approaches exist where multiples are not treated as annoying noise that is to be eliminated but regarded as signal that has travelled the subsurface many times more than a primary und thus can also give us information about the geology (Helbig and Brouwer, 1993).

A very interesting advance in the direction of non-conventional multiple suppression using Hopfield ANNs is due to Calderòn-Macias et al. (1997).

3. Neural Networks as Non-Linear Filters

Typically, in classic inverse theory it is assumed that we possess relatively accurate knowledge of the physical model underlying a certain problem. This knowledge is expressed as an operator $\mathcal{A}$, and can be written in matrix form. If this operator is applied to a model vector $\vec{x}$, it roughly reproduces the data vector $\vec{d}$ consisting of actually measured data:

$$\mathcal{A}\,\vec{x} \approx \vec{d} \tag{1}$$

In a classic inverse problem we know the measured data $\vec{d}$ and we also know the operator $\mathcal{A}$, which could be, for example, a normal move-out (NMO) - operator [Calderòn-Macias (1997) shows a neural net approach to NMO correction], or a filter to predict and subtract multiples. What we want is a good model for the data $\vec{d}$: the model parameters $\vec{x}$. eq.(1) can be inverted in a formal manner for $\vec{x}$,

$$\vec{x} \approx \mathcal{A}^{-1}\vec{d}. \tag{2}$$

This will never be an equality, since there is always noise present in the data. But by minimizing the least squares error $||\mathcal{A}\vec{x} - \vec{d}||^2$ we can iteratively find the best approximation of our model to the measured data.

This is the situation when the mapping rule from the model space to the data space is known and can be expressed as a matrix operator $\mathcal{A}$. However, in many cases this relationship is *not* known or only known empirically, thus cannot be expressed explicitly in mathematical form. In this situation classic inverse theory as described above fails, and the neural networks appear on stage.

The action of a neural network can be described by the same equation as classic inverse theory (eq.(1)). Here, we do not know the operator $\mathcal{A}$, but we have a set of examples in the form of data from an input space x and corresponding data from an output space d. These examples should be representative of the mapping from input to output space. The task of the neural network is then to find this relationship (the operator $\mathcal{A}$) by learning from training examples. $\vec{x}$ is one of many input data vectors of the training set, and $\vec{d}$ is the corresponding desired output vector. The operator $\mathcal{A}$ now denotes the weight matrix of the neural network, which has to be optimized in order to guarantee not only correct mapping of the training data, but also generalization to data not included during training. This is accomplished by a learning rule often based on gradient descent. One of the numerous advantages of the neural net approach, for example, is that it is not restricted to the L2-norm, so that the error measure can be adapted to the specific problem.

Normally a neural network consists of three or four layers - an input layer, one or two hidden layers, and an output layer - as shown in Figure 2. The layers are fully connected with weights. Therefore the weight matrix is more complex in this situation than in eq.(1). The network can be expressed as

$$\tilde{\mathcal{B}}\,[\,\tilde{\mathcal{A}}\,\vec{x}\,] = \vec{o} \tag{3}$$

$$\text{or} \quad f[\,\mathcal{B}\,f[\,\mathcal{A}\,\vec{x}\,]] = \vec{o} \tag{4}$$

where $\mathcal{A}$ is the weight matrix associated with the connections from the input layer to the hidden layer, $\mathcal{B}$ is the weight matrix associated with the connections from the hidden to the output layer, $\vec{x}$ is the input, $\vec{o}$ the output of the network, and f is a nonlinear function (shown in Figure 2). The matrices $\tilde{\mathcal{A}}$ and $\tilde{\mathcal{B}}$ are related to the original matrices $\mathcal{A}$ and $\mathcal{B}$ via the non-linearity f,

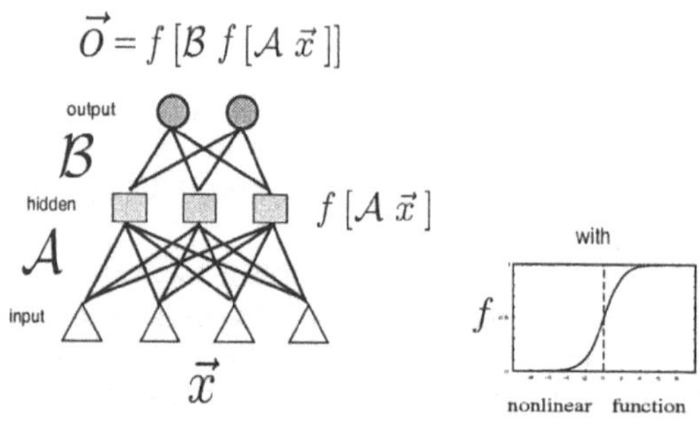

Figure 2. The three-layer backpropagation neural network with the input vector $\vec{x}$, the output vector $\vec{o}$, the weight matrix $\mathcal{A}$ associated with the connections from the input layer to the hidden layer, the weight matrix $\mathcal{B}$ associated with the connections from the hidden layer to the output layer, and a non-linear function f.

Instead of using the standard backpropagation learning algorithm we employed the Rprop (resilient propagation) algorithm (Riedmiller and Braun, 1993) which shows considerably faster convergence. In contrast to other gradient-descent algorithms, Rprop does not use the magnitude of the gradient, but only its sign. It begins with an initial small update value, and then increases this value by a factor greater than 1 (typical value: 1.2), if the current gradient has the same direction (sign) as the previous gradient, but decreases this value by a factor smaller than 1 (typical value: 0.5), if the gradient has the opposite direction (sign). This update then is added to the weight, if the gradient is negative (progression in the positive direction towards the minimum), and subtracted from the weight, if the gradient is positive (progression in the negative direction towards the minimum). The principle is shown in Figure 3.

4. Multiple Suppression Trace by Trace

To investigate the behavior of the neural network and - for comparison - the Wiener filter on the attenuation of multiple reflections in synthetic data, we generated a synthetic subsurface model and computed a number of seismograms. The synthetic model, having

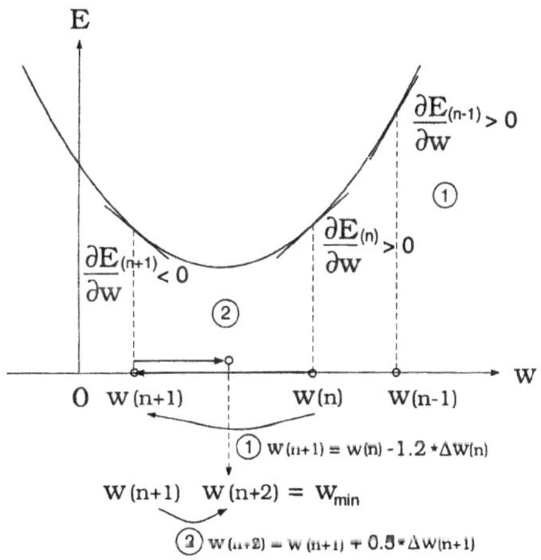

Figure 3. Principle of the Rprop-algorithm: ① The gradients of the steps n-1 and n are both positive, so the last update $\Delta w(n)$ is increased by the factor 1.2 and subtracted (since the current gradient is positive) from the current weight $w(n)$. ② The gradients of the steps n and n+1 have different signs, so the last update $\Delta w(n+1)$ is decreased by the factor 0.5 and added (since the current gradient is negative) to the current weight $w(n+1)$. E is the total error.

TABLE 1. Synthetic model for trace-by-trace multiple suppression.

Reflector	*Time (Depth)*	*Refl. Coeff.*
Water Surface	0	-1
Water Bottom	9	[-0.5 : -0.3]
1. Deep Reflector	22 - 42	[0.6 : 0.8]
2. Deep Reflector	55 - 75	[-0.6 : -0.4]
3. Deep Reflector	76 - 96	[0.7 : 0.9]

a total depth of 100 time/depth[1] units, is made up of three deep reflectors with a depth varying ± 10 time units and a variability of 0.2 in the reflection coefficients. The water surface reflection coefficient is -1. The sea bottom is located at a fixed depth of 9 time units with the reflection coefficient varying in the range [-0.5 : -0.3]. Table 1 shows the entire model. The relatively high reflectivities were chosen in order to see clearly what the neural net, or the Wiener filter, respectively, does to the signal.

By propagating a vertically incident plane wave through different reflector configurations (i.e. depth and reflection coefficient are chosen randomly within their respective ranges), a set of 200 different zero-offset seismograms was calculated, convolved with a Gabor-wavelet, perturbed by various amounts of noise and finally split into one set of 100 training patterns and another set of 100 test patterns. In Figure 4 at the bottom one example of such a seismic trace, containing free-surface as well as internal multiples, is shown (the signal-to-noise ratio is S/N = 2.0 in this example).

As depicted in Figure 4, the neural net input is the seismic trace containing all kinds of multiple reflections and noise , and the desired output is the seismic trace with only the primary reflection events, which are known from the given synthetic model. According to the problem, the MLP consists of 100 input neurons, between 10 and 30 hidden neurons, and 100 output neurons. The training time varied according to the contamination of the seismograms with noise. For the noise-free training set about 100 to 200 epochs of training were sufficient, whereas for the very noisy data sets (e.g. S/N = 0.6) the network needed up to 1000 such epochs.

The neural net output gives the arrival times as well as the reflection strengths of the desired primary reflections. However, the amplitude characteristic of the seismic wavelet is destroyed. Thus, this method is not applicable if in a subsequent processing step true amplitudes of the seismic trace are required, although the reflection strengths of the individual reflectors are reproduced quite reliably.

The set containing the 100 test patterns was not only presented to the trained neural net but also to a Wiener prediction filter, with prediction distance $\alpha = 8$, and of length 30. This parameter choice proved to yield the best results for this data set.

Figure 5 shows the result for a given trace for neural net testing on the left, and for

[1] In seismics "depth" often is expressed in terms of the two-way travel time of the seismic signal

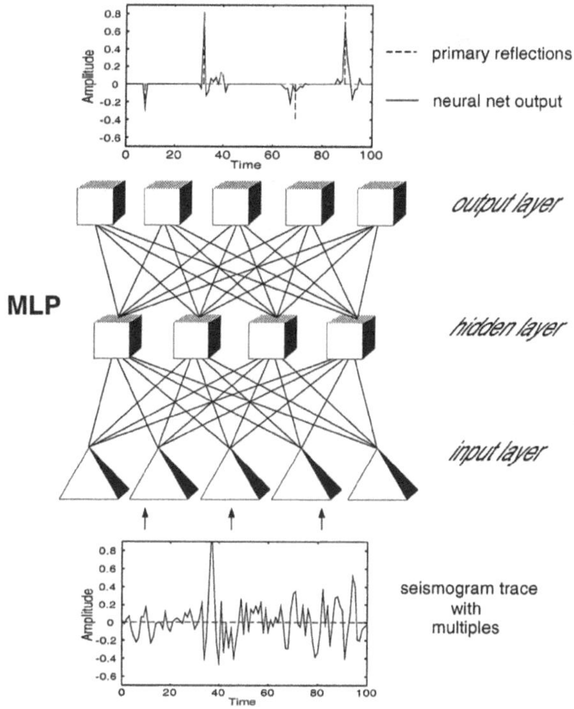

Figure 4. **Training of the neural net. Presentation of the training set consisting of 100 seismic traces containing a variety of multiple reflections and different levels of noise at the input layer, and simultaneously providing the desired output i.e. the arrival time and reflection strength of the primary reflections.**

Wiener filtering on the right. The figure shows one example out of 100 test traces with signal-to-noise ratios ranging from the noise free case (trace 2) to a S/N ratio of 0.6 (trace 5). Trace 1 is the desired output, i.e. the actual reflectivity for this example.

The performance of the neural net and of the Wiener filter for the whole test set is shown graphically in Figure 6. It shows the percentage of correctly detected reflectors. The percentage of correct neural net detection is plotted versus the percentage of correct Wiener filter detection. The 45-degree line separates the area where the neural net performance is better (white area) from the area where the Wiener filter yields higher detection rates (gray area). On the two diagrams, significantly more dots lie in the white area, demonstrating that the neural net shows an overall better performance. The left part of Figure 6 shows the performance with respect to the different amounts of noise in the data. It can be seen that with increasing amount of noise, the performance of both the neural net and the Wiener filter generally becomes worse. However, the noise factor is not as crucial for reliable reflector detection as the depth of the respective reflector. The

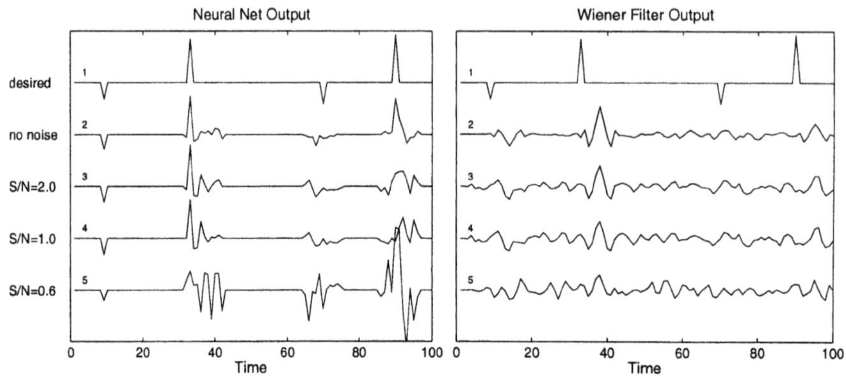

Figure 5. **Comparison between the neural net performance (on the left), and the Wiener filter (predictive deconvolution) performance (on the right), for synthetic seismogram deconvolution with various levels of noise. Only 1 out of 100 examples is shown for each signal-to-noise ratio. Trace 1 is the true reflectivity, and traces 2-5 are the neural net or Wiener filter outputs.**

right part of Figure 6 depicts the performance with respect to the four different reflectors. The water bottom (o), for example, is always detected to 100% by the neural net, while the Wiener filter only scores 100% for the noise-free case (68%, 51%, and 37 %, respectively for increasing noise).

To create the plots of Figure 6, each output trace was divided into windows centered at the true reflection event and only the neurons with the highest output in the corresponding windows were taken into account. The diagram displays how often (in %) the neural net activated the correct neuron (i.e. within ± 1 deviation from the correct value). Since the water bottom does not change depth, it is detected with a score of 100% by the neural net. This provides a criterion to assess the proper functioning of the algorithm. The Wiener filter, on the other hand, does a better job only for the shallowest deep reflector ($\times$). The reason is that this reflector has a very high reflection coefficient and thus the reflection event is not distorted much by multiples. The second deep reflector ($\square$) is the hardest to detect correctly for both the neural net as well as for the Wiener filter because the signal here directly interferes with a strong multiple. But even so around 60% of the neural net picks lie within a range of ± 1 time samples around the true value, whereas the Wiener filter performance only lies between 18% and 35%. For the third deep reflector (∇) the neural net shows detection rates similar to those for the second deep reflector. The Wiener filter also performs better than for the second reflector, possibly because there is not much interference with multiple energy.

For the noisy data sets the neural net performance, as well as the Wiener filter performance, decreases with increasing amounts of noise. It has to be noted that the score for the Wiener filter always has its peak value at a deviation of five samples from the correct value (see the traces in the right panel of Figure 5). The reason for this is the difficulty in picking the onset of a seismic wavelet, which the Wiener filter tries to restore in the

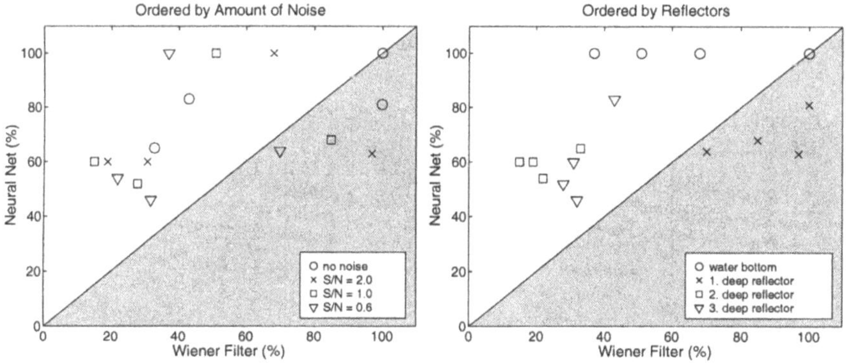

Figure 6. The neural net performance plotted against the Wiener filter (predictive deconvolution) performance in percentage of correctly detected events. Left: detection percentage ordered by noise content of the seismic data. Right: detection percentage ordered by reflector depth. The 45-degree line separates the area where the neural net performance is higher (white area) from the area where the Wiener filter yields superior detection rates (grey area).

filter process. Therefore we decided to pick the peak value of the wavelet, which is at a constant offset of five samples. The output from predictive deconvolution has to be processed further with a spiking deconvolution operator in order to achieve even sharper events.

Due to its high reflection coefficient, the first deep reflector is detected quite reliably by the Wiener filter. However, in the presence of noise it fails almost completely for reflectors at greater depth, whereas the neural net still shows a clear trend for correct detection. It is hard to give an objective measure of comparison between the results of the neural net on the one hand and the Wiener filter on the other, because in seismics the success of a processing step is often judged by visual inspection. Criteria that involve input data for success assessment are biased by the underlying assumptions of the algorithm and thus cannot really provide an independent measure (an example is the residual wavelet). Thus, if a method reveals subsurface structure or information that was previously hidden or distorted, it is rated as successful, even if the overall change in the data is very small.

Table 2 shows the correlation coefficients between the desired and actual output for the neural net and the Wiener filter. The tabulated coefficients represent average values for the entire test set of 100 traces.

5. Multiple Suppression with Neural Net Ensembles

In real seismic data processing we not only deal with zero-offset seismograms from vertically incident plane waves (as in the previous section). We normally record many shot

TABLE 2. Correlation coefficients between the desired and actual output for the neural net and Wiener filter. The tabulated coefficients represent average values for the entire test set of 100 traces.

	Neural Network	*Wiener Filter*
No Noise	0.67	0.49
$S/N = 2.0$	0.36	0.35
$S/N = 1.0$	0.28	0.28
$S/N = 0.6$	0.27	0.25

gathers (as described in section 2) containing many traces for receiver offsets from 100 m up to a few km. A typical shot gather is shown in Figure 7, right. If we shoot a whole profile we obtain such a shot gather, say every 50 m. We now show the suppression of multiple reflections in marine seismic sections for an entire profile by training an ensemble of neural networks. We demonstrate the method on a synthetic data set and train different neural nets for different offsets in the seismograms. This corresponds to performing neural net deconvolution in the common-offset domain. After deconvolving all common-offset gathers, the data are resorted into common-midpoint gathers and then stacked.

To test the performance of neural network deconvolution on synthetic data, we created a subsurface model, consisting of several reflectors. The sea bottom dips from 200 m to 300 m in depth, and the deeper interfaces have the form shown in Figure 9, left. The total depth of the model is about 4 km. This 2-D model was then assumed to be horizontally stratified locally (within 4-km intervals). This was done in order to use the reflectivity method (Fuchs and Mueller, 1971), which requires 1-D geometry for computing the reflection seismograms. In this way we calculated a set of 100 reflection seismograms (CMP gathers), containing free-surface multiples, internal multiples and

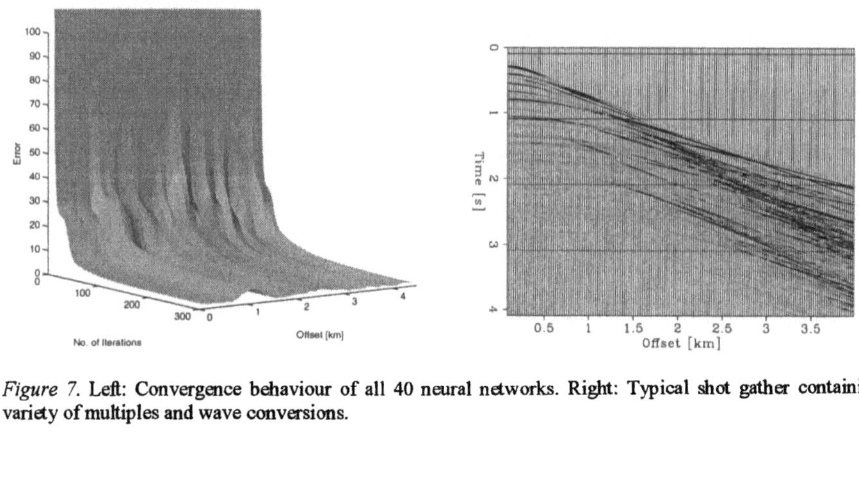

Figure 7. Left: Convergence behaviour of all 40 neural networks. Right: Typical shot gather containing a variety of multiples and wave conversions.

converted waves, which were then resorted into common-offset (CO) gathers.

The reflectivity method can switch on or off the various kinds of multiples and/or conversions. It is therefore a practical tool for testing neural network multiple suppression, since it permits us to produce input data (full wave field) and desired output data (primaries only) with the same algorithm. The section with the smallest offset containing all multiples and conversions is shown in Figure 8, left. This is the input for a given neural network. The stack of all CMP gathers containing only primary reflections, which is the desired result for the deconvolution efforts, is shown in Figure 9, left. Figure 8, right shows a stack of the data after normal move-out (NMO) correction with the known velocity model . If the section in Figure 8, left is compared with this stacked section, we observe that many multiple reflections have already been attenuated. However, it was presumed that we know the subsurface velocity model. In the very common case when we do not have a correct velocity model available, the NMO-stack yields far worse results. If we have some information about the subsurface from well-logs in the region, the neural net method produces better results when we do not know the exact 2-D velocity model for the whole profile.

For suppressing the multiple energy in the data we designed an ensemble of neural networks to perform deconvolution in the common offset domain. This seems to make more sense than processing in the CMP domain, where the neural net would have to average over all offsets. Since the structure of the information varies greatly from near to far offset, a neural net would have to adapt to a large pattern space, which leads to poorer results. This latter way of processing would be analogous to the application of a single prediction filter to all offsets in the conventional case.

After resampling the data from a 4 ms to a 16 ms time interval, and from a spatial interval of 25 m to one of 100 m we obtained 40 sections, each containing 100 traces. These data were split into a training set containing every $4th$ trace of each section and a test set containing the remaining traces. This would correspond to a real world situation where we have well-log information from boreholes every 4 km, with the aim to interpolate the subsurface structure from the seismic data between the boreholes. Although such well-spacing exists in some oil fields, in general the number of boreholes is very small, since wells are expensive to drill. Our experiments suggest that it is not necessary to have more than a few boreholes available. We trained a neural network with seismograms that were computed from artificial variations of real well-log data and achieved good results. Of course, the well-log variations should mimic the geologic situation in the investigation area, since a reflector that does not appear in the training data is unlikely to be detected.

On the other hand, the neural net is not a mere interpolator between well-log data in the traditional sense. It learns to extract the subsurface information from seismic data, and additionally makes use of the available geologic data from borehole measurements.

For the different offsets in the CMP gathers we trained different neural networks - one for each offset. The Rprop algorithm (Riedmiller and Braun, 1993, section 3) proved to be most efficient for this task. We tested different network configurations, of which a three-layer network with one hidden layer produced the best results.The stack of the neural net output is shown in Figure 9, right.

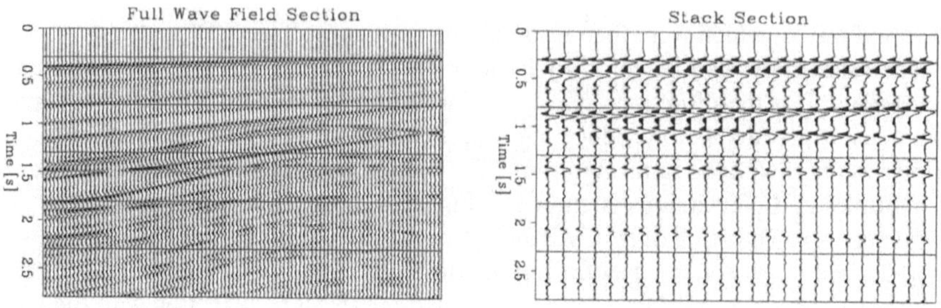

Figure 8. **Left: Input for the neural network (common-offset gather for the nearest offset, containing the full wave field). Right: NMO-stack of the input data using the known velocity model.**

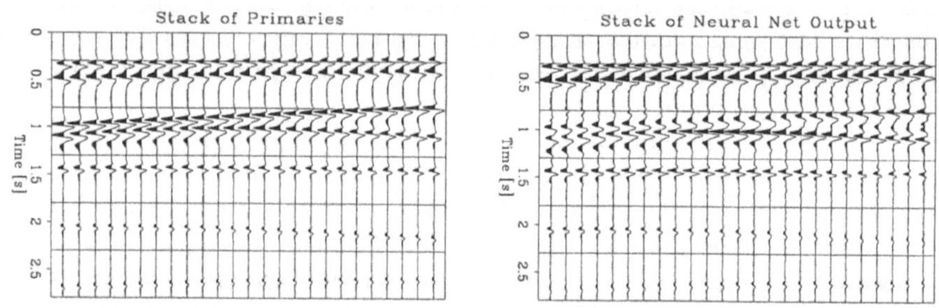

Figure 9. **Left: Stack of the CMP gathers containing primaries only (desired output). Right: Stack of the neural network deconvolved CMP gathers.**

Several deep reflectors were detected, and multiple energy was attenuated significantly. The convergence behavior of all networks for the different offsets is shown in Figure 7, left: The training error is plotted versus the number of iterations and the offset (in the CMP gathers). Convergence normally was reached already after 100 iterations, but for offsets around 1 km the error stayed at a higher level. It could be, that for this offset many reflection hyperbolae intersect, so that the information provided to the neural net is ambiguous (compare with the shot gather in Figure 7, right).

This example should demonstrate that the neural network can detect primary reflection events in data highly contaminated by multiple energy. Even the shape of completely hidden deep reflectors was revealed, given some information from well-logs. Generally, the present method cannot work without any subsurface information from sonic or density logs. The borehole data represent the constraints for inversion of the data, while the seismic data is used to infer the structure between the wells.

6. Combined Domain Multiple Suppression

As an alternative approach we now describe our Combined Domain Multiple Suppression method. All multiple attenuation techniques based on filtering have one thing in common: they remain in one parameter domain and depend on the assumption that the transformation of the data into this domain separates primary from multiple reflections (e.g. $f - k$ filtering). Our idea was to combine the different discriminatory powers in the various domains and use meaningful attributes from each data space. In other words, if a multiple cannot be separated from a primary in the velocity spectrum, it might be possible in the f-k domain or vice-versa. Feeding the neural net with selected attributes from as many different domains as possible leaves the decision with the neural net regarding which combination of attributes corresponds to a primary or a multiple.

We computed the following attributes : instantaneous amplitude (window including 4 neighboring traces) of the first trace in each seismic section, time of the window center, instantaneous amplitude of the envelope of the first trace, maximum value of the velocity spectrum (semblance) for each time sample, location of the maximum in the velocity spectrum for each time sample (i.e. the velocity corresponding to the highest peak in the spectrum), a horizontal window from the velocity spectrum for each time sample, instantaneous phase, and instantaneous frequency of the first trace. The principles of our approach are shown in Figure 10. All attributes were normalized in such a way that each attribute had zero mean and standard deviation 1.

We picked all seven reflectors from our synthetic data set (described in section 5) containing the sections with only the primaries, and used this information as the desired output for the training. Then we split the whole data set into a training and a test set, and trained a network with 58 input neurons (for the 58 attribute values), 8 to 20 hidden neurons and one output neuron. The single-node output of the net can be configured to deliver a one for the presence of a primary event and a zero for its absence. However, to obtain a feeling for network classification reliability, we allowed a continuous output instead of the hard-limited zero or one. The results for one example of a shot gather are shown in Figure 11.

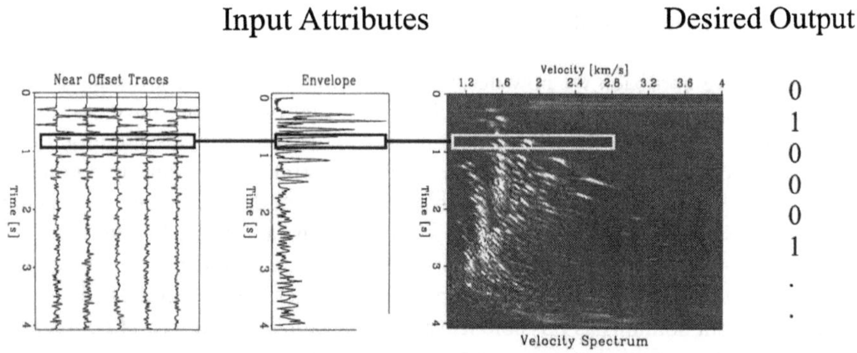

Figure 10. Combined Domain Multiple Suppression. The input consists of a number of attributes from different domains. Shown here are the instantaneous amplitude of 5 near-offset traces, the envelope of the first trace, and the velocity spectrum from which different attributes have been computed.

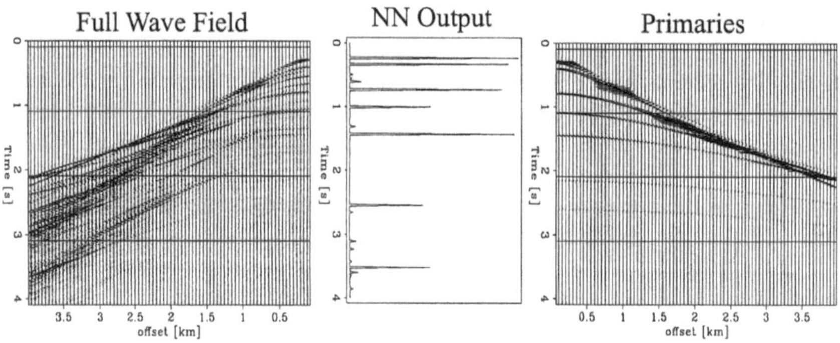

Figure 11. Results for one typical shot gather for Combined Domain Multiple Suppression. The seismogram on the left shows the raw data containing all multiples and converted waves. The seismogram on the right shows the desired output containing primaries only. In the center, the corresponding classification result of the neural network is displayed. It reliably detects the upper five reflectors and the lowest reflector, but fails on the sixth reflector. The lowermost spike is a misclassification of a multiple.

7. Conclusions

Neural networks are able to recover the main features of the desired primary reflections from prestack seismic data. The results suggest that the neural net approach to deconvolution or multiple removal is promising, especially since it can easily handle non-linear data interrelations. The quality of the result varies, since the neural net generalizes from the input for relatively few seismograms, and attempts to remove multiple energy on the basis of empirically learned rules. In the case of zero-offset data as well as for entire seismic sections, the neural net method proved that it could reveal the desired information even in the case of data heavily corrupted by noise. However, performance decreases with decreasing signal-to-noise ratio. By using not only the seismic data as input to the neural net, but also a number of selected attributes from other parameter domains, it is possible to improve discriminatory power and unveil information deeply hidden in the obscurity of the raw seismograms.

References

Backus, M. M., 1959, Water reverberations - their nature and elimination: Geophysics, **XXIV**, 233–261.

Berkhout, A. J., and Verschuur, D. J., 1997, Estimation of multiple scattering by iterative inversion, part I: theoretical considerations: Geophysics, **62**, 1586–1595.

Berryhill, J. R., and Kim, Y. C., 1986, Deep-water peg legs and multiples: emulation and suppression: Geophysics, **51**, 2177–2184.

Calderòn-Macias, C., Sen, M. K., and Stoffa, P. L., 1997, Hopfield neural networks, and mean field annealing for seismic deconvolution and multiple attenuation: Geophysics, **62**, 992–1002.

Calderòn-Macias, C., 1997, Artificial neural systems for interpretation and inversion of seismic data: Ph.D. thesis, University of Texas at Austin.

Carrion, P. M., 1986, A layer-stripping technique for suppression of water-bottom multiple reflections: Geophysical Prospecting, **34**, 330–342.

Fuchs, K., and Mueller, G., 1971, Computation of synthetic seismograms with the reflectivity method and comparison with observations: Geophys. J. Roy. Astr. Soc., **23**.

Helbig, K., and Brouwer, J. H., 1993, Surface multiples as a tool of calibration in shallow avo studies: 55th Mtg. Eur. Assoc. Expl Geophys., Abstracts, Session:A040.

Kneib, G., and Bardan, V., 1997, 3d targeted multiple attenuation: Geophysical Prospecting, **45**, 701–714.

Peacock, K. L., and Treitel, S., 1969, Predictive deconvolution: theory and practice: Geophysics, **34**, 155–169.

Riedmiller, M., and Braun, H., 1993, A direct adaptive method for faster backpropagation learning. The rprop algorithm: IEEE International Conference on Neural Networks, San Francisco, CA, March 28-April 1, 1993, IEEE.

Robinson, E. A., and Treitel, S., 1980, Geophysical signal analysis: Prentice-Hall, Inc.

Schneider, W. A., Prince, E. R. J., and Giles, B. F., 1965, A new data-processing technique for multiple attenuation exploiting differential normal moveout: Geophysics, **30**, 348–362.

Taner, M. T., 1980, Long period sea-floor multiples and their suppression: Geophysical Prospecting, **28**, 30–48.

Verschuur, D. J., and Berkhout, A. J., 1997, Estimation of multiple scattering by iterative inversion, part II: practical aspects and examples: Geophysics, **62**, 1596–1611.

Weglein, A. B., Carvalho, P. M., and Stolt, R. H., 1992, Nonlinear inverse scattering for multiple suppression: application to real data. part I: 62nd Ann. Internat. Mtg., Soc. Expl. Geophys., Expanded Abstracts, 1093–1095.

Wiggins, J. W., 1988, Attenuation of complex water-bottom multiples by wave-equation-based prediction and subtraction: Geophysics, **53**, 1527–1539.

Yilmaz, O., 1987, Seismic data processing: Society of Exploration Geophysicists, Tulsa, OK.

APPLICATION OF ARTIFICIAL NEURAL NETWORKS TO SEISMIC WAVEFORM INVERSION

QIAODENG HE and HUI ZHOU

Department of Geophysics,
Changchun University of Science and Technology,
Changchun 130026,
PEOPLES REPUBLIC OF CHINA
E-Mail: heliu@public.cc.jl.cn

Abstract

A three-layer feedforward neural network is described, which has been applied to geophysical parameter inversion. The number of elements in the input layer is equal to the number of recorded samples. During training, the global minimum of the energy function is determined using a decimal-encoded genetic algorithm. It is necessary to use weights and thresholds of neurons as a gene group. To determine the weights and thresholds corresponding to the global minimum of the energy function, a large search range is used initially, which is then progressively reduced in order to accelerate convergence. The network was trained using both 5 and 10 numerically modeled records of the vertical component. After convergence, the network was tested using a randomly generated three-layer transverse isotropic model. The inversion results are very encouraging.

1. Introduction

In recent years, the study of artificial neural networks (ANNs) has progressed significantly, and applications may now be found in many fields including pattern recognition, signal processing, decision making, optimization, and knowledge engineering, to mention but a few (Ji Fan et al, 1991). Significant advances have also been made in theoretical studies of ANN paradigms, training etc (Haykin, 1999).

The application of ANNs in applied geophysics began in the late 1980's, when they were mainly employed in pattern recognition. Huang et al. (1989) made use of the Hopfield network model to the recognition of 'bright spots'. Liu et al. (1989) applied an ANN to automate seismic trace event tracking. McCormack (1990) and

W. Sandham and M. Leggett (eds.),
Geophysical Applications of Artificial Neural Networks and Fuzzy Logic, 89-101.

Veezhinathan and Wagner (1990), employed ANNs for editing seismic traces. Murat and Rudman (1992) and Veezhinathan and Wagner (1990) studied the automated picking of seismic first arrivals using an ANN. Raiche (1991) applied ANN pattern recognition to geophysical inversion, whilst Cheng and Zhu (1995) investigated oil and gas prediction using a Kohonen ANN.

During the process of ANN pattern recognition, the network inputs are the samples to be recognized, for example the seismic first arrivals, and the network output represents the judgment, which is usually 'yes' or 'no', usually represented by '1' and '0'. Applying this type of recognition to geophysical inversion is very limited, because geophysical parameters e.g. the layer velocity, are continuous variables. In the feasibility study of Calderon-Macias and Sen (1993), the interpretation parameter is the output of the neural network, which is a continuous value. Michaels (1994) modeled the one-dimensional inelastic wave equation using circuits and pointed out that it is possible to obtain velocity parameter inversion with these circuits. Roth and Tarantola (1994) studied seismic waveform inversion using an ANN. In this case, the thickness of each layered model was known and assumed to be the same, and the unknown was the velocity to be determined.

Calderon-Macias and Sen (1993) and Röth and Tarantola (1994) investigated the inversion of geophysical parameters using an ANN, which employed a gradient-descent method for the network training. During training, the squared sum difference between the expected output and the actual output values was defined as the objective function, i.e. the network energy function. Unfortunately, the energy function exhibits multiple minima. In order to search for the global minimum of the energy function, network-training methods based on simulated annealing and genetic algorithms, were employed.

In this chapter, an ANN, trained using a genetic algorithm, is applied to seismic waveform inversion. The primary issues which have been addressed, include adapting the genetic algorithm for network training, determining the range of connecting weights to employ, investigating network convergence acceleration methods, and evaluating characteristics of the network when applied to waveform inversion.

Due to the large number of network input nodes, weight coefficients and threshold values required for waveform inversion, the conventional binary-coded genetic algorithm is not suitable for PCs unless considerable memory is available. A decimal-coded genetic algorithm was therefore investigated which reduces the memory requirements and enables realization of the genetic algorithm using conventional PCs.

Seismic waveform inversion is a very complicated nonlinear problem. During the training and association process, the ANN trained with the genetic algorithm will not linearize the objective function, associated media model and the wave function of the seismic data. The method which was finally implemented, is a nonlinear inversion process.

2. Artificial Neural Network Model

There are various topological structures, or paradigms, for an ANN. A three-layer feedforward network was chosen for the work reported here (Figure 1), since it can approximate any continuous function with arbitrary accuracy (Ji Fan et al, 1991).

Let the network input be x_i (i=1,2,...,N). Then the input at each hidden layer node may be expressed as follows,

$$A_j^h = \sum_{i=1}^{N} w_{ij}^h x_i - \theta_j^h \tag{1}$$

where superscript h denotes the hidden layer, A_j is the input of the jth neuron, w_{ij} is the connecting weight between the ith input unit and the jth hidden unit, θ_j is the threshold value, which can be considered as a connecting weight corresponding to an input value of -1, and i_j is the output of the jth unit.

The output at the hidden layer is,

$$i_j = f^h(A_j^h) \tag{2}$$

where

$$f(A) = \left(1 + e^{-A}\right)^{-1} \tag{3}$$

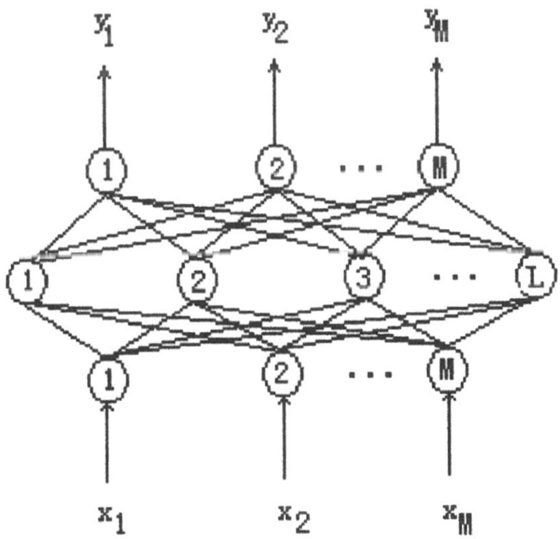

Figure 1. The three-layer feedforward network.

Similarly, the output node can be written as,

$$A_k^o = \sum_{i=1}^{L} w_{kj}^o i_j - \theta_k^o \tag{4}$$

and

$$y_k = f_k^o(A_k^o) \tag{5}$$

in which superscript *o* denotes the output layer.

The error is defined as,

$$E = \frac{1}{2}\sum_{k=1}^{M}(d_k - y_k)^2 \tag{6}$$

in which d_k is the expectation value of the kth unit. Equation (5) is the energy function or the objective function of the inversion problem.

3. Decimal-Encoded Genetic Algorithm

Usually a genetic algorithm represents unknown parameters using binary-encoding. Such binary-encoded genetic algorithms are difficult to realize in computers with inadequate memory, especially PCs, when the unknowns are too many.

A decimal-encoded method is proposed here to overcome the memory problems for PCs when using binary-encoding. The decimal-encoding method makes use of random values between 0 and 1 to represent multiple parameters of the model. This method saves considerable computer memory and is suited to multiple parameter optimization problems.

Let p_i be the variable to be solved ($p_i \in [a_i, b_i]$), and q_i the corresponding parameter ($q_i \in [0,1]$). The relationship between p_i and q_i may be expressed as follows,

$$q_i = \frac{p_i - a_i}{b_i - a_i} \tag{7}$$

When the resolution (or sample rate) d_i of p_i is known, the resolution of q_i is dq_i,

$$dq_i = \frac{d_i}{b_i - a_i} \tag{8}$$

q_i may be transformed to q_{ic} using decimal-encoding, comparable to the resolution of p_i, expressed as follows,

$$q_{ic} = \left[\frac{q}{dq_i}\right] dq_i \tag{9}$$

in which [•] denotes an integer operation.

Let the decoding rule for transforming the decimal-encoded q_{ic} to an actual model parameter be,

$$p_i = q_{ic}(b_i - a_i) + a_i \tag{10}$$

Observing eqs.(7)–(10), it can be seen that each model parameter is represented by a single value between 0 and 1, which is a simple encode and decode method. Because each model parameter requires only one decimal value, the decimal-encoding method can save considerable computer memory when compared to the conventional binary-encoding method.

Expressing model parameters using the decimal-encoded genetic algorithm, is a similar process to the binary-encoded algorithm. However, to determine the initial gene group, some random numbers are generated and converted to decimal codes with a specific resolution given by eq.(9). The selection operation of the genetic algorithm is the same as that for binary-encoding. The crossover operation is carried out through the probability p_c. The codes of two chromosomes to be crossed, are exchanged immediately after the cross-position is randomly determined, and two new codes are generated randomly to replace the previous codes in the cross-position. The parameters in the cross-position are varied after the cross-operation, which is similar to that for the binary-encoded genetic algorithm. The mutation process mutates a position randomly, with probability p_m. The mutated code is generated randomly. It differs slightly from the variation process in the binary-encoded genetic algorithm, but is essentially the same.

4. Genetic Algorithm Training Strategy

Due to the existence of the nonlinear hidden layer (i.e. the middle layer) in the multi-layered neural network (Figure 1), many local minima exist in the error function or energy function. Thus, the gradient training method cannot guarantee finding a global minimum. Stochastic training methods were therefore adapted.

Stochastic training methods adjust variables of the network, associated with the stochastic process, probability and energy. They optimize the objective function of the network, which means that the actual output closely approaches the expected output. The stochastic training methods include simulated annealing (Rothman, 1985) and genetic algorithms (Sambridge et al., 1991). Variables (connecting weights and threshold values) change randomly during the stochastic training process, and the energy function is determined after these random variations occur.

The genetic algorithm is used during the training process. Inputs and outputs, together with the topological structure of the network, are known. The unknowns include the connecting weights and threshold values of the network, and these are used as the gene group. The initial status for the evolution is the stochastic association of genes. Moreover, the gene groups may be formed randomly. The selection, crossover and mutation operations act on the gene groups. The connecting weights and neuron threshold values are iteratively obtained.

The genetic algorithm coding can only be done when the upper and lower variable limits are known. However, the connecting weights and thresholds are not known beforehand, and even their ranges of values are very difficult to predict. To determine the weights and thresholds corresponding to the global minimum of the energy function, a large search range is carried out, which is then progressively reduced to accelerate convergence.

5. Examples of Seismic Waveform Inversion

Assume there are I seismic traces, each of length J. Hence, the number of neurons required for the network input layer $N=I.J$, which is equal to the total number of input samples. The number of hidden nodes in hidden layer L is arbitrary, but it cannot be too small or large. The number of nodes in the output layer is unknown and to be determined, however it is assumed to be a value M (the number of parameters).

From eq.(3), it can be seen that the range of values of $f(A)$ is [0,1]. However, in seismic waveform inversion, the inverting parameters are the elastic parameters and the thickness of the layered media, and the variation of these is not between [0,1]. Therefore, the values of the input and the desired output must be normalized.

Normalization of seismic traces is carried out by normalizing all the seismic traces for training and testing to [-1,1], using the same normalizing factor. Due to the dimension of each desired output being different, they should be normalized individually. For example, normalization of the elastic parameters is carried out simply by dividing by the maximum value. Similarly, normalization of the thickness parameters is carried out by dividing by the maximum thickness value.

After normalization, the known model and its seismic data are applied at the input and output of the neural network for supervised training.

5.1. FIVE-SAMPLE TRAINING

Using the elastic parameters and the thickness data for a three-layered medium, ten models were produced using the random method, in which the interval of the elastic parameter (c_{33}/ρ) was $0.5\text{x}10^6\ \text{m}^2/\text{s}^2$ and the interval thickness was 10m (Table 1). The vertical components of the 1-D zero-offset seismograms for these ten models were computed using the pseudo-spectral method, based on the Hartley transform (Zhou and He, 1995).

TABLE 1. Ten theoretical models (*) and the associated network results ($c_{33}/\rho \times 10^6\ m^2/s^2$).

Layer		1		2		3	
		c_{33}/ρ	h(m)	c_{33}/ρ	h(m)	c_{33}/ρ	h(m)
Samples							
1	*	9.5	200.0	6.0	200.0	11.0	
		9.7	200.0	6.2	186.0	10.9	
2	*	9.0	200.0	6.5	190.0	11.0	
		9.3	200.2	6.4	183.8	11.0	
3	*	11.5	190.0	6.0	170.0	10.0	
		11.3	186.3	6.0	171.6	10.2	
4	*	10.5	190.0	6.0	160.0	11.5	
		10.5	186.7	5.8	162.8	10.8	
5	*	11.0	200.0	6.0	190.0	10.0	
		10.7	200.8	6.1	191.3	10.6	
6	*	12.0	210.0	6.0	200.0	11.0	
		10.7	201.3	6.3	191.4	10.5	
7	*	11.0	210.0	6.5	160.0	10.0	
		7.4	147.0	1.5	159.3	11.6	
8	*	12.0	200.0	6.5	160.0	10.0	
		11.1	195.6	6.6	180.7	10.3	
9	*	11.5	210.0	6.5	170.0	11.0	
		9.0	163.9	1.9	174.7	11.5	
10	*	10.5	190.0	6.0	190.0	10.5	
		10.7	201.6	6.4	191.4	10.5	

After individual normalization, the first five seismograms and their corresponding model parameters were used as the inputs and desired outputs, respectively, of the neural network for training. Initially, the upper and lower limits for the connecting weights and threshold values were 15 and -15, respectively. The number of iterations used during training was 3500. Results are shown in Table 1, and the error curve, in Figure 2.

RMS values of the error between the model parameters produced by the neural network and the theoretical model parameters, are shown in Figure 3. In order to study the regular pattern of convergence for the neural network, more iterations were used.

Figure 2 shows the error curve for the object function. It displays a high rate of decrease up to 300 iterations, but a very low rate between 300 and 2900 iterations. Beyond 3000 iterations there is a moderate rate of decrease.

Figure 3 shows the error between the desired output of the testing samples and the actual output, after training is terminated (samples 6-10). Errors also exist among the 5 training samples (samples 1-5). The errors of these samples are different with the error of the 3rd sample being the least. There are quite large errors among the 5 testing samples and the errors of the 6th, 8th and 10th are relatively small, with those of the 7th and 9th being quite large. In Table 1, the associative results of the 5 training samples are very close to the synthetic values, but there are still some differences.

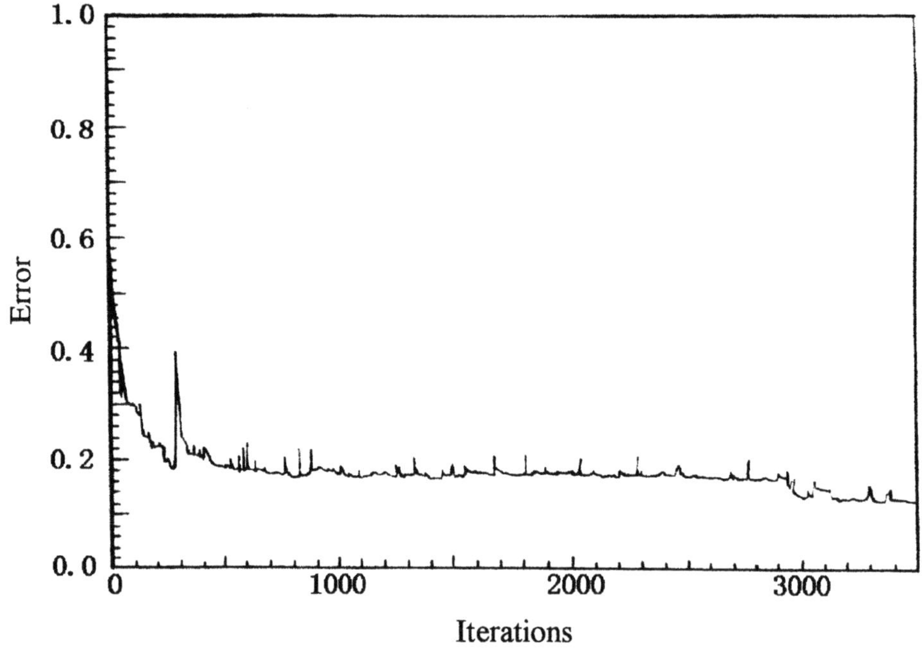

Figure 2. Error curve for the object function.

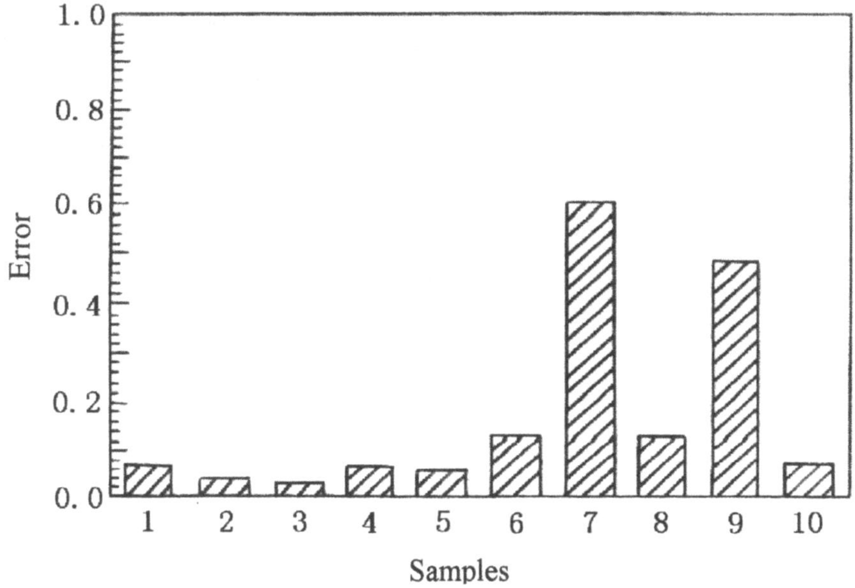

Figure 3. Errors between the association network and the theoretical values. Samples 1-5 are used for training, and samples 6-10 for testing.

In order to improve the association accuracy of the network, re-training was required. It can be seen from Figure 2 that the error function of the network decreases very slowly, using the present training methodology. However, due to the large range of values and high resolution, some connecting weights and threshold values are still not optimized. The upper and lower threshold values were therefore gradually decreased during further network training.

During re-training, the search range of the connecting weights and the threshold values were defined as,

$$w_{ij} \in [w_{ijo} - w_L, w_{ijo} + w_H] \tag{11}$$

where w_{ijo} is the optimum connecting weight or threshold value obtained at the end of the previous training (the connecting weight and threshold value corresponding to the minimum value of the object function in the group); w_L and w_H are positive. Equation (11) indicates that the new search is carried out using w_{ijo} as the center of the range, and $w_L + w_H$ as the width of the range.

Using this approach, the errors (as evidenced in Figure 4) have decreased significantly, and are now considerably smaller than those in Figure 2. Errors reduced quickly during early iteration times, and slowly during later iteration times.

The optimum connecting weights and threshold values of the network obtained by re-training, were used to associate the network and the theoretical models of the associating samples marked "*" in Table 1. The results are shown in Figure 5.

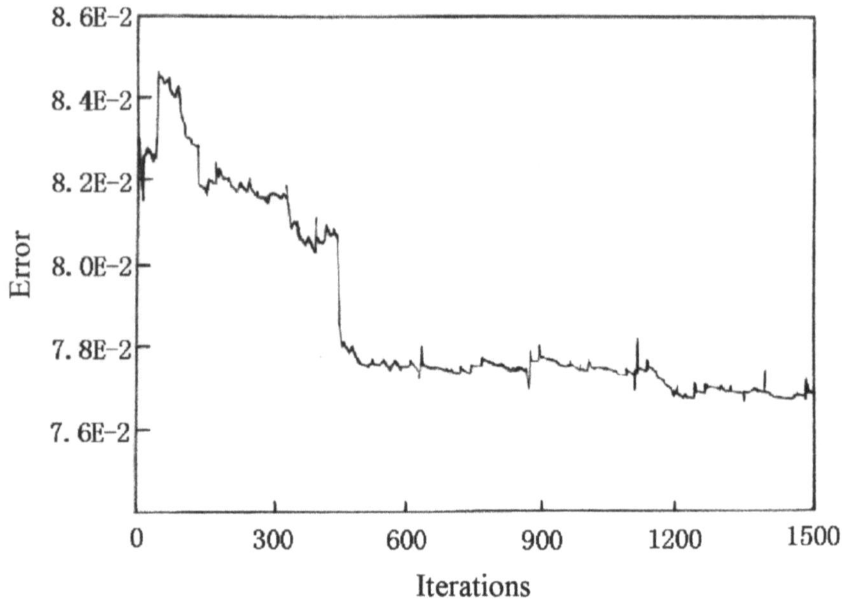

Figure 4. Error curve for the object function after re-training.

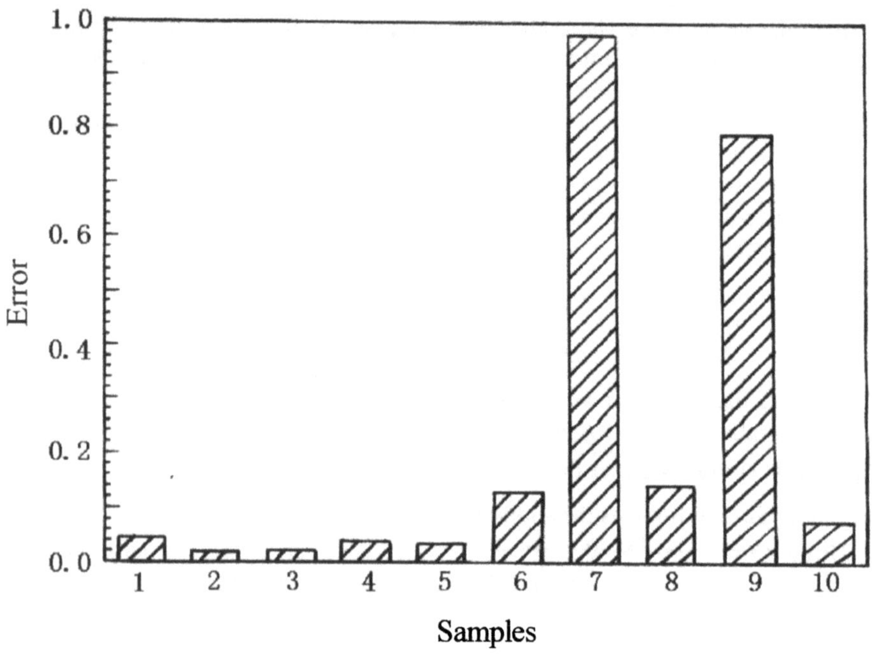

Figure 5. Errors between the re-trained association network and the theoretical values. Samples 1-5 are used for training, and samples 6-10 for testing.

Comparing Figures 3 and 5, it can be seen that the errors of the samples used for training, have decreased significantly. Figure 5 also shows that the associating results of the five training samples are very close to the theoretical values.

For instance, the associating results of the 2nd and the 3rd models are identical to the theoretical values. The errors of the five testing samples had the tendency to increase with the decrease of the training sample errors. Thus, the errors between the associating result of testing samples and the theoretical values must also be increasing, especially for the 7th and the 9th samples. Even though the difference between the associating results of the 6th, 8th and 10th samples and the theoretical results are not unacceptable. If the associating results of the 6th, 8th and 10th samples can be considered successful, then the associating success rate of untrained samples is 60%.

Highly trained neural networks have a very high recognition capability. As for the untrained samples, the network makes a much more restrictive judgment as to whether they belong to the same class of training samples, during the associating process. Therefore, the fact that the training sample errors reduce when errors for the untrained samples increase, is strictly justified.

Artificial neural networks may be regarded as a simplified simulation of the biological neural system. Their main feature is that they are capable of recognizing patterns or samples which are assimilated through learning, and they have a highly accurate associating rate for such patterns.

In the case of untrained samples on the other hand, the network makes a judgment as to which group the samples belong, by experience. However, whether this is correct or not, depends on the level of training and the number of training samples. In order to improve the accuracy of the network, the number of training samples was increased from 5 to 10, as described in the following section.

5.2. TEN-SAMPLE TRAINING

In order to determine the relation between the number of training samples and the success rate of association, the forward records and model parameters (normalized) of 10 theoretical models were used to train the network, and the normalized records of 5 models were used to test the association of the network. After an appropriate number of training iterations, the optimum connecting weights and threshold values of the network were obtained, which were then used for association. The differences between the association results and real situation, are shown in Figure 6.

Referring to Figure 6, it can be seen that the errors for most of the training samples (1-10) are less than 0.10, with only the errors of the 1st and 8th samples being greater than 0.10. For the testing samples, the maximum error is approximately 0.23. This value is only 2.5 times the average error (0.09) of the ten training samples. Using only 5 training samples (see Figure 5), the average error is less than 0.02 and the maximum error of the testing samples is 0.99 (the ratio of both is about 50). Therefore, the accuracy of the associating results for the neural network using 10 training samples, is relatively higher than the network using 5 training samples. Hence, increasing the number of training samples definitely improves the associating capability of the

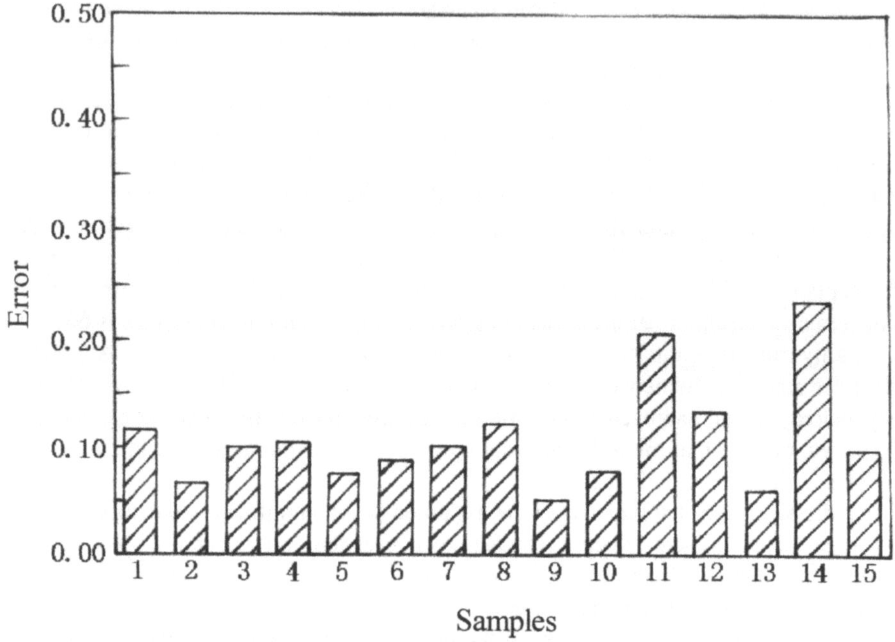

Figure 6. Errors between the re-trained association network and the theoretical values, for 10-sample training. Samples 1-10 are used for training, and samples 11-15 for testing.

network, as expected.

6. Conclusions

Artificial neural networks have been applied to the problem of geophysical inversion, with particular consideration given to convergence and speed of convergence. A three-layer feedforward neural network has been described, with the number of elements in the input layer equal to the number of recorded samples.

During training, the global minimum of the energy function was determined using a decimal-encoded genetic algorithm. It was necessary to use weights and thresholds of neurons as a gene group. To determine the weights and thresholds corresponding to the global minimum of the energy function, a large search range was used initially, which was then progressively reduced in order to accelerate convergence.

The method was tested using a randomly generated three-layer transverse isotropic model. The network was trained using numerically modeled records of the vertical component. After convergence, the network was subjected to an association process. The inversion results are very encouraging, and the method provides a basis for further study with potential applications in other geophysical inversion domains.

References

Calderon-Macias, C., and Sen., M. K., 1993, Geophysical interpretation by artificial neural systems: a feasibility study: 63rd Ann. Internat. Mtg., Soc. Expl. Geophys., Expanded Abstracts, 254~257.

Cheng, Z., and Zhu, G., 1995, Application of Kohonen network to the lateral prediction of oil and gas: Geophys. Prosp. Ptr., 34, 53~56.

Haykin, S., 1999, Neural networks – a comprehensive foundation: Prentice-Hall, Inc.

Huang, K. Y., Liu, W. H., and Chang, I. C., 1989, Hopfield model of neural networks for detection of bright spots: 59th Ann. Internat. Mtg., Soc. Expl. Geophys., Expanded Abstracts, 444~446.

Ji Fan, Fan Junbo, and Tan Yongdong, 1991, Neural network and application of neural computer: Westsouth Transportation University Publishing House.

Liu, X., Xue, P., and Li, P., 1989, Neural network method for tracing seismic events: 59th Ann. Internat. Mtg., Soc. Expl. Geophys., Expanded Abstracts, 716~718.

Michaels, P., 1994, Recurrent neural network representation of inelastic wave equation and full-waveform inversion without local minima: 64th Ann. Internat. Mtg., Soc. Expl. Geophys., Expanded Abstracts, 22~25.

Murat, M. E., and Rudman, A. J., 1992, Automated first arrival picking: a neural network approach: Geophys. Prosp., 40, 587~604.

McCormack, M. D., 1990, Seismic trace editing and first break picking using neural networks: 60th Ann. Internat. Mtg., Soc. Expl. Geophys., Expanded Abstracts, 321~324.

Raiche, A., 1991, A pattern recognition approach to geophysical inversion using neural nets: Geophys. J. Internat., 105, 629~648.

Röth, G., and Tarantola, A., 1994, Inversion of seismic waveforms using neural networks: 64th Ann. Internat. Mtg., Soc. Expl. Geophys., Expanded Abstracts, 788~791.

Rothman, D. H., 1985, Nonlinear inversion, statistical mechanics and residual statics estimation: Geophysics, 50, 2784~2796.

Sambridge, M. S., Tarantola, A., and Kennett, B. L. N., 1991, An alternative strategy for non-linear inversion of seismic waveforms: Geophys. Prosp., 39, 723~736.

Veezhinathan, J., and Wagner, D., 1990, A neural network approach to first-break picking: Proc. Internat. Joint Conf. Neural Networks, San Diego CA, 1, 235~240.

Zhou, H., and He, Q., 1995, Seismic wave modeling in heterogeneous medium by Hartley transform: Oil Geophys. Prosp., 30, 593~601.

References

SEISMIC PRINCIPAL COMPONENTS ANALYSIS USING NEURAL NETWORKS

KOU-YUAN HUANG

Department of Computer and Information Science,
National Chiao Tung University,
Hsinchu,
TAIWAN
E-Mail: kyhuang@cis.nctu.edu.tw

Abstract

A neural network is described which uses an unsupervised generalized Hebbian algorithm (GHA) to find the principal eigenvectors of the covariance matrix for different types of seismogram. Principal components analysis (PCA) using the GHA network enables the extraction of information regarding seismic reflections and uniform neighboring traces. The seismic data analyzed are seismic traces with 20, 25, and 30 Hz Ricker wavelets. The GHA network is also applied to analyze (a) fault, reflection and diffraction patterns after NMO correction, (b) bright spot patterns, and (c) a real seismogram from the Mississippi Canyon. The properties of high amplitude, low frequency, and polarity reversal are observed from projections on the principal eigenvectors. The GHA network also provides significant seismic data compression.

1. Introduction

Principal components analysis (PCA), also known as the Karhunen-Loeve transformation, has been investigated and used in many applications (Hotelling, 1933; Chien and Fu, 1967; Fukunaga and Koontz, 1969; Pelat, 1974; Jain, 1976, 1977; Hagen, 1982; Jones, 1985). Given a set of random data **X** with dimension N and $\mathbf{M}=E[\mathbf{X}]=\mathbf{0}$, the correlation matrix (covariance matrix) $\mathbf{Q}=E[(\mathbf{X}-\mathbf{M})(\mathbf{X}-\mathbf{M})^T]=E[\mathbf{X}\mathbf{X}^T]$ can be computed. The eigenvalues and the corresponding eigenvectors can be found from **Q**. The principal eigenvectors point in the principal directions of the distribution of data.

In the 1980s, Hagen (1982) and Jones (1985) applied PCA to a seismic data set. Hagen (1982) considered the input data vectors in the vertical trace direction, and

W. Sandham and M. Leggett (eds.),
Geophysical Applications of Artificial Neural Networks and Fuzzy Logic, 103-122.

computed the principal components to evaluate the subtle character changes of porosity in neighboring uniform seismic traces. Jones (1985) considered the input data vectors in the horizontal direction, and likewise computed the principal components to separate diffraction and reflection patterns from seismic data after normal moveout (NMO) correction.

Several neural network algorithms have been proposed for PCA (Oja and Karhunen, 1980; Oja, 1982, 1985; Sanger, 1989a, b; Baldi and Hornik, 1989; Krogh and Hertz, 1990; Kung, 1993; Bannour and Azimi-Sadjadi, 1995; Peper and Noda, 1996). Oja's learning rule is capable of finding one principal eigenvector of a covariance matrix (Oja and Karhunen, 1980; Oja, 1982, 1985). However, Sanger's unsupervised generalized Hebbian learning algorithm (GHA) is capable of finding many principal eigenvectors of a covariance matrix in decreasing eigenvalue order (Sanger, 1989a, b).

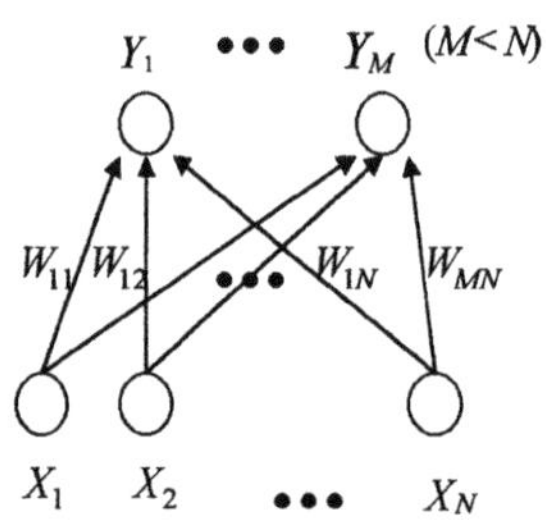

$$Y_i = \sum_{j=1}^{N} W_{ij} X_j, \ i=1, \dots, M.$$

Figure 1. Sanger's neural network.

Here, the neural network with Sanger's unsupervised GHA (Sanger, 1989a) is adopted to find the principal eigenvectors of a covariance matrix using different varieties of seismograms. The neural network is shown in Figure 1. The input data are fed into the neural network iteratively to determine the principal eigenvectors. The advantage of learning is the ability to find the principal eigenvectors incrementally from data as they become available. It is not necessary to compute the covariance matrix from the complete data set in advance, so the eigenvectors can bypass this be derived from the data directly. While applying the traditional and laborious method of eigenvector analysis to seismic data with 512 input dimensions, the size of the covariance matrix (512x512) may exceed some computers' memory limitation. These eigenvectors cannot be readily solved using the traditional method. However, using neural networks with GHA can decrease the computational requirements and storage for a small number of output eigenvectors. For example, using the notation in Section 2, if there are 5 output eigenvectors, then $\mathbf{YX}^T$ will have only 5x512=2,560 elements, and $\mathbf{YY}^T$ will have only 25 elements. The required memory is significantly less. Sanger notes, "when the number of inputs is large and the number of required outputs is small, GHA provides a practical and useful procedure for finding eigenvectors"

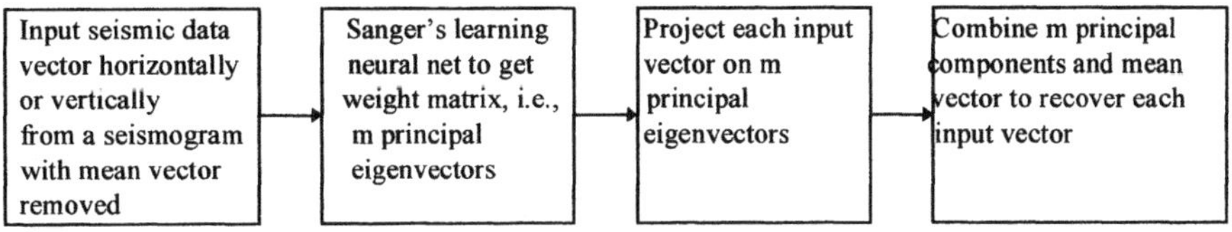

Figure 2. Schematic of Sanger's learning network for seismic PCA.

(Sanger, 1989a). Seismic data tend to involve large data sets with significantly large dimensions. Thus, information regarding a small number of reflections and uniform traces can considerably improve the seismic interpretation. With these advantages, PCA using GHA should be a better method for calculating and analyzing seismic data.

Based on Sanger's GHA learning rule for the seismic PCA, Figure 2 shows the processing steps using neural networks with the input data vectors in either horizontal or vertical directions. The data analyzed are seismic traces with 20, 25, and 30 Hz Ricker wavelets. Also, the PCA using the neural network is applied to analyze faults, reflection and diffraction patterns after NMO correction, bright spot patterns, and a real seismogram of the Mississippi Canyon.

2. Neural Network for Generalized Hebbian Learning

The neural network with Sanger's GHA possesses a single layer (see Figure 1). As stated, the input N-dimensional column vector is $\mathbf{X}$, $E[\mathbf{X}]=\mathbf{0}$, $\mathbf{W}$ is a MxN weight matrix, and the output M-dimensional column vector is $\mathbf{Y}=\mathbf{WX}$ with $M<N$, i.e. $Y_i = \sum_{j=1}^{N} W_{ij} X_j$, i=1, ... , M. The input data are fed into the network iteratively to determine the principal eigenvectors of the correlation (covariance) matrix $\mathbf{Q}=E[(\mathbf{X}-\mathbf{M})(\mathbf{X}-\mathbf{M})^{\mathrm{T}}]=E[\mathbf{XX}^{\mathrm{T}}]$. The weight matrix $\mathbf{W}$ must be updated for each iteration step.

Sanger (1989a) proved that the GHA neural network with a probability of one, converges from random initial weights to finding the principal eigenvectors of the input covariance matrix in decreasing eigenvalue order. The GHA learning rule is given by,

$$W_{ij}(t+1) = W_{ij}(t) + \gamma(t) Y_i \left(X_j - \sum_{k=1}^{i} Y_k W_{kj} \right)$$

where j is the input index and i is the output index, or in matrix form as,

$$\Delta \mathbf{W}(t) = \gamma(t)(\mathbf{YX}^{\mathrm{T}} - \mathrm{LT}[\mathbf{YY}^{\mathrm{T}}]\mathbf{W}(t))$$

where $\mathbf{X}=[X_1, X_2, \ldots, X_N]^{\mathrm{T}}$, $\mathbf{Y}=[Y_1, Y_2, \ldots, Y_M]^{\mathrm{T}}$, and LT[.] denotes a lower triangular matrix, i.e. all elements above the diagonal of its matrix argument are zeros. $\lim_{t\to\infty} \gamma(t) = 0$ and $\sum_{t=0}^{t=\infty} \gamma(t) = \infty$. The i-th eigenvector $\mathbf{e}_i$ is W_{ij}, j=1, ... , N, and the corresponding eigenvalues satisfy $\lambda_1 > \lambda_2 > \ldots > \lambda_i > \ldots > \lambda_M$.

The "projection values" of the input data vector **X** onto unit eigenvectors $\mathbf{e}_i$ are $a_i = \mathbf{X}^T \mathbf{e}_i = \mathbf{e}_i^T \mathbf{X}$, i=1, ... , M. $a_i \mathbf{e}_i$ is defined as the "projection vector" or "principal component" of input data **X** onto $\mathbf{e}_i$. Small numbers of a_i and $a_i \mathbf{e}_i$ are used in the interpretations of seismic principal components.

For the seismic experiments reported here, the definition of convergence is that if all the absolute values of W_{ij} $(t+1)$- W_{ij} (t) are less than a constant (0.0001), then the iterations terminate. For the examples in Sanger (1989a), he chose $\gamma(t)$ empirically at a fixed value between 0.1 and 0.01, which leads to exceptional convergence. Here, Sanger's empirical values on $\gamma(t)$ are adopted whereby $\gamma(t)$ is held fixed at a value between 0.1 and 0.01. From our seismic experiments, the number of iterations increase as the value of $\gamma(t)$ decreases. Also if $\gamma(t)$ decreases linearly from 0.1 to 0.01 in other experiments, then it remains a small fixed value after several more iterations. The formula design is:

$\gamma(t)$ = -0.1*(# of iteration - 1000)/1000, when the number of iterations is less than or equal to 900,

$\gamma(t)$ = 0.01, when the number of iterations is greater than 900.

In order to have a criterion for comparison, the mean-squared error (MSE) and the normalized mean-squared error (NMSE) (Sanger, 1989a) are calculated. The NMSE is the ratio of the mean-squared error (MSE) to the data variance. For a seismic intensity I at position m, n with the mean removed, and a recovered intensity $\hat{I}$,

$$\text{NMSE}=E[(I_{m,n}-\hat{I}_{m,n})^2]/E[I^2_{m,n}]$$

3. Data Compression

For data compression, the p largest eigenvalues and their corresponding p eigenvectors are selected, and the smallest $(M\text{-}p)$ eigenvalues are discarded from the data representation. The data representation for **X** becomes $\mathbf{X}'=\sum_{i=1}^{p} a_i \mathbf{e}_i$. The mean-squared error between random vector **X** and **X'** is the summation of the discarded M-p eigenvalues (Chien and Fu, 1967). This dimensionality reduction technique leads to the following: (1) considerable memory is saved in the data representation, and (2) notable, effective features are selected for pattern recognition.

4. PCA in Seismic Ricker Wavelet Analyses

Initially, the basic seismic PCA experiments involved different layers with different Ricker wavelets. The PCA applications have been extended here to simulate seismograms and real seismic data. Table 1 lists the results for each seismic experiment for Figures 3 to 16.

TABLE 1. Experimental results of Figures 3 to 16 for different learning rates $\gamma(t)$. Nomenclature: dimension of input vector (#1), number of input vectors (#2), number of steps to convergence (#3), mean-squared error (MSE), and normalized mean-squared error (NMSE).

Convergence steps #3, MSE, NMSE	**Learning rate $\gamma(t)$ as a decreasing function**	**Learning rate $\gamma(t)$ as a constant (0.045)**
Figure 3. #1=2. #2=128. σ^2=.00182582	#3= 469 MSE=.00000136 NMSE=.00074489	#3= 697 MSE=.00000170 NMSE=.00093188
Figure 4. #1=2. #2=128. σ^2=.00509162	#3= 237 MSE=.00000011 NMSE=.00002089	#3= 432 MSE=.00000021 NMSE=.00004053
Figure 5. #1=3. #2=128. σ^2=.001168	#3=811 MSE=.000004662 NMSE=.003989943	#3=1573 MSE=.000001545 NMSE=.001322128
Figure 6. #1=128. #2=3. σ^2=.001168	#3= 654 MSE=.00001847 NMSE=.01580673	#3=1104 MSE=.00001356 NMSE=.01160910
Figure 7. #1=3. #2=128. σ^2=.000961	#3= 864 MSE=.000006476 NMSE=.006740044	#3=3216 MSE=.000000281 NMSE=.000292096
Figure 8. #1=128. #2=3. σ^2=.000961	#3= 815 MSE=.00002695 NMSE=.02804647	#3= 1710 MSE=.00000679 NMSE=.00706588
Figure 9. #1=32. #2=128. σ^2=.004551	#3= 253 MSE=.002908 NMSE=.639098	#3= 445 MSE=.002908 NMSE=.639040
Figure 10. #1=128. #2=32. σ^2=.004551	#3= 556 MSE=.002675 NMSE=.587913	#3= 854 MSE=.002675 NMSE=.587908
Figure 11. #1=64. #2=512. σ^2=.0044	#3= 432 MSE=.0035 NMSE=.8017	#3= 768 MSE=.0035 NMSE=.8013
Figure 12. #1=512. #2=64. σ^2=.0044	#3= 255 MSE=.0034 NMSE=.7773	#3= 394 MSE=.0034 NMSE=.7768
Figure 13. #1=64. #2=512. σ^2=.004438	#3= 439 MSE=.003502 NMSE=.789089	#3= 432 MSE=.003501 NMSE=.788915
Figure 14. #1=512. #2=64. σ^2=.004438	#3= 237 MSE=.003511 NMSE=.791054	#3= 322 MSE=.003504 NMSE=.789480
Figure 15. #1=64. #2=512. σ^2=.000651	#3= 70 MSE=.000212 NMSE=.325803	#3= 127 MSE=.000212 NMSE=.325218
Figure 16. #1=512. #2=64. σ^2=.000651	#3= 80 MSE=.000166 NMSE=.255123	#3= 137 MSE=.000165 NMSE=.253560

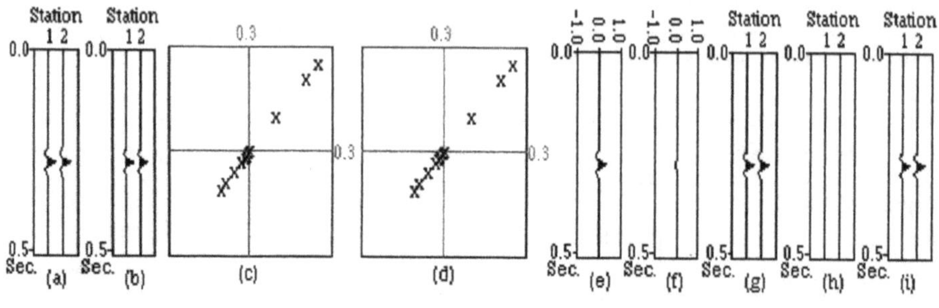

Figure 3. **(a) Seismic data (20 Hz Ricker wavelet). (b) Seismic data (with the horizontal mean vector removed). (c) Distribution from (a). (d) Distribution from (b). (e) and (f) 1st and 2nd projection values, respectively. (g) and (h) 1st and 2nd components, respectively. (i) Sum of the 2 components and the mean vector.**

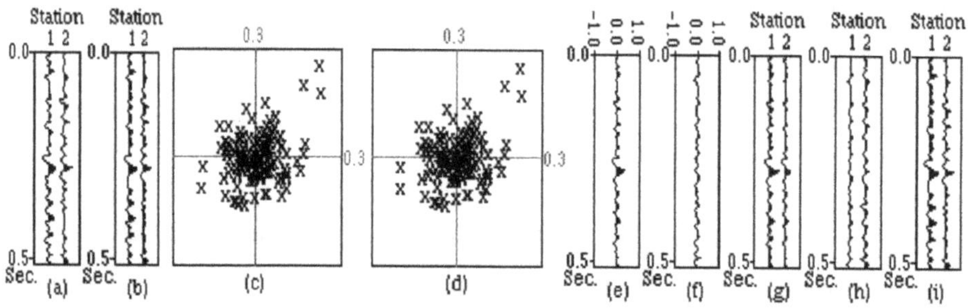

Figure 4. **(a) Seismic data (20 Hz Ricker wavelet). (b) Seismic Data (with the horizontal mean vector removed). (c) Distribution from (a). (d) Distribution from (b). (e) and (f) 1st and 2nd projection values, respectively. (g) and (h) 1st and 2nd components, respectively. (i) Sum of the 2 components and the mean vector.**

4.1. 2-D PCA OF RICKER WAVELETS IN ONE SEISMIC LAYER USING INPUT IN THE HORIZONTAL DIRECTION.

Figure 3(a) shows one layer containing 20 Hz, zero-phase Ricker wavelets with a reflection coefficient of -0.25 in two seismic traces. The sampling rate is 0.004 sec. Using horizontal input data vectors, the data vector type of 20 Hz Ricker wavelets at the layer is $[c, c]^T$, and is in one line with high correlation in Figure 3(c), the 2-D scatter diagram from Figure 3(a). Figure 3(b) consists of the traces with the horizontal mean vector removed. Figure 3(d) is the 2-D scatter diagram of the data from Figure 3(b). The neural network can find the first eigenvector corresponding to the layer's Ricker wavelets. Figure 3(e) is the projection value on the first eigenvector for each data vector. Figure 3(f) is the projection value on the second eigenvector for each data vector. Figure 3(g) is the first component. Figure 3(h) is the second component. Figure 3(i) is the recovered portion from the first and the second components plus the mean vector, $\mathbf{X'} = \sum_{i=1}^{2} a_i \mathbf{e}_i + \mathbf{M}$. The sum of several components plus the mean vector in the later analyses of Figures 4 - 17 verifies this description.

Figure 4(a) shows the addition of 10-56 Hz Gaussian white band noise (mean=0, standard deviation=0.1) to Figure 3(a). Referring to Figure 4, it can be seen that the neural network still finds the first eigenvector, even though noise is present.

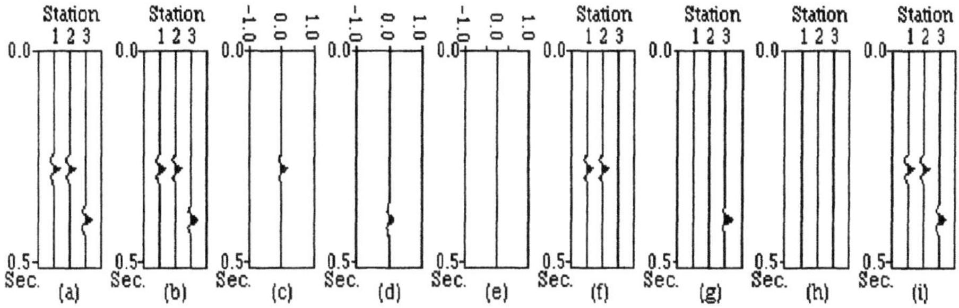

Figure 5. (a) Seismic data (20 Hz Ricker wavelet). (b) Data (with the horizontal mean vector removed). (c), (d), and (e) 1st, 2nd, 3rd projection values, respectively. (f), (g), and (h) 1st, 2nd, 3rd components, respectively. (i) Sum of the three components and the mean vector.

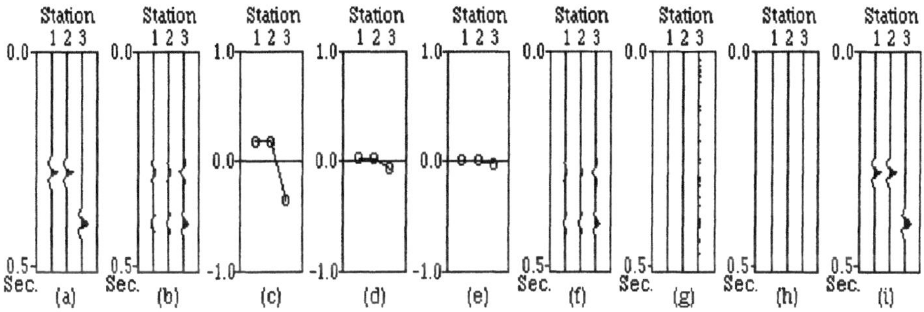

Figure 6. (a) Seismic data (20 Hz Ricker wavelet). (b) Data (with the vertical mean vector removed). (c), (d), and (e) 1st, 2nd, 3rd projection values, respectively. (f), (g), and (h) 1st, 2nd, 3rd components, respectively. (i) Sum of the three components and the mean vector.

4.2. PCA FOR TWO SEISMIC LAYERS

4.2.1. *Input in the Horizontal Direction*

In Figure 5(a), the 20 Hz zero-phase Ricker wavelets are on 3 traces. One horizontal layer is on traces 1 and 2 and the other layer is on trace 3. Input data are selected in the horizontal direction; so the data are in 3-D. The input data vector type at the first layer is $[c, c, 0]^T$, the input data vector type at the second layer is $[0, 0, c]^T$. Figures 5(c), (d), and (e) show the three projection values. The projection components for the first and second eigenvectors enable the first and second layers to be recovered, as shown in Figures 5(f) and (g), respectively. From the distances between the origin $[0, 0, 0]^T$ and these 3-D data vectors, the distribution of the data in the first layer is longer than that of the second layer; so the largest component corresponds to the first layer. Elongated data distributions can thus generate the first eigenvector.

4.2.2. *Input in the Vertical Direction*

Figure 6(a) shows the same signal as Figure 5(a). Input data vectors are selected in the vertical direction. The dimension is 128. One trace reflects one sample. The sample number 3 is too small, and the mean vector is not **0**. Figure 6(b) consists of traces with the mean vector removed. After removal of the mean vector, the wavelet morphologies are altered. Using PCA and adding the mean vector, the wavelets of the layers may be recovered, as shown in Figure 6(i).

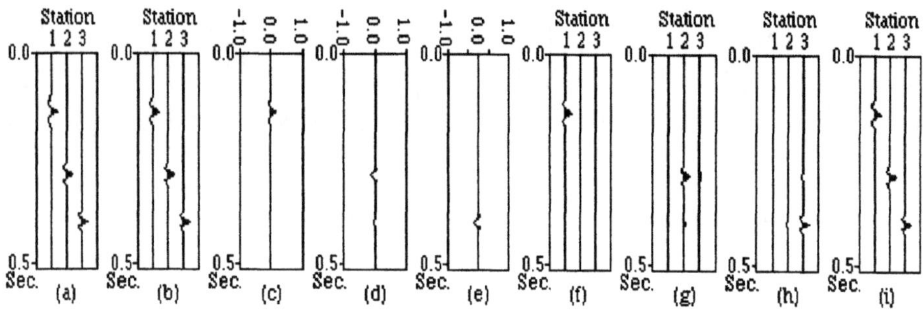

Figure 7. (a) Seismic data (20, 25, 30 Hz Ricker wavelets). (b) Data (with the horizontal mean vector removed). (c), (d), and (e) 1st, 2nd, 3rd projection values, respectively. (f), (g), and (h) 1st, 2nd, 3rd components, respectively. (i) Sum of the three components and the mean vector.

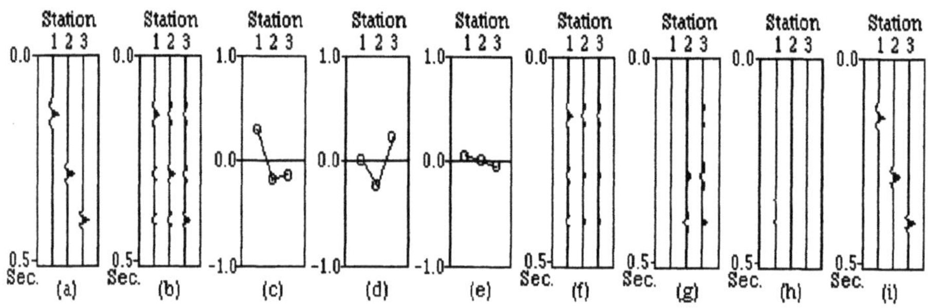

Figure 8. (a) Seismic data (20, 25, 30 Hz Ricker wavelets). (b) Data (with the vertical mean vector removed). (c), (d), and (e) 1st, 2nd, 3rd projection values, respectively. (f), (g), and (h) 1st, 2nd, 3rd components, respectively. (i) Sum of the three components and the mean vector.

4.3. ANALYSIS OF RICKER WAVELETS WITH THREE SEISMIC LAYERS

4.3.1. *Input in the Horizontal Direction*

In Figure 7(a), the 20 Hz, 25 Hz, and 30 Hz zero-phase Ricker wavelets are on the first, second, and third traces, respectively. They have the same maximum amplitude (0.25). The difference between them is the duration, i.e. number of points at the three Ricker wavelets. Input data are selected in the horizontal direction. The input data vector types are $[c, 0, 0]^T$ at the first layer, $[0, d, 0]^T$ at the second layer, and [0, 0,

$e]^T$ at the third layer. The first layer has the largest number of points in 3-D space corresponding to the 20 Hz Ricker wavelet. Figures 7(c), (d), and (e) show the three projection values. The largest component corresponds to the first layer with the 20 Hz Ricker wavelet. Figures 7(f), (g) and (h) indicate that the projection components on the first, second and third eigenvectors, recover the first, second and third layers, with the 20, 25 and 30 Hz Ricker wavelets, respectively. Hence, if the layer with the Ricker wavelet in the same amplitude range can contribute more points, such as the 20 Hz Ricker wavelet, then the data direction is significant, and the eigenvector can dominate.

4.3.2. *Input in the Vertical Direction*
Figure 8(a) is the same signal as shown in Figure 7(a). Input data are selected in the vertical direction. The mean vector is not **0**. Figure 8(b) produces traces with the mean vector removed. After removal, the wavelets are again modified. In Figure 8(i), using PCA and adding the mean vector, the wavelets can be recovered.

5. PCA for Simulated and Real Seismograms

In this section, the neural network based PCA is applied to the analysis of simulated and real seismograms. In order to compare the performance, a traditional numerical method of PCA is carried out for the simulated bright spot seismogram.

5.1. PCA FOR FAULTS

5.1.1. *Input in the Horizontal Direction*
The seismogram in Figure 9(a) depicts a fault. The left-hand side incorporates 24 uniform traces, and the right hand side has 8 uniform traces. The seismic trace comprises 20 Hz zero-phase Ricker wavelets with a reflection coefficient of -0.2, 4 ms sampling interval, and 10-56 Hz Gaussian white band noisc (mcan=0, standard deviation=0.1). Using the input data vectors in the horizontal direction, the data type of the layer at the left side is $[c, c, c, ..., 0, 0, ..., 0]^{W_{ij}}$ and the data type of the layer at right side is $[0, 0, ..., c, c, ..., c]^T$. Thc two sides of the fault correspond to two principal eigenvectors. Figure 9(b) is the data with the mean vector removed. Figure 9(c) displays the projection values of each horizontal data vector on 3 eigenvectors. Figures 9(d) and (e) reveal the projection component of each horizontal data vector on the first and second eigenvectors, corresponding to the first and second layers, respectively. Figure 9(f) shows the summation of three principal components and the mean vector. The two sides of the fault correspond to two principal eigenvectors. The 24 uniform wavelets on the left-hand side reflect the largest projection value on the first eigenvector; and the 8 uniform wavelets on the right hand side correspond to the largest (negative) projection value on the second eigenvector. The distances between data vectors $[c, c, c, ..., c, 0, ..., 0]^T$ and the origin are greater than the distances

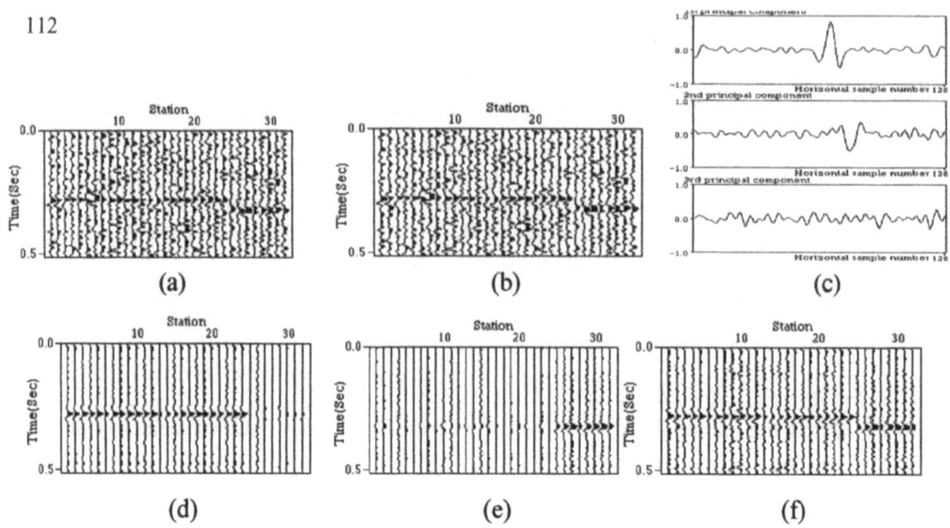

Figure 9. (a) A fault. (b) Data (with the horizontal mean vector removed). (c) 3 projection values. (d) 1st component. (e) 2nd component. (f) Sum of the three components and the mean vector.

between data vectors $[0, 0, ..., 0, c, ..., c]^T$ and the origin for the horizontal data vector. Elongated data distributions generate the first eigenvector, which is the same reason for the results in the previous experiment.

5.1.2. *Input in the Vertical Direction*

Input data were also selected in the vertical direction. One trace reflects one sample. The mean vector is again not **0**. Figure 10(a) exhibits traces with the mean vector removed. After removal, the wavelets on the first layer are once again modified. From Figure 10(e), using PCA and adding the mean vector, the wavelets are recovered.

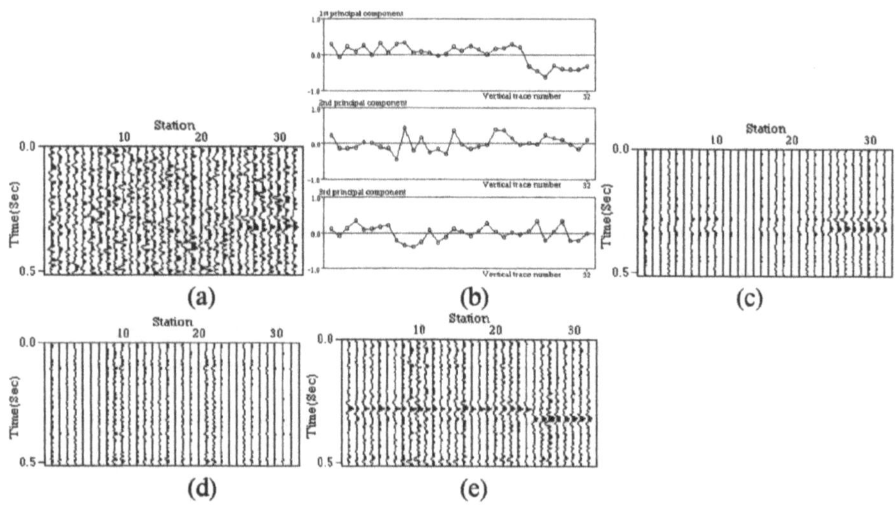

Figure 10. (a) Data (with the vertical mean vector removed). (b) Three projection values. (c) 1st component. (d) 2nd component. (e) Sum of the three components and the mean vector.

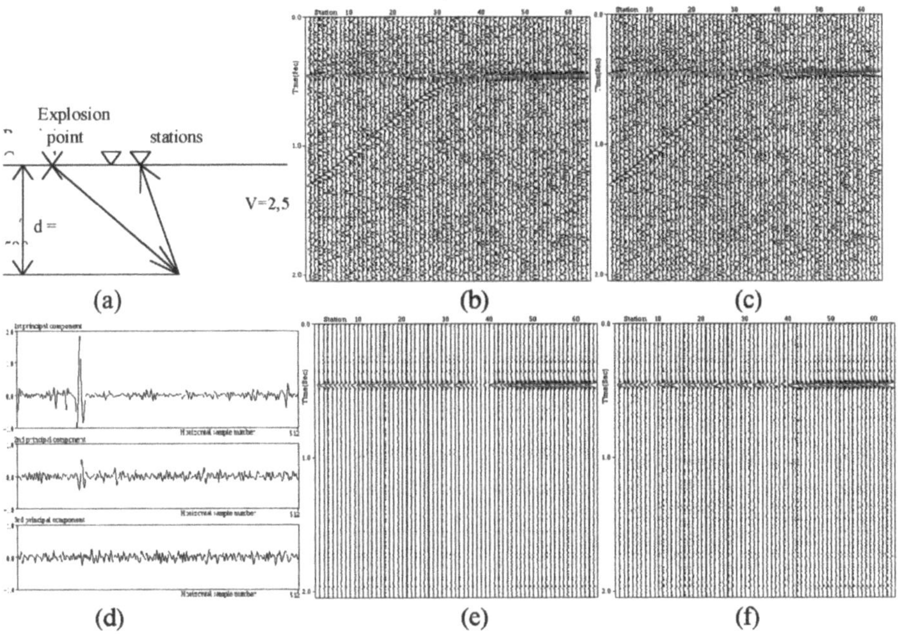

Figure 11. (a) Geological model. (b) Reflection and diffraction patterns. (c) Data (with the horizontal mean vector removed). (d) Three projection values. (e) 1st component. (f) Sum of the three components and the mean vector.

5.2. PCA FOR DIFFRACTION PATTERNS

The geological model for this case is shown in Figure 11(a). The depth of the layer is 500 m, the seismic P-wave velocity is 2,500 m/s, and the receiving station interval is 50 m. The synthetic seismogram after normal moveout correction (NMO) consists of the reflection and diffraction patterns in Figure 11(b). The source signal is a 20 Hz zero-phase Ricker wavelet, and the reflection coefficient of 0.2. Gaussian white band 10-56 Hz noise is maintained in the seismogram.

5.2.1. *Input in the Horizontal Direction*

Using the input data vectors in the horizontal direction, the data type is $[c, c, ..., c, 2c, 2c, ..., 2c]^T$ for the whole uniform layer. Figure 11(e) reveals that the diffraction seismic pattern is filtered from the horizontal reflection layer, in the projection component of each horizontal input data vector on the first eigenvector.

5.2.2. *Input in the Vertical Direction*

Input data are selected in the vertical direction. One trace consists of one sample. The mean vector is as usual not $\mathbf{0}$. Figure 12(a) exhibits the traces with the mean vector removed. The Ricker wavelet morphologies on the right hand side of the layer are kept the same as before, but the amplitudes are reduced.

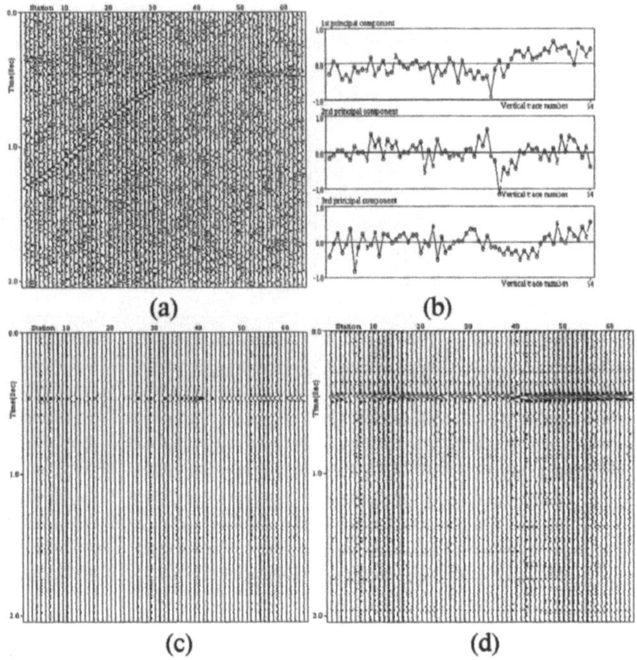

Figure 12. (a) Data (with the vertical mean vector removed). (b) Three projection values. (c) 1st component. (d) Sum of the three components and the mean vector.

The projection value on the first eigenvector in Figure 12(b) shows the uniform property on the right hand side of the traces. In Figure 12(c) the diffraction pattern is filtered using one principal component.

5.3. PCA FOR BRIGHT SPOTS

5.3.1. *Input in the Horizontal Direction*

Figure 13(a) displays a structure of bright spots that indicate the gas and oil sand zones with a large (negative) reflection coefficient of -0.29 at the top of the gas sand zone (Dobrin, 1976). The seismogram has 64 traces with 512 samples per trace. Using the input data vectors in the horizontal direction to compute the 5 principal eigenvectors of the covariance matrix, Figure 13(c) reveals that the projection value of each horizontal vector on the first eigenvector shows the high amplitude content and polarity reversal of the wavelets. Figure 13(d) depicts the projection of each horizontal data vector on the first eigenvector. The central part of the first layer at 0.3 s and the central flat part of the bright spot structure, a total of 4 layers, appear in the projection component on the first eigenvector. However, the whole first layer at 0.3 s does not correspond to the first eigenvector. The reason is that the number of data vectors in the central flat three layers of the bright spot structure is three times more than that in the one layer at 0.3 s.

Figure 13(i) shows the recovered seismogram using five principal components. Since there are non-flat layers in the bright spot pattern, it is necessary to choose more principal components for the recovered seismogram.

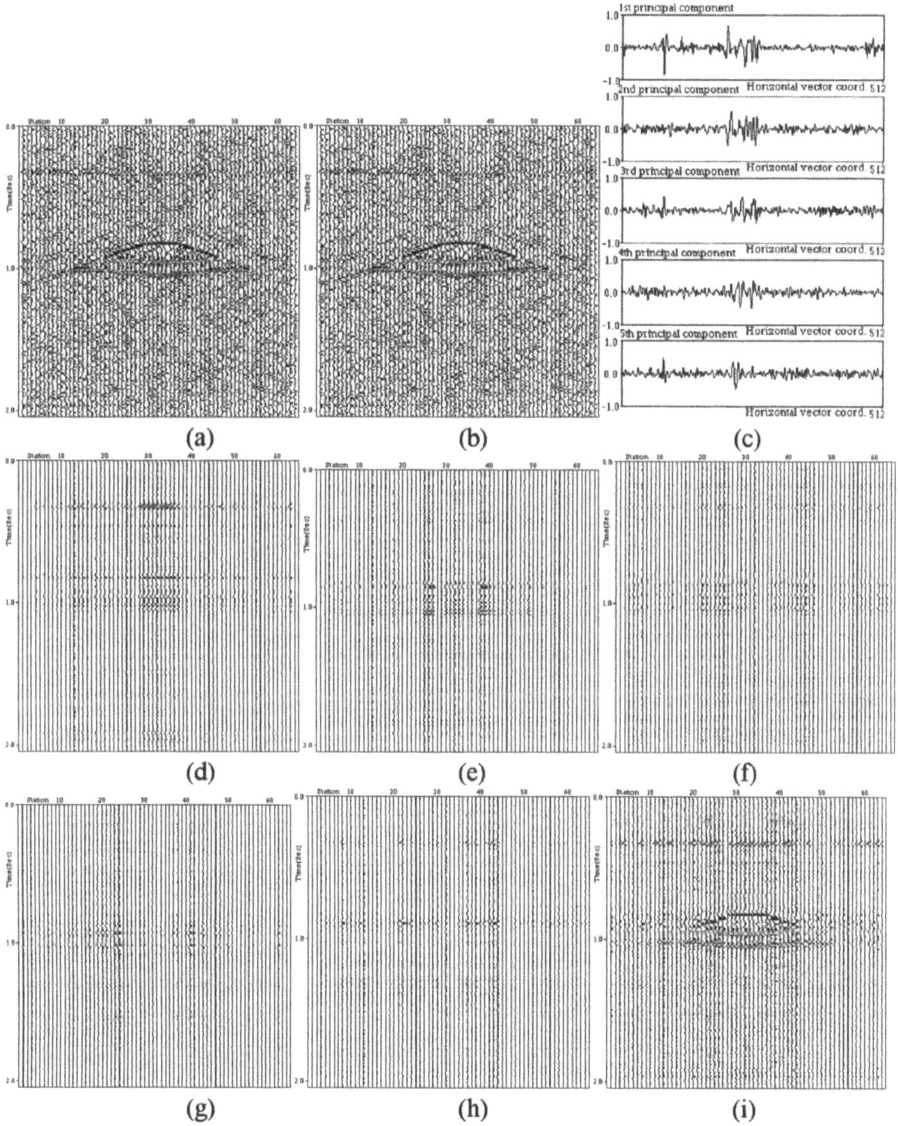

Figure 13. (a) Bright spot pattern. (b) Data (with the horizontal mean vector removed). (c) Five projection values. (d) 1st component. (e) 2nd component. (f) 3rd component. (g) 4th component. (h) 5th component. (i) Sum of the five components and the mean vector.

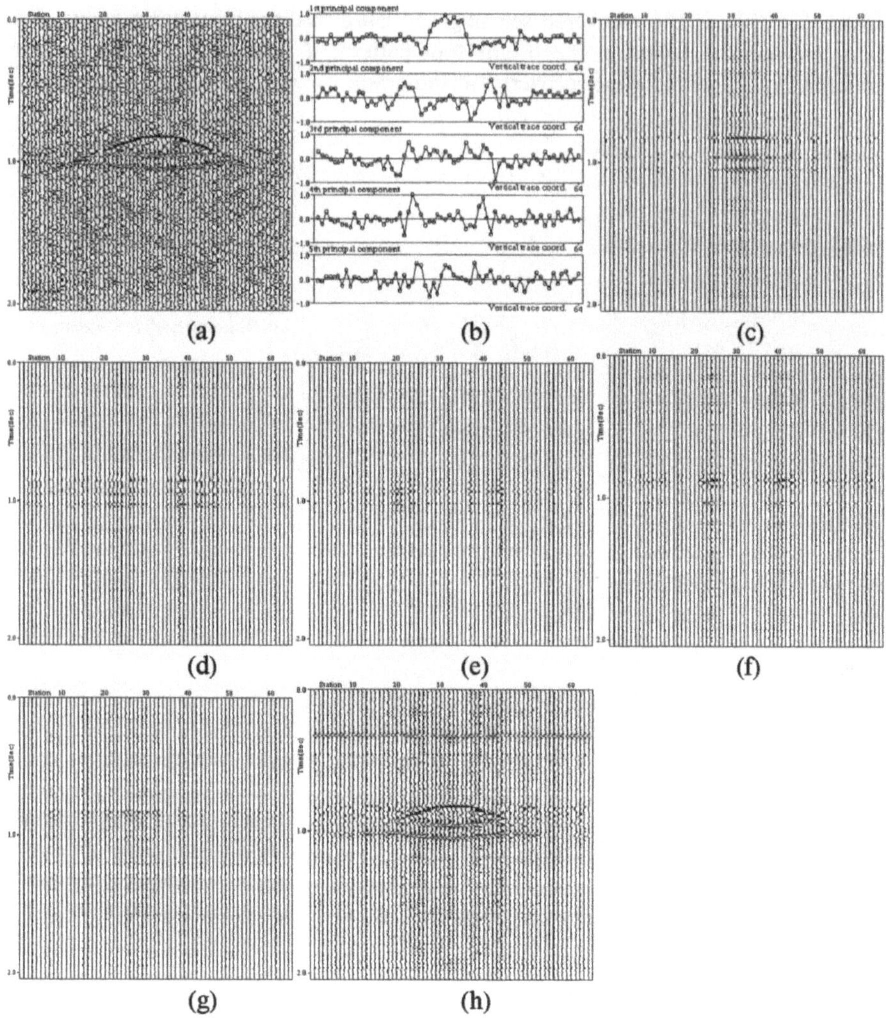

Figure 14. (a) Data (with the vertical mean vector removed). (b) Five projection values. (c) 1st component. (d) 2nd component. (e) 3rd component. (f) 4th component. (g) 5th component. (h) Sum of the five components and the mean vector.

5.3.2. *Input in the Vertical Direction*

Input data vectors are selected in the vertical direction. Each trace is again one sample. In the first principal projection values of Figure 14(b), the traces in the central part show a uniform property, and the neighboring traces at the two sides reveal the polarity reversal.

5.4. PCA FOR A REAL SEISMOGRAM FROM THE MISSISSIPPI CANYON

5.4.1. *Input in the Horizontal Direction*

A real seismogram from the Mississippi Canyon is shown in Figure 15(a). Figure 15(c) displays the 5 projection values. The central, left and right parts, strongly reflect the projections of the first, second and third eigenvectors, in Figures 15(d), (e) and (f), respectively. The structure can be decomposed into three significant principal components. Figure 15(g) shows the recovered data using three principal components.

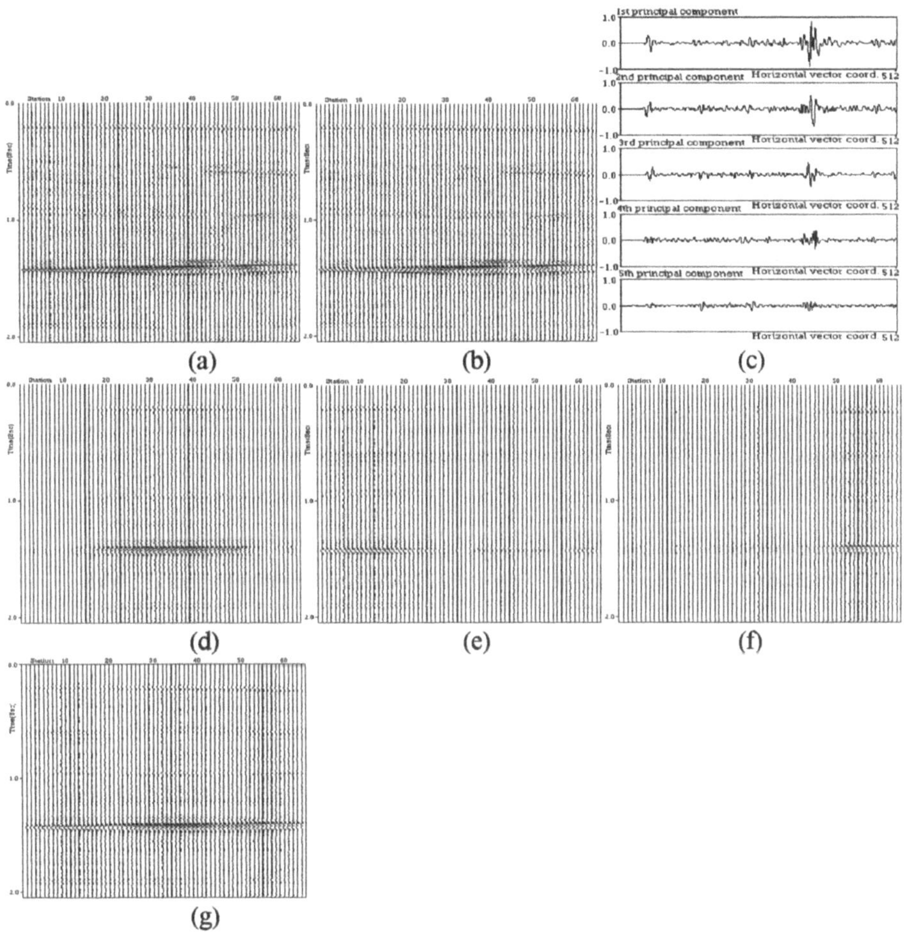

Figure 15. (a) Seismogram from the Mississippi Canyon. (b) Data (with the horizontal mean vector removed). (c) Five projection values. (d) 1st component. (e) 2nd component. (f) 3rd component. (g) Sum of the three components and the mean vector.

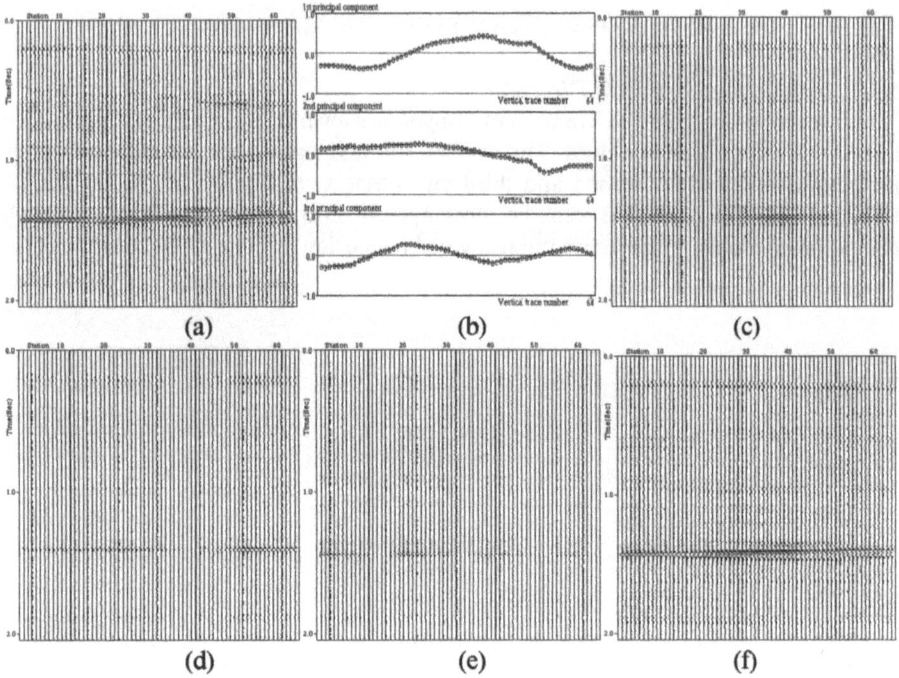

Figure 16. (a) Data (with the vertical mean vector removed). (b) 3 projection values. (c) 1st component. (d) 2nd component. (e) 3rd component. (f) Sum of the 3 components and the mean vector.

5.4.2. *Input in the Vertical Direction*

The input data vectors are selected in the vertical direction in Figure 16(a). Figure 16(b) presents the three projection values, in which the first projection value shows the high amplitude and polarity reversal at the central part and both sides. The seismic traces at the central part of the seismogram show a uniform property. Figure 16(c) exhibits the projection component of each trace on the first eigenvector. Figure 16(f) shows the recovered data using three principal components.

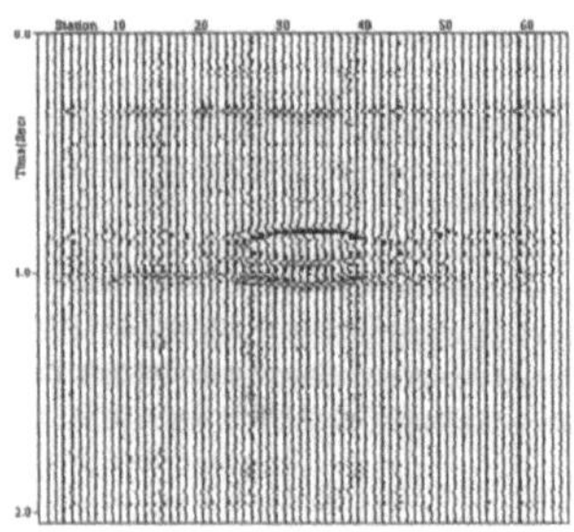

Figure 17. Input vector in the horizontal direction. Sum of the five components and the mean vector by the traditional numerical method.

5.5. TRADITIONAL PCA FOR BRIGHT SPOT SEISMOGRAMS

5.5.1. *Input in the Horizontal Direction*

Traditional PCA was applied to the bright spots in Figure 13(a). The input data vectors were selected in the horizontal direction, and the dimension was 64. The first step involves the determination of the 64x64 covariance matrix from the complete data set. The second step is the acquisition of the five principal eigenvectors using the traditional Power method of numerical analysis, which can also detect the several dominant eigenvectors (Carnahan, et al., 1969). The third step is to find the five principal components and the recovered seismogram. Figure 17 exhibits the recovered data using five principal components. In order to avoid the problem of large dimensions in DOS-based computer memory, the computation is carried out sequentially. The MSE is .003551 and the NMSE is .800152. Both are close to the results in Table 1 for Figure 13.

5.5.2. *Input in the Vertical Direction*

The input data vectors are selected in the vertical direction in Figure 13(a). However, the 512x512 dimension of the covariance matrix causes serious memory problems in DOS-based computers. Hence, it is not feasible to obtain the covariance matrix and the eigenvectors.

6. Discussion

(1) The bright spot seismogram in Figure 13(a) has a magnitude between -0.4036 and 0.4622. The real seismogram in Figure 15(a) has a magnitude between -0.2075 and 0.2316. The dimension of the input data vector can reach 512. Although the dimensionality of the seismic data is quite large, and the magnitude of the seismic data is extremely small, the information regarding the seismic layers and the uniform seismic traces, can be extracted by PCA using GHA.

(2) For an input vector with dimension N, the traditional method uses memory of order $O(N^2)$ because of the covariance matrix. However, the neural network based on Sanger's learning rule, uses memory of the order $O(N)$ to determine the eigenvectors. For an input seismic data vector in the vertical direction with 512 dimensions, the covariance matrix is 512x512. The memory for real numbers is 1,048,576 bytes (=4x512x512) and exceeds the restriction of 64K bytes for DOS-based computers. In the present study, our computation requirements using the traditional method, greatly exceed the computation power of a DOS-based computer. However, using the iterative approach involved with Sanger's learning rule, the eigenvectors can be found using Microsoft FORTRAN on a DOS-based computer without memory problems. In summary, the computation of eigenvectors for seismic data becomes feasible using neural networks based on Sanger's learning rule in a DOS-based computer.

(3) Seismic data may be compressed significantly. For a seismogram with 64 traces and 512 samples per trace, there are 32,768 samples. If input data are selected in the

vertical direction to find three principal eigenvectors, each eigenvector has 512 dimensions. Each seismic trace has 3 projection values; so there are 3x64 samples for the projection values of 64 traces. One mean vector (512 dimensions), 3 eigenvectors with 3x512 samples, and projection values of 64 traces are stored. The total stored data is 512+3x512+3x64=2,240 samples. This indicates a 32,768/2,240=14.6 fold information redundancy.

If input data vectors are selected in the horizontal direction to find three principal eigenvectors, each eigenvector has 64 dimensions. Each seismic horizontal data vector has 3 projection values, so there are 3x512 samples for the projection values of 512 vectors. One mean vector (64 dimensions), 3 eigenvectors with 3x64 samples, and the projection values of 512 vectors are stored. The total stored data is 64+3x64+3x512=1,792 samples. This indicates a 32,768/1,792=18.3 fold information redundancy. If the seismic layers are mainly flat, then the recovered seismogram is acceptable if several principal components are employed. From the data representation point of view, the principal components network provides a useful technique for seismic data compression.

(4) The initial values of the coefficients W_{ij} of the weight matrix **W** may bias the final direction of the first eigenvector towards a positive or negative direction. To overcome this problem, the initial weight coefficients W_{ij} are set to random values. After convergence, the first principal projection value is checked. If the projection value is positive, then the set of eigenvectors is accepted. Otherwise, the initial weight coefficients W_{ij} are changed until the first principal projection value is positive.

(5) If the data distribution in 2-D is highly correlated along a line, then the first eigenvector can be computed from the network, but the remaining second eigenvector cannot be derived to the correct value. The same occurs for the *N*-dimensional case.

(6) How many eigenvectors can be chosen? In the case of the real seismogram, the first to the fifth principal projection values of each data are checked. The eigenvectors with significantly large projection values are selected. Alternatively, all the data can be projected on each eigenvector, and the variance of the data on each eigenvector can be calculated. The variances are the eigenvalues λ_i, and $\lambda_1 > \lambda_2 > \ldots > \lambda_i > \ldots > \lambda_M$ (Chien and Fu, 1967). Eigenvectors with larger variances can be chosen and the eigenvectors with smaller variances may be neglected.

(7) If there are few layers in a seismogram, E[**X**]= **0** is easily obtained in the horizontal input direction. Usually E[**X**]= **0** is not easily obtained in the vertical input direction because of the layer effect. In the vertical input direction, if E[**X**] is not equal to **0**, the recovered components will not match the original seismogram until the mean vector is added.

(8) The principal components from the original seismogram may be subtracted, then the remainder consists of the non-uniform property patterns: slope patterns, curve patterns such as diffraction patterns, and noise.

7. Conclusions

(1) Seismic data with a uniform or similar property have high correlation. PCA can improve seismic interpretation, and can demonstrate how many uniform classes there are and where their locations are in the seismogram, for both input directions, which are not easy to identify by experience. The real seismogram from the Mississippi Canyon is an excellent example.

(2) Using the input data vectors in the horizontal direction, the principal components reveal the uniform property of wavelets from the same horizontal layer. The projection values on the first eigenvector display the high amplitude content and polarity reversal for wavelets at different layers. These properties appear in the analyses of the seismogram of the bright spot pattern and real seismogram from the Mississippi Canyon. From simulations of the Ricker wavelets for different layers, the projection of the first eigenvector also shows the long duration (low frequency) content of the wavelet.

(3) Using the input data vectors in the vertical direction, the principal components can show the uniform property of the neighboring seismic traces. The projection on the first eigenvector reveals the uniform and reverse properties of the traces. In the analyses of the seismogram for the bright spot pattern and the real seismogram from the Mississippi Canyon, the traces in the central part show a similar waveform morphology, and the neighboring traces at two sides show polarity reversal.

(4) The seismic diffraction pattern after NMO may be separated from the horizontal reflection layer, using the first principal component for the input data in both horizontal and vertical directions.

(5) The two sides of a fault correspond to two principal eigenvectors in both directions.

(6) PCA can be used for seismic data compression.

Acknowledgements

The author would like to thank Kevin M. Barry of Teledyne Exploration for providing real seismic data, and Mr. I. H. Huang for his pre-programming work. This investigation was supported by the National Science Council, Taiwan, under NSC-85-2213-E-009-122 and NSC-88-2213-E-009-135.

References

Baldi, P., and Hornik, K., 1989, Neural network and principal component analysis: Learning from examples without local minima: Neural Networks, **2**, 53-58.

Bannour, S., and Azimi-Sadjadi, M. R., 1995, Principal component extraction using recursive least squares learning: IEEE Trans. Neural Networks, **6**, 457-469.

Carnahan, B., Luther, H. A., and Wilkes, J. O., 1969, Applied numerical methods: John Wiley & Sons, Inc., New York.

Chien, Y. T., and Fu, K. S., 1967, On the generalized Karhunen-Loeve expansion: IEEE Trans. Information Theory, **13**, 518-520.

Dobrin, M. B., 1976, Introduction to geophysical prospecting: 3rd ed., New York, McGraw-Hill, ch. 10.

Fukunaga, K., and Koontz, W. L. G., 1969, Application of the Karhunen-Loeve expansion to feature selection and ordering: IEEE Trans. Comput., **19**, 311-318.

Hagen, D. C., 1982, The application of principal components analysis to seismic data sets: Geoexploration, **20**, 93-111.

Hotelling, H., 1933, Analysis of a complex statistical variables into principal components: J. Educ. Psychol., **24**, 417-441 and 498-520.

Jain, A. K., 1976, A fast Karhunen-Loeve transform for a class of random processes: IEEE Trans. Commun., **24**, 1023-1029.

—— 1977, A fast Karhunen-Loeve transform for digital restoration of images by white and colored noise: IEEE Trans. Comput., **26**, 560-571.

Jones, I. F., 1985, Applications of the Karhunen-Loeve transform in reflection seismology: Ph.D. thesis, The University of British Columbia, Vancouver, Canada.

Karhunen, J., and Oja, E., 1981, Optimal adaptive compression for high-dimensional data: Proceedings 2nd Scand. Conf. on Image Analysis, Helsinki, Finland, 152-157.

—— 1982, New methods for stochastic approximation of truncated Karhunen-Loeve expansions: Proceedings 6th Int. Conf. on Pattern Recognition, Munich, FRG, 550-553.

Krogh, A., and Hertz, J., 1990, A. Hebbian learning of principal components, *in* Eckmiller, R., Hartmann, G. and Hauske, G., Ed., Parallel Processing in Neural Systems and Computers: Amsterdam, Elsevier, 183-186.

Kung, S. Y., 1993, Digital neural networks: Prentice Hall, ch.8.

Oja, E., 1982, A simplified neuron model as a principal component analyzer: Journal of Mathematics and Biology, **15**, 267-273.

—— 1985, On stochastic approximation of the eigenvectors and eigenvalues of the expectation of a random matrix: Journal of Mathematical Analysis and Applications, **106**, 69-84.

Oja, E., and Karhunen, J., 1980, Recursive construction of Karhunen-Loeve expansions for pattern recognition purposes: Proceedings 5th Int. Conf. on Pattern Recognition, Miami Beach, Fl., 1215-1218.

Pelat, D., 1974, Karhunen-Loeve series expansion: a new approach for studying astrophysical data: Astron. and Astrophys, **33**, 321-329.

Peper, F., and Noda, H., 1996, A symmetric linear neural network that learns principal components and their variances: IEEE Trans. Neural Networks, **7**, 1042-1047.

Sanger, T. D., 1989a, Optimal unsupervised learning in a single-layer linear feedforward neural network: Neural Networks, **2**, 459-473.

—— 1989b, An optimality principle for unsupervised learning, *in* Touretzky, D. S., Ed., Advances in neural information processing systems, Vol. I: San Mateo, CA: Morgan Kaufmann, 11-19.

SECTION B

LITHOLOGY, WELL LOGS, PROSPECTIVITY MAPPING AND RESERVOIR CHARACTERISATION

FUZZY CLASSIFICATION FOR LITHOLOGY DETERMINATION FROM WELL LOGS

ANTOINE TOUMANI

DMT GmbH,
GeoTec Division, Geo-Engineering,
Am Technologiepark 1, 45307 Essen,
GERMANY
E-Mail: toumani@dmt.de

Abstract

A new hybrid fuzzy classification method has been employed successfully for the discrimination of lithology from well logs. It is based on the correlation of log responses from key boreholes having a known lithology, with the logs from neighbouring boreholes where only well logs are available. The method combines a model-based supervised classification and a model-free unsupervised clustering into a single classification. Both supervised/unsupervised parts are linked together by the fuzzy membership of the log data within the determined lithology. Fuzzy membership of the data serves also as a quality measure of the classification results, and provides valuable information concerning the reliability of the model. The classification approach has been used successfully for determining Upper Carboniferous lithologies from open hole logs in a multi-well study.

1. Introduction

One of the goals of well logging is to derive an accurate interpreted lithology column, and consequently avoid the need for expensive coring. To achieve a good correspondence between log-derived and core-recorded lithologies, maximum information needs to be extracted from the log data. However, it is not always straightforward to translate log responses in terms of geologically meaningful lithofacies based on physical, chemical and biological characters. Different tools such as the graphical representation of log data in the form of cross plots, and multivariate statistical data analysis, have been employed to determine lithology from well logs (Serra, 1986). New information processing technologies such as expert systems and

W. Sandham and M. Leggett (eds.),
Geophysical Applications of Artificial Neural Networks and Fuzzy Logic, 125-142.

artificial neural computing have also been applied towards achieving this goal (Hoffman et al., 1988; Westfal & Bornholdt, 1996). In this chapter two efficient methods of data analysis have been investigated. A vague description of the data using fuzzy logic has been embedded in the statistical methods of classification.

Since the introduction of fuzzy sets by Zadeh (1965) as a generalization of ordinary sets, many researchers recognised their potential for solving pattern recognition problems. Fuzzy clustering began with Ruspini (1969). In his "next neighbour" clustering criterion, the grouping of data is based on their proximity to a training sample (i.e. labelled data: data with known memberships into reference classes). Dunn (1974) and Bezdek (1981) developed the well-known fuzzy *c*-means clustering algorithm, which minimizes the scatter of data around cluster prototypes. Generalization of the fuzzy *c*-means algorithm to resemble maximum likelihood estimates of mixture densities, by using different norm metrics, one for each cluster, has been investigated by Gustafson and Kessel (1979). The above clustering algorithms are subject to the condition that the memberships of a data point across clusters are positive numbers and sum to one. This probabilistic constraint is relaxed by Krishnapuram and Keller (1993), through adaptation of the possibilistic interpretation of membership into the framework of the fuzzy *c*-means clustering algorithm, thereby avoiding the dependency of membership assignment on the number of clusters (i.e. relative membership). Pattern recognition algorithms for detecting not only volume clusters, but also "thin shells" like curves and surfaces, have recently been developed. The reader is referred to the tutorial presentation of Bezdek (1996) for a review of fuzzy classification algorithms and their current status.

Changes of the log responses for the same geological formation, acquired in different locations, have to be considered in the classification process. These changes cause problems for the supervised classification of new log data using a model based on information obtained in the key borehole. Theoretical developments for such hybrid consideration, which are synonymous with "imperfect or partial supervision" of data classification, are very few. Existing algorithms (Pedryc, 1985; Stefanou, 1985; Lashgari, 1991; Bensaid et al., 1996) use model-representative training samples to supervise the clustering of the data.

In this chapter, a new approach to partial supervision of data clustering is presented, which fits a normal distribution model in order to recognize the true distributions (i.e. clusters) of data under investigation. The theoretical background of the clustering technique which has been developed is described, and its operation is illustrated using examples which provide insight into the method.

2. The Approach

The basis of many statistical and heuristical transformations of well logs into lithology is the concept of "electrofacies", which is defined as the set of log responses which characterize a bed and permit it to be distinguished from other beds (Serra & Abott, 1982). The idea is to represent a set of *p*-log readings at a given depth level as a point in p-dimensional space. An electrofacies is then materialized as a cluster in this space, that is, an area of relatively high concentration of points, since the log responses are similar. The task is to attach to each measurement level, the cluster to which it belongs. From a statistical perspective, two approaches may be considered:

The first approach involves supervised classification. The electrofacies (i.e. lithology model) are pre-specified in the form of a multivariate normal distribution (Delfiner et al., 1984). The only problem which remains is to assign the log data level-by-level to the appropriate electrofacies, using a Bayes classifier. The advantage is that the classification is fast. In addition, it provides a measure of confidence in the classification results. Supervised classification has the disadvantage that the results depend on the reliability of the lithology model.

The second approach is to determine clusters from the log data (Wolff & Pelissier-Combescure, 1982). This is called clustering or unsupervised classification. It has the advantage that the data are simultaneously considered, and consequently their inherent structures are revealed. However, clustering has the disadvantage that derived clusters must be interpreted each time clustering is carried out. This is a time-consuming iterative procedure. Only in fuzzy clustering, where the membership of the data in the clusters is not crisp, is a measure of confidence in clustering results provided, and an overlap of clusters effectively handled. Covariance-clustering, such as maximum likelihood estimation of mixture densities under a Gaussian assumption, is well suited for recognising different shapes of clusters. Many log analysts have used both methods for lithology determination from well logs (Serra et al., 1985; De La Cruz & Takizawa, 1985; Stowe & Hock, 1988).

In this chapter, the supervised and unsupervised fuzzy covariance classification are combined into a single classification approach (Figure 1). The hybrid fuzzy classification makes use of pre-defined electrofacies as well as clusters residing in the data itself, in order that the similarity between known electrofacies and data derived clusters can be detected despite their different geometries.

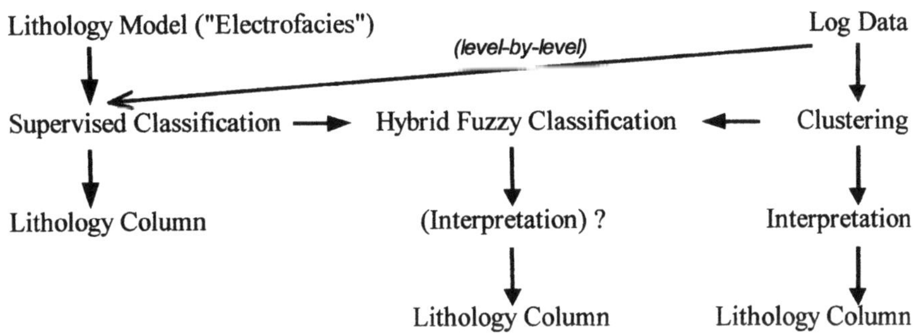

Figure 1. Lithology determination from well logs.

Whether new interpretations of derived clusters have to be made, is dependent on how much the yielded clusters differ from the pre-specified clusters or electrofacies. By utilising known electrofacies, the hybrid fuzzy classification method has the advantage of clustering the data faster than in the unsupervised case. Another advantage is that by deriving membership of log data in combined criteria (data cluster & model cluster), information regarding reliability of the model is obtained. If the quality of data membership is insufficient, classification has to proceed with a higher influence of the log data.

3. Hybrid Fuzzy Classification

A cluster is characterized by a cloud of points showing high concentration. An optimal partition of a data set tends to yield well-separated compact clusters, if the data truly contain these structures. The total scatter of a data set around its average is constant, and does not depend on how the data set is partitioned into clusters. Since this total scatter is equal to the sum of the within-cluster scatter and between-cluster scatter, it is sufficient to optimize one of the two scatters. The minimization of the within-cluster scatter leads automatically to maximization of between-cluster scatter. Most clustering methods using an objective function are based on this idea and its mathematical realization (Duda & Hart, 1973).

The minimization of within-cluster scatter is used as the clustering criterion, because of its computational feasibility in real situations. The within-cluster scatter in hybrid fuzzy classification, is defined as the sum of the scatter of the data around the centres of data derived clusters, and the scatter of the data around the centres of pre-specified clusters (Toumani, 1995):

$$J(U,V,A) = (1-R) \underbrace{\sum_{j=1}^{n} \sum_{i=1}^{c} u_{ij}^{m} d_{ij}^{2}}_{\textit{Unsupervised}} + R \underbrace{\sum_{j=1}^{n} \sum_{i=1}^{c} u_{ij}^{m} d'^{2}_{ij}}_{\textit{Supervised}} \rightarrow \min \qquad (1)$$

with:

$$d_{ij}^{2} = \left\| x_j - v_i \right\|_{A_i}^{2} = (x_j - v_i)^T A_i \, (x_j - v_i)$$

$$d'^{2}_{ij} = \left\| x_j - v_i' \right\|_{A_i'}^{2} = (x_j - v_i')^T A_i' (x_j - v_i')$$

where J is the objective mapping function into R (set of positive real numbers),

n is the number of data,

c is the number of clusters,

p is the number of variables,

$X = (x_1, x_2, \ldots, x_n)$ with $x_i \in R^p$ is the data,

$V = (v_1, v_2, \ldots, v_c)$ with $v_i \in R^p$ is the prototype of i-th data derived cluster,

$V' = (v'_1, v'_2, \ldots, v'_c)$ with $v'_i \in R^p$ is the given prototype of i-th model cluster,

d^2_{ij} is the quadratic distance of x_j from v_i, normed by p x p symmetric positive definite matrix A_i,

d'^2_{ij} is the quadratic distance of x_j from v'_i, normed by the given p x p symmetric positive-definite matrix A'_i,

$(x_j - v_i)^T$ is the transpose of $(x_j - v_i)$,

$R \in [0,1]$ is the mixing parameter,

$m \in [1,\infty)$ is a weighting exponent called a fuzzifier,

U is c x n fuzzy partition matrix of X in c-clusters, whose elements ,u_{ij} (membership of j-th data point in the i-th cluster couple), satisfy the following conditions:

$$u_{ij} \in [0,1] \ \forall \ (\textit{for all}) \ i,j \ ; \quad \sum_{i=1}^{c} u_{ij} = 1 \ \forall j \ ; \quad 0 < \sum_{j=1}^{n} u_{ij} < n \ \forall \ i$$

It is important to note in this hybrid classification scheme, that the i-th model and the i-th data derived cluster have to be considered as a couple. This coupling is achieved by the membership of data into cluster couples (eq.(1)). The unsupervised part is known as fuzzy c-means clustering (Bezdek, 1981).

The influence of the data on the scatter within the cluster couples is weighted by the m-th power of their membership of those cluster couples. The higher the value of exponent m, the fuzzier the partition of the data in the clusters, and this results in a higher overlap between the clusters. Investigations on the exponent m under well-defined conditions have shown that a value of two provides reliable classification results (Choe & Jordan, 1992). The mixing parameter, R, regulates the influence of both parts (supervised and unsupervised) on the classification. A high value of R means a high influence of the pre-specified cluster and vice-versa. The membership, u_{ij}, is constrained, such that it is restricted to values *between* 0 and 1 instead of 0 *or* 1, as in hard clustering. The sum of the membership of a data point in the c-clusters is 1. No cluster is allowed to be empty and to have a cardinality of n, which is the number of data points. The norm matrices, A'_i, A_i, $i=1,..,c$, describe the scatter shapes of the given and derived clusters, respectively.

The minimization of J (see eq.(1)), subject to the normalization condition (i.e. the sum of memberships of a data point in the c-clusters is 1), provides the following solution for V (Bezdek, 1981) and U (Toumani, 1995):

$$v_i(k) = \frac{\sum_{j=1}^{n} (u_{ij})^m x_j(k)}{\sum_{j=1}^{n} (u_{ij})^m}, k = 1,2,\ldots p \tag{2}$$

$$u_{ij} = \frac{1}{\sum_{k=1}^{c} \left[\frac{(1-R) d_{ij}^2 + R\, d'^2_{ij}}{(1-R) d_{ik}^2 + R\, d'^2_{ik}} \right]^{1/m-1}} \tag{3}$$

$$u_{ij} = 1, \forall x_j = v_i = v_i' \, (u_{kj} = 0, \forall k \neq i)$$

By minimizing J with respect to A, each A_i is constrained by requiring the determinant of A_i, $det(A_i)$, to be a fixed positive number during the iterative minimization of J, since the unsupervised part in eq.(1) may be made as small as necessary by simply making A_i, $i = 1,..,c$, less positive definite ($det(A_i)$ close to zero). Allowing A_i to vary while keeping its determinant fixed, thus corresponds to seeking an optimal cluster shape fitting the data to a fixed cluster volume. The local minima of J with respect to A_i, with determinant, $\rho_i = det(A_i) > 0$, is constant, can be achieved only if (Gustafson & Kessel, 1979):

$$A_i = [\rho_i \det(Sf_i)]^{(1/p)} (Sf_i^{-1}), \quad 1 \le i \le c,$$

where

$$Sf_i = \sum_{j=1}^{n} (u_{ij})^m (x_j - v_i)(x_j - v_i)^T$$

is the fuzzy scatter matrix of the i-th cluster in the unsupervised part, whose inverse is Sf_i^{-1}.

If $R=0$ (unsupervised case), then the clustering criterion and the membership assignment will be identical with fuzzy c-means clustering and its solution, respectively. For $R = 1$ (supervised case), the membership assignment will be identical with the ones of "next prototype" classifier (Keller et al., 1985). In the supervised classification only one iteration is required, since only an assignment of the data to the given reference cluster will be done. The smaller the value of R, the more the entire data set will be considered, and this leads to a higher number of iterations if the data have different cluster geometries compared to those of the given model. This property of the hybrid classification is important, since in the supervised classification the data will be assigned point for point to the appropriate cluster without consideration of the data's inherent structure, which can only be detected if the entire data set is analysed. A practical aspect of the hybrid fuzzy classification, is that statistical parameters of the derived clusters (averages and scatter matrices) in previously carried out fuzzy c-means clustering, can serve as a model to classify new data without manual intervention, since in both parts of the classification scheme the same clustering criterion and the same norm for distance are used. Therefore a high compatibility between the supervised and unsupervised components is ensured.

The assumption of a normal distribution of cluster elements could be a restriction for general use of the hybrid classification approach, compared to other hybrid classification techniques (Pedryc, 1985; Stefanou, 1985; Lashgari, 1991; Bensaid et al., 1996), where model-based/model-free training samples for the supervision of clustering can be used. However, heuristic construction of training samples (i.e. membership assignment) is constrained in practice to a small number of variables, since visualisation of inter-relationships between the variables is limited to a low-dimensional space. Conversely, a training sample resulting from a previously cluster analysis, is conditioned in its distribution by the used clustering algorithm, criterion and norm metric (Bezdek, 1981). The lack of methods generating reliable membership functions from training data remains a serious problem in many fuzzy set applications (Krishnapuram & Keller, 1993). For lithology and porosity determination from well logs, the normal distribution of log responses of geological formations is supported in principal, not only for sedimentary rocks (Delfiner et al., 1984; Quirein et al., 1986; Mitchell & Nelson, 1988), but also for magmatic and metamorphic rocks (Büttgenbach, 1990). The uncertainties of the tool measurements and supposed tool responses can therefore be handled effectively, using the simplified parameterization of the normal distribution and the well-known statistical approaches for estimation of the distribution's parameters.

Krishnapuram and Keller's concept (1993), to relax the probabilistic constraint of the membership assignment in order to facilitate a possible interpretation of the memberships in sense of evidence of its typicality, is integrated in the hybrid classification formula (eq.(1)). In addition to minimization of within-cluster scatter, maximization of the membership of cluster elements is also required. The solution of the membership is the same as their solution for possible versions of fuzzy c-means, if the quadratic distance, d^2_{ij}, is exchanged by the hybrid equivalents, $(1\text{-}R)\, d^2_{ij} + R\, d'^2_{ij}$. However, in the following examples, the membership is calculated according to eq.(3), i.e. constrained membership.

4. Application

The hybrid fuzzy classification was used to determine lithologies from well logs of exploration wells in the Upper Carboniferous. In Figure 2 the standard logging programme for coal exploration used by coal mining companies in Germany is shown. Four different logs are used in the classification. Sonic and density are coal significant measurements, whereas gamma ray and resistivity measurements are ambiguous with respect to coal, since clean sandstones have similar log responses. Coal has high resistivity and acoustic traveltimes, and low density and gamma ray values. Sandstones are distinguished from silt/mudstones by low gamma ray levels and traveltimes, and high resistivity values. The log response of density and traveltime is very similar for fossil plant remains, compared to coal with dirt bands (Schmitz, 1983).

Cluster analyses are carried out in a key well, and the clusters are carefully calibrated with the corresponding core record to define the electrofacies. The result of fuzzy *c*-means clustering, including membership of the log data into the appropriate lithology, is shown in Figure 2 for a 45m depth interval. Rock symbols are defined in Legend 1. The main lithologies - sandstone, silt/mudstone and coal - are easily recognized from the log data. Two types of sandstone can be traced from well logs: argillaceous compact and porous sandstone. Only the differentiation between sand-rich siltstone on one side and clay-rich silt/mudstone on the other side, can be achieved from the available well logs. The occurrence of the high velocity sideritic layers in silt/mudstone makes it more difficult to distinguish between the silt/mudstone group members. The membership values of well logs into silt/mudstone are lower than those into sandstone and coal, because of the high overlap in log responses of the silt/mudstone lithologies.

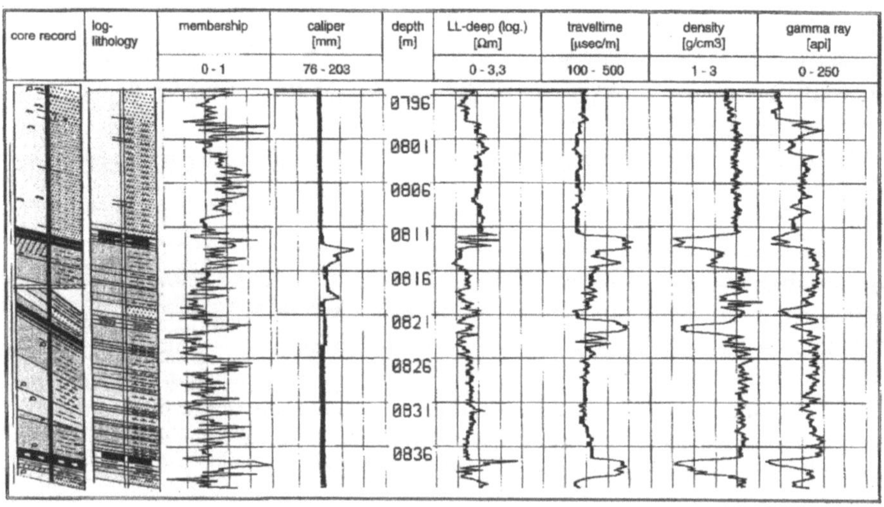

Figure 2. Log-derived lithology compared with core record (key well).

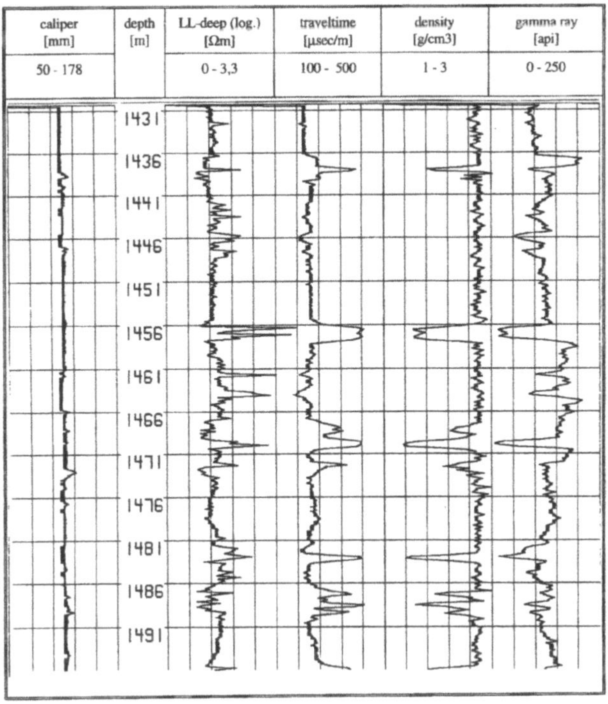

Figure 3a. Acquired well logs from a coal exploration well (borehole no. 1).

The pre-defined electrofacies from the key well are used in the hybrid classification of logs from new wells. The classification results of the well logs displayed in Figure 3a are shown in Figure 3b for different mixing parameters, *R*.

In the supervised case where *R* is equal to one, the classification result is poor. Only the differentiation between coal, silt/mudstone and the upper thick sandstone bed is reliable. With a higher influence of log data, a reliable distinction between clay-rich silt/mudstone and sandy/sand-rich siltstone can be achieved. It is important to note the improvement of the membership of the data in the electrofacies is dependent on the increase of the log data influence. Between the second (S2) and third seam (S3) no differentiation between clay-rich silt/mudstone and sandy siltstone is possible. This is caused by the different gamma ray levels in the top and bottom of the second seam. The fossil plant remains are recognised in some intervals as coal with dirt.

Figure 4 shows the difference in the scatter between three lithologies for different influences of pre-specified electrofacies in the two-dimensional gamma ray-density space. The border of the cluster is twice the standard deviation. A high overlap between sandstone and siltstone is observed by the high influence of electrofacies of the key well.

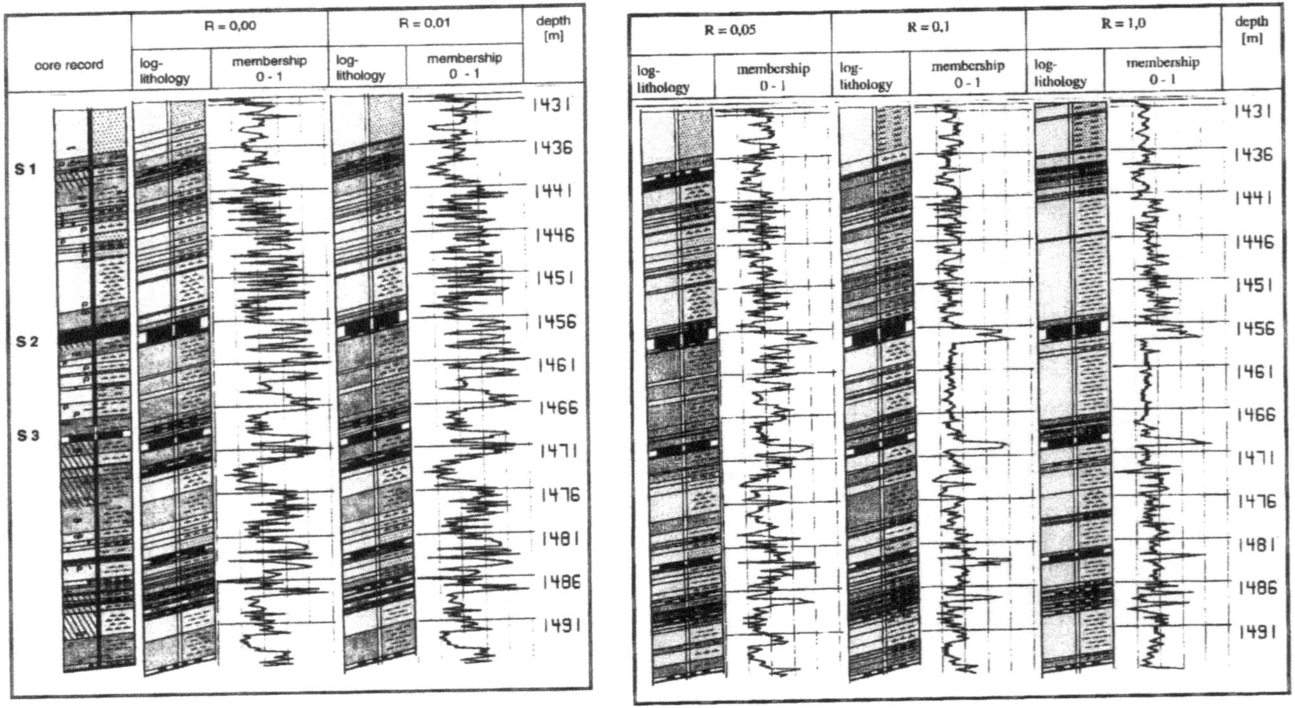

Figure 3b. Dependence of classification results on mixing Parameter *R* (borehole no. 1).

The overlap and scatter is minimized by increasing the influence of log data. Only a small improvement in minimization of the scatter can be achieved by values of *R* lower than 0.05. This observation is consistent with the number of required iterations for different values of *R* (Table 1). Due to stabilization of the clusters, the number of additional iterations needed by increasing the influence of the log data, will then be smaller for *R*=0.05.

Fuzziness measures of the clusters (Bezdek, 1981; Xuanli & Gerardo, 1991), which are mainly used in the problems of cluster validity (determination of the optimal number of clusters), also provide valuable information regarding the reliability of the classification results. A validity function measures the overall "compactness" of the partition found by a particular clustering method. It is then often presumed that a large change in slope or large jump in the validity functions indicate levels at which natural groups exist in the data. It may be physically plausible to expect "good" clusters at more than one value of *c* (number of clusters) as the data are separated into successively finer substructures. A significant problem with this strategy is that many objective functions are monotonic in *c*, which tends to obscure pathological behaviour unless the change is quite radical. For most applications however, the range of variation of *c* is previously known.

In Figure 5 three fuzziness measures are displayed for different values of mixing parameter *R*. A Maximum of the measures *B* and *G* as well as minimum of *S*, suggest compact well-separated clusters. By decreasing *R*, the measures *S*, *B* and *G* are optimized continuously up to *R*=0.05. *S* reaches its optimum value when approaching the value *R*=0.05, *G* by the value *R*=0.01 and *B* by the value *R*=0. Reliable classification results may then be achieved with values of *R* lower than 0.05. The difference of the three lithologies between the key well (reference) and the investigated well (*R*=0) are clearly visible (Figure 4). The density of sandstone and siltstone in the new well is rather high and has a low scatter, caused by carbonization, while diagenesis. The high scatter and positive covariance between gamma ray and density in coal has no lithological reason. Because the resolution of the tools is constrained with respect to the thickness of the coal seams, and because of the large difference in log responses between coal and surrounding strata, averaged log responses are often registered in the direct neighbourhood of the coal seam.

TABLE 1. Number of required iterations for different values of *R* (borehole no. 1).

R	Iterations
1	1
0.5	3
0.1	5
0.05	19
0.01	21
0.00	24

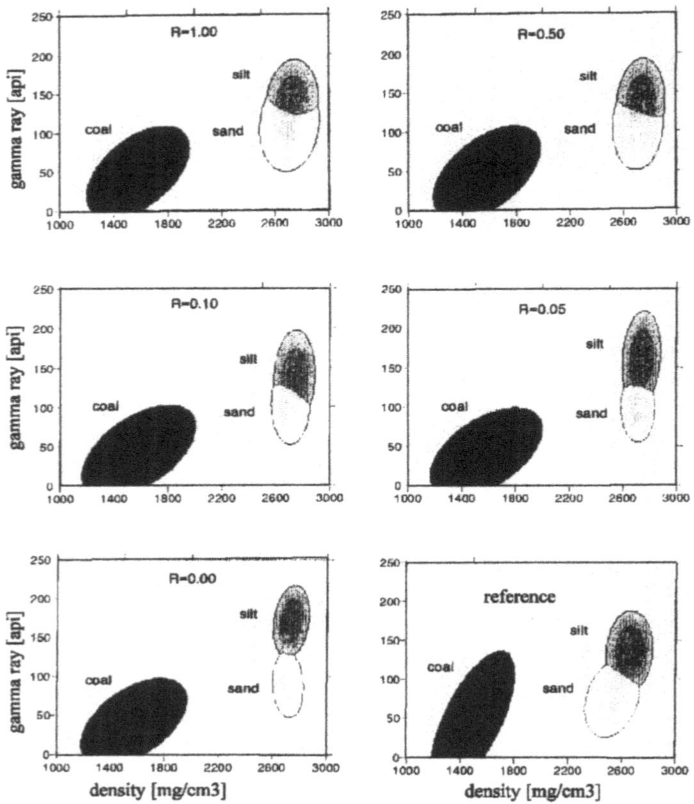

Figure 4. Protype and scatter for different lithologies with respect to mixing parameter *R* (borehole no. 1).

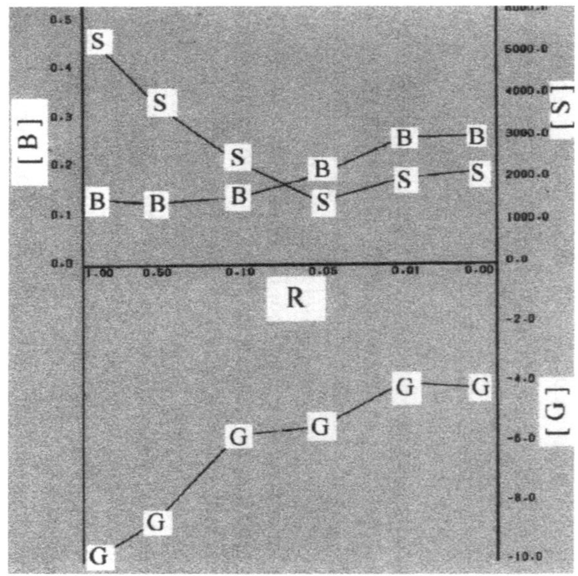

Figure 5. Fuzziness measures with respect to mixing parameter *R* (borehole no. 1).

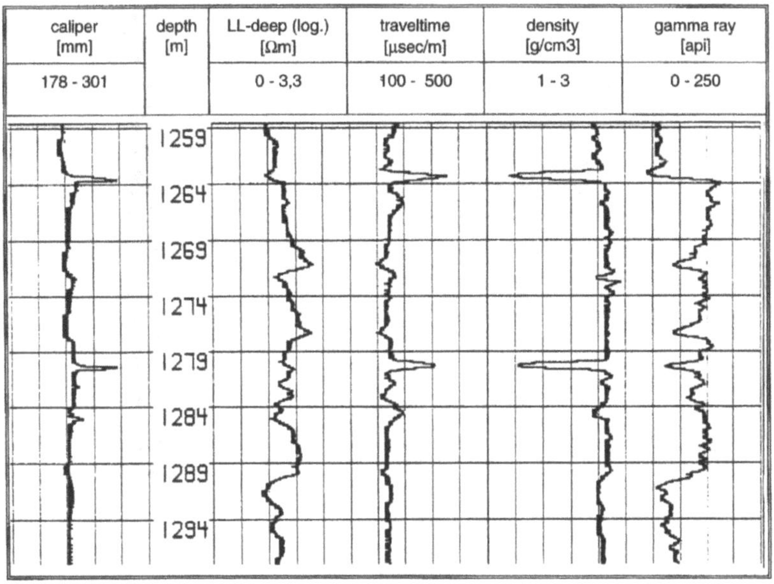

Figure 6a. Acquired well logs from a coal exploration well (borehole no. 2).

The method was tested in another borehole, which was 70 km distant from the key well compared to 5 km distant in the first example. The data is shown in Figure 6a. Due to breakouts, which may be seen from caliper data, no reliable resistivity response was recorded. In the case of supervised classification (Figure 6b), the siltstone is recognised as argillaceous sandstone. The mis-classification of silt/mudstone as argillaceous sandstone is caused by the higher compaction of the formation in this well. Because sandstone normally has lower traveltime values than siltstones, the difference in gamma ray values between the compact sandstone of the key well and the silt/mudstone in the investigated well, is overcome by the similarity in their traveltime values. The well logs in the investigated borehole were acquired using a sampling rate one-fifth that of the sampling rate for logs from the key well. Due to the low sampling rate and the small thickness of coal with dirt bands, the lithologies of coal and coal with dirt could not be distinguished from each other. Therefore, coal shows different within-cluster scattering behaviour compared to that of the surrounding lithologies (Figure 7): the scatter increases by increasing the influence of the data. The increase of coal scatter is directly related to disappearance of coal with dirt lithology at the floors of coal seams from the classification results (Figure 6b). With higher influence of the data, higher reliability in the classification results is observed. In this example, a higher influence of log data than in the previous examples was required.

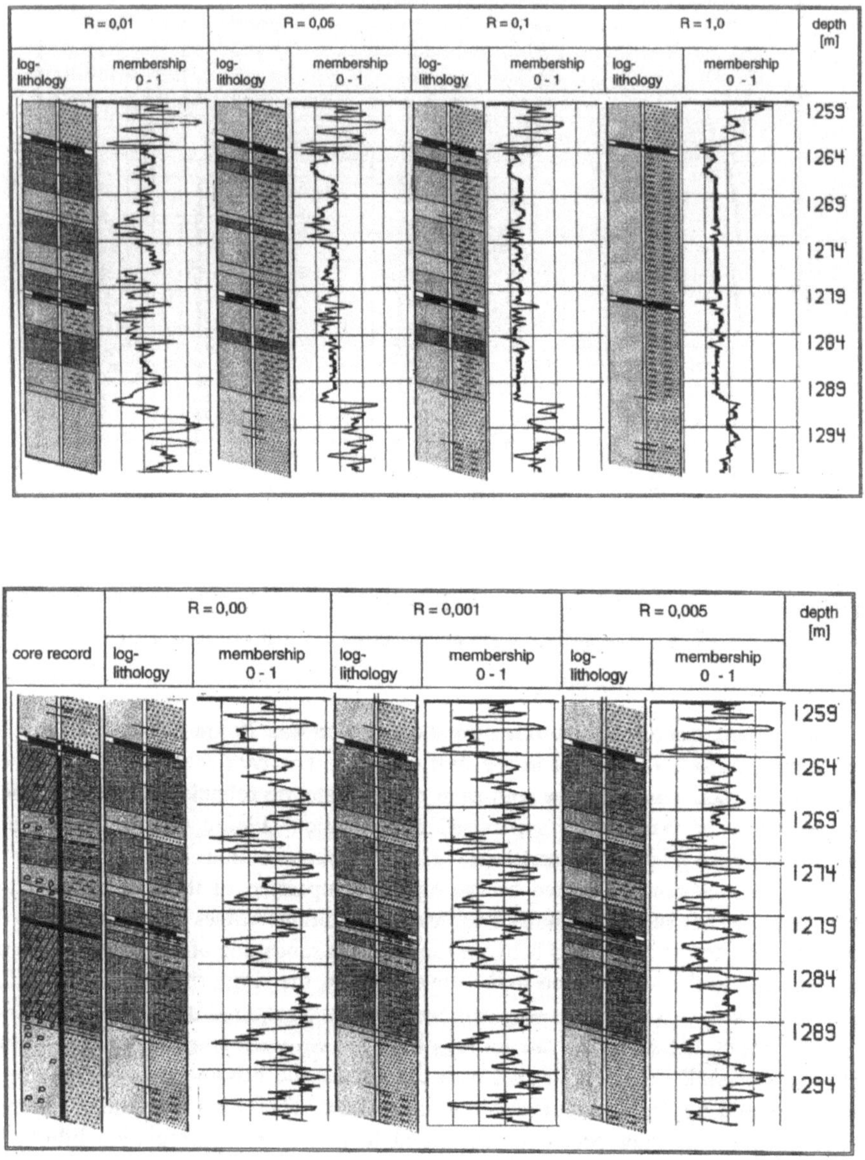

Figure 6b. Dependence of classification results on mixing parameter *R* (borehole no. 2).

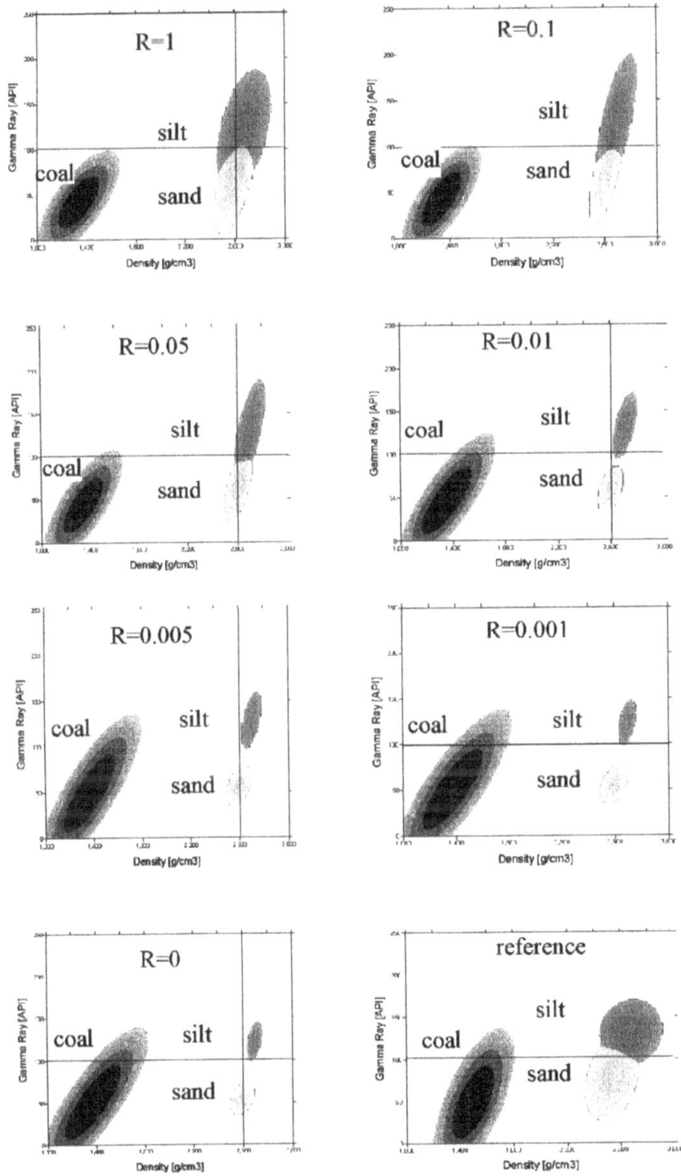

Figure 7. **Prototype and scatter for different lithologies with respect to mixing parameter R (borehole no.2).**

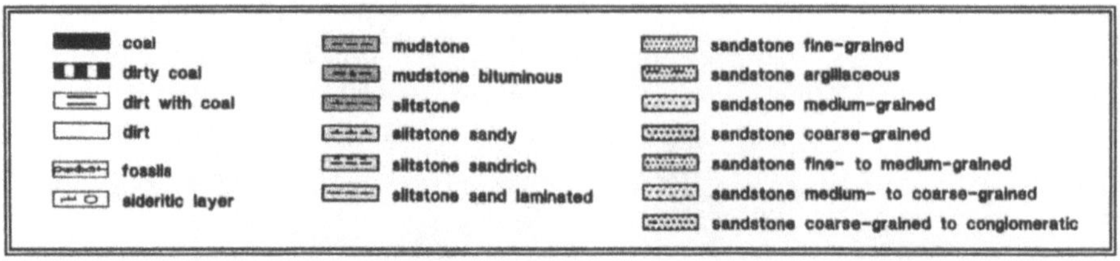

Legend 1. Symbols of rocks and their constituents.

The examples show that reliable classification results can only be achieved for low values of the mixing parameter, *R*. A value for *R* of 0.5 does not imply equal influence of both supervised and unsupervised components on the classification results, since clusters derived exclusively from the data are not yet determined. How far the influence of the model should be reduced to obtain reliable classification results, is dependent on how representative the model is for the data structures under investigation.

It should be mentioned that post-processing (contextual classification) has been applied to classification for the examples shown, in order to ensure depth consistency of the results. In the heuristic classification which has been employed, a lithology section with thickness smaller than the given threshold is re-classified and merged into the lithology above or below according to its affinity. To preserve the coal seam structure, re-classification is carried out only on the coal surrounding lithologies.

5. Conclusions

A new computing technique for lithology determination from wireline logs is presented. The hybrid fuzzy classification method combines the widely used methods of supervised and unsupervised classification, improving their advantages and overcoming their disadvantages. The method is very suitable for multiple well lithological studies, where diagenetic changes in the lithologies are expected and have to be adequately treated. The definition of electrofacies can be improved automatically by tracing changes in geometry of the multi-dimensional log space. The membership of data within electrofacies (both pre-specified and data-derived) is fundamental for a decision regarding the reliability of the classification results and the model. Fuzziness measures are also successfully employed to estimate the spatial validity of the model used in the classification process. Improvement of the classification results is suggested if the possibilistic version of the introduced hybrid classification is used, since determination of absolute (i.e. not relative) membership of the data into clusters will lead to better quantification of those clusters.

Lithology determination using fuzzy classification is helpful in formation evaluation, where mineral composition, porosity and saturation are computed using quantitative evaluation of logs and pre-defined formation models (formation constituents and their log responses). Because of the constraint of the number of formation constituents according to the available well logs, simple log analysis programs assume the same model throughout the interval to be interpreted. The model selection has to be done manually and is usually based on visual inspection of the results. No aids are provided for the selection and/or combination of the models. Since the memberships of every depth level in different lithologies (i.e. formation models) are provided, as a result of fuzzy classification, and since the membership values sum to one, an automatic selection of formation model (appropriate electrofacies) or combination of models according to their probability (memberships) is given. Such automatic combination makes it practical to use a very explicit model for each facies by evaluation of reservoirs with complex lithologies.

Acknowledgements

I wish to express my thanks to DMT for allowing publication of this chapter, and to my colleagues who helped me to prepare it.

References

Bensaid, A.M., Hall, L.O., Bezdek, J.C., and Clarke, L.P., 1996, Partially supervised clustering for image segmentation: Pattern Recognition, **29**, 859-871.

Bezdek, J.C., 1981, Pattern recognition with fuzzy objective function algorithms: Plenum Press.

Bezdek, J.C., 1996, A sampler of non-neural fuzzy models for clustering and classification: EUFIT '96, 4th European Congress on Intelligent Technique and Soft Computing, September 2, Tutorial Presentation, Aachen, Germany.

Büttgenbach, T., 1990, Statistische Untersuchungen der Geschwindigkeits – Dichte Relation in kristallinen Gesteinen auf der Basis einer Datenbank: Ph.D. Dissertation, H77, Universität zu Köln, Germany.

Choe, H., and Jordan, J.B., 1992, On the optimal choice of parameters in a fuzzy c-means algorithm: IEEE International Conference on Fuzzy Systems, San Diego, 349-354.

De La Cruz, M.A., and Takizawa, H., 1985, Automatic facies analysis in the Arab formations, El Bunduq field, offshore Abu Dahbi/Qatar: 60th Annual Technical Conference and Exhibition of the SPE, Las Vegas, NV September, 22-25.

Delfiner, P.C., Peyret, O., and Serra, O., 1984, Automatic determination of lithology from well logs: 59th Annual Technical Conference and Exhibition of the SPE of AIME, Houston, Texas, September, 16-19.

Duda, R.O., and Hart, P.E., 1973, Pattern classification and scene analysis: John Wiley & Sons.

Dunn, J.C., 1974, A fuzzy relative of the ISODATA process and its use in detecting compact, well separated clusters: J. Cybern., **3**, 32-57.

Gustafson, D.E., and Kessel, W.C., 1979, Fuzzy clustering with a fuzzy covariance matrix, *in* Fu, K.S., Ed., Proc. IEEE-CDC, **2,** IEEE Press, Piscataway, New Jersey, 761-766.

Hoffman, L.J.B., Hoogerbruge, P.J., and Lomas, A.T., 1988, LOGIX, a knowledge-based system for petrophysical evaluation: SPWLA 29th Annual Logging Symposium, June 5-8, R 1- 13.

Keller, J.M., Gray, M.R., and Givens, J.A., 1985, A Fuzzy K-nearest neighbour algorithm: IEEE Transactions on Systems, Man, and Cybernetics, **SMC-15**, July/August, 580-585.

Krishnapuram, R., and Keller, J.M., 1993, A possibilistic approach to clustering: IEEE Transactions on Fuzzy Systems, **1**, 98-110.

Lashgari, B., 1991, Fuzzy classification with application to geophysical data, *in* Aminzadeh, F., and Simaan, M., Eds., Expert systems in exploration, Geophysical Development Series, **3**, Society of Exploration Geophysicists.

Mitchell, W.K., and Nelson, R.J., 1988, A practical approach to statistical log analysis: SPWLA 29th Annual Logging Symposium, June 5-8, 1-19.

Pedryc, W., 1985, Algorithm of fuzzy clustering with partially supervision: Pattern Recognition Lett., **3**, 13-20.

Quirein, J., Kimminau, S., La Vigne, J., Singer, J., and Wendel, F., 1986, A coherent framework for developing and applying multiple formation evaluation models: SPWLA 27th Annual Logging Symposium , June 9-13, S 1-19.

Ruspini, E., 1969, A new approach to clustering: Information and Control, **15**, 22-32.

Schmitz, D., 1983, Die Interpretation der verschiedenen Gesteinstypen des Oberkarbons (Westfal A-C) aus Kernbohrungen des Ruhrreviers nach geophysikalischen Bohrlochmessungen: Ph.D. Dissertation, Technische Universität Hannover, Germany.

Serra, O., and Abott, H.T., 1982, The contribution of logging data to sedimentology and stratigraphy: Trans. of SPE of AIME, Feb., 117-131.

Serra, O., Delfiner, P., and Levert, J.C., 1985, Lithology determination from well-logs: Case study: SPWLA 26th Annual Logging Symposium, June 17-20, WW 1-19.

Serra, O., 1986, Fundamentals of well-log interpretation: **2**, The Interpretation of Logging Data, Elsevier publication.

Stephanou, H.E., 1985, Imperfectly supervised classification using fuzzy prototypes: IEEE Proceedings of Computer Vision and Pattern Recognition Conf., San Francisco, 144-148.

Stowe, I., and Hock, M., 1988, Facies analysis and diagenesis from well logs in the Zechstein carbonates of northern Germany: SPWLA 29th Annual Logging Symposium, June 5-8, HH 1-25.

Toumani, A., 1995, Fuzzy Klassifikation von geophysikalischen Bohrlochmessungen zur Lithologie-bestimmung am Beispiel des Oberkarbons: Ph.D. Dissertation, Universität zu Köln, H102, Germany.

Westphal, H., and Bornholdt, S., 1996, Lithofacies prediction from wireline logs with genetic algorithms and neural networks: Zeitschrift der deutschen geologischen Gesellschaft, Stuttgart, Germany, 455-474.

Wolff, M., and Pelissier-Combescure, J., 1982, FACIOLOG - automatic electrofacies determination: SPWLA 23rd Annual Logging Symposium, July 6-9, F 1-23.

Xuanli, L.X., and Gerardo, B., 1991, A validity measure for fuzzy clustering: IEEE Transactions on Pattern Analysis and Machine Intelligence, **13**, 841-847.

Zadeh, L.A., 1965, Fuzzy sets: Information and Control, **8**, 338-353.

RESERVOIR PROPERTY ESTIMATION USING THE SEISMIC WAVEFORM AND FEEDFORWORD NEURAL NETWORKS

PING AN[¥], WOOIL M. MOON[†] and FOTIS KALANTZIS[‡]

¥ *Schlumberger-GeoQuest,*
5599 San Felipe, Suite 1700,
Houston, TX 77056-2722,
USA
E-Mail: pan@houston.geoquest.slb.com

† *Department of Geological Sciences,*
The University of Manitoba,
Winnipeg,
CANADA
E-Mail: wmoon@cc.umanitoba.ca

‡ *Geophysical Specialist (PGeoph.),*
Calgary,
Alberta,
CANADA
E-Mail: fkalantzis@home.com

Abstract

Feedforward neural networks are used to estimate reservoir properties. The neural networks are trained using known reservoir properties from well-log data, and the seismic waveform at the well locations. The trained neural networks are then applied to data from the complete seismic survey to generate a map of predicted reservoir properties. The theoretical analysis and tests with synthetic models, show that the neural networks are adaptive to coherent noise, and that random noise in the training samples can increase the robustness of the trained neural networks. This approach has been applied to a mature oil field in Northern Alberta, Canada, to explore the Devonian reef edge oil by estimating the product of porosity and net pay thickness. The resulting prediction map was used to select new well locations and design horizontal well trajectories. Four wells were drilled based on the prediction, and all were successful. This increased oil field production by about 20%.

W. Sandham and M. Leggett (eds.),
Geophysical Applications of Artificial Neural Networks and Fuzzy Logic, 143-156.

1. Introduction

One of the major applications of feedforward neural networks (FNNs) is mapping function approximation. Recent studies have shown that feedforward neural networks are, under very general conditions of the hidden unit activation function, universal approximators provided that sufficient hidden units are available (Hornik, 1991). These mapping function approximators are data driven and model free, which means that they may be established based on known input and output data pairs, with no pre-defined mathematical models required for the functional relation between input and output.

Reservoir properties at well locations, such as porosity and net pay thickness, can be derived reliably from core samples or well-log measurements. These properties vary laterally from one well to another. Normally, it is very difficult to predict the reservoir properties away from the wells, especially near the edge of an oil field. Seismic data, particularly 3-D surveys, contain valuable information about the lateral variation of reservoir properties. When there are wells inside the seismic coverage, it is natural to infer the reservoir property between the wells by interpreting the seismic data and using the reservoir property at well locations as spatial control points. In this case, a training sample set of input and output data pairs can be collected. The input is the seismic data around the reservoir interval at the well locations. The output is the reservoir property derived from the well data. Assuming that there exists a functional or statistical relationship between the seismic data and the reservoir property, FNNs can be applied to establish a model of the relationship using the training sample set. This model can then be used to predict the reservoir properties away from the wells (An and Moon, 1993).

A seismic data set contains information about subsurface geology and noise. The information induced by the change of reservoir properties is usually less obvious than that from other geological factors, such as reflections induced by the change of geological formation. From the point of view of directly predicting reservoir properties using the seismic waveform, the reflections near the reservoir may be treated as strong coherent noise. Obviously, one of the important issues regarding direct use of the seismic waveform to predict reservoir properties, is whether the FNNs are adaptive to coherent noise. Seismic data also contain random noise. A trained neural network must be robust enough to predict the change of reservoir properties accurately when random noise exists.

Preliminary studies show that FNNs, if trained properly, can identify small changes of reservoir properties even when the seismic data contain strong coherent and random background noise (An, 1994).

This chapter reviews, theoretically, the mapping function approximation using FNNs and coherent and random noise associated with this approach. Test results are shown using synthetic seismic data contaminated with strong sine (coherent) and random noise. A real example of a reservoir characterization project using this approach is given, and the advantages and pitfalls of the approach are discussed.

2. Neural Network Mapping Function Approximation

An FNN can be represented as a mapping function that is defined by a set of weights and connection parameters. Let $(\mathbf{x},\mathbf{y})$ be the input-output vector pairs, where $\mathbf{x}=[x_1,x_2,...,x_l]^T$ is an input vector and $\mathbf{y}=[y_1,y_2,...,y_m]^T$ is an output vector. A mapping function $\mathbf{G}$ between $\mathbf{x}$ and $\mathbf{y}$ can be approximated using an FNN with as few as one hidden layer such that,

$$\mathbf{y}=\hat{\mathbf{G}}(\mathbf{W},\mathbf{x})+\mathbf{e} \tag{1}$$

where $\hat{\mathbf{G}}$ is an approximation of $\mathbf{G}$; $\mathbf{W}$ represents the weights and network specifications; $\mathbf{e}=[e_1,e_2,...,e_m]^T$ is an error vector. When the activation functions are continuous, bounded and non-constant, by adjusting $\mathbf{W}$, $\mathbf{y}$ can be approximated by $\hat{\mathbf{G}}(\mathbf{W},\mathbf{x})$ to any degree of accuracy, provided that sufficient hidden units are available (Hornik, 1991). The mapping function $\hat{\mathbf{G}}(\mathbf{W},\mathbf{x})$ can be approximated from the sample data.

Consider a set of training samples,

$$\mathbf{S}_N=\{(\mathbf{x}_1,\mathbf{y}_1),(\mathbf{x}_2,\mathbf{y}_2),\cdots,(\mathbf{x}_N,\mathbf{y}_N)\} \tag{2}$$

where $(\mathbf{x}_i,\mathbf{y}_i)$ $(i=1,2,...,N)$ is a known input-output vector pair. $\mathbf{y}_i$ contains reservoir properties from the well-log data, and $\mathbf{x}_i$ can be a seismic trace or seismic attribute. A training algorithm, such as backpropagation (Rumelhart and Williams, 1986) - one of the most widely used and investigated algorithms - can be used to minimize the error term,

$$\mathbf{e}(\mathbf{W})=\frac{1}{2}\sum_{p=1}^{N}|\mathbf{y}_p-\hat{\mathbf{G}}(\mathbf{W},\mathbf{x}_p)|^2 \tag{3}$$

with respect to the training sample set $\mathbf{S}_N$. $|\cdot|$ denotes the Euclidean norm of a vector, and $\mathbf{y}_p$ is the desired target output vector when input vector $\mathbf{x}_p$ is presented to the network.

The approximation $\hat{\mathbf{G}}$ is localized and relies on the training sample set $\mathbf{S}_N$. The trained neural network can be applied only to the same area and play where the training samples are collected.

3. Coherent Noise in Neural Approximations

The difference in seismic response resulting from the variation of reservoir properties, is usually small if it is compared with reflections near the target layer. These reflections can usually be observed over a large area and are very useful for geological structure interpretation. However, in the context of identifying the small lateral variation of reservoir properties, these reflections may be treated as coherent noise, especially those immediately above the target layer. Hence, it is an essential requirement that FNNs should be able to calibrate themselves with the strong coherent noise and accurately identify the small lateral reservoir variations from the seismic data. This noise can be simulated by an identical background noise.

Let $\mathbf{x}' = \mathbf{x} + \mathbf{c}$ where $\mathbf{c} = [c_1, c_2, ..., c_l]^T$ is a noise signal vector. The training sample set then becomes,

$$\mathbf{S}'_N = \{(\mathbf{x}'_1, \mathbf{y}_1), (\mathbf{x}'_2, \mathbf{y}_2), \cdots, (\mathbf{x}'_N, \mathbf{y}_N)\} \tag{4}$$

where $\mathbf{x}'_i = \mathbf{x}_i + \mathbf{c}$ $(i = 1,\ 2, ..., N)$. The error to be minimized becomes,

$$\mathbf{e}(\mathbf{W}) = \frac{1}{2} \sum_{p=1}^{N} |\mathbf{y}_p - \hat{\mathbf{G}}(\mathbf{W}, \mathbf{x}'_p)|^2 \tag{5}$$

The neural network then adjusts **W** during the training process so that a new mapping function can be obtained. The new mapping function accepts input seismic trace $\mathbf{x}'_i$ $(i = 1,\ 2, ..., N)$, and delivers the same output $\mathbf{y}_i$ $(i = 1,\ 2, ..., N)$ as if **c** does not exist.

4. Effects of Random Noise in the Sample Data Set

Let $\mathbf{r} = [r_1, r_2, ..., r_l]^T$ be a vector of random noise variables independent of $\mathbf{x}$. The variation of the network output caused by the disturbance of $\mathbf{r}$ is,

$$\partial\, \mathbf{y} = \mathbf{G}(\mathbf{W}, \mathbf{x} + \mathbf{r}) - \mathbf{G}(\mathbf{W}, \mathbf{x}) \tag{6}$$

A sensitivity function can be defined as,

$$T(\mathbf{W}) = \sum_{p} \left\langle |\partial \mathbf{y}_p|^2 \right\rangle \Big/ \left\langle |\mathbf{r}|^2 \right\rangle \tag{7}$$

where $\langle \cdot \rangle$ is the statistical expectation of the argument. It is obvious that the smaller the $T(\mathbf{W})$, the more resistant the network is to random noise. Therefore, both the error [eq.(3)] and the sensitivity function [eq.(7)] should be minimized simultaneously.

If $\mathbf{r}$ satisfies,

$$\langle \mathbf{r} \rangle = \underset{\approx}{\mathbf{O}} \quad \text{and} \quad \left\langle \mathbf{r}\mathbf{r}^T \right\rangle = \sigma_r{}^2 \underset{\approx}{\mathbf{I}}, \tag{8}$$

where $\underset{\approx}{\mathbf{O}}$ is the zero matrix and $\underset{\approx}{\mathbf{I}}$ is the identity matrix. An alternative method to minimize both eq.(3) and eq.(7) is to introduce a vector of random noise variables $\mathbf{n} = [n_1, n_2, ..., n_l]^T$ into the training sample set (Matsuoka, 1992), which satisfies,

$$\langle \mathbf{n} \rangle = \underset{\approx}{\mathbf{O}} \quad \text{and} \quad \left\langle \mathbf{n}\mathbf{n}^T \right\rangle = \sigma_n{}^2 \underset{\approx}{\mathbf{I}} \tag{9}$$

The seismic image contains random noise and so do the training samples selected from the seismic data. When the random noise in the seismic data satisfies eq.(9), the random noise in the training samples selected from the seismic data should satisfy eq.(9), also. The training sample set $\mathbf{S}_N$ defined by eq.(2) becomes,

$$\mathbf{S}_N = \{(\mathbf{x}_1 + \mathbf{n}, \mathbf{y}_1), (\mathbf{x}_2 + \mathbf{n}, \mathbf{y}_2), \cdots, (\mathbf{x}_N + \mathbf{n}, \mathbf{y}_N)\} \tag{10}$$

The error is then,

$$\mathbf{e}(\mathbf{W}) = \frac{1}{2} \sum_{p=1}^{N} | \mathbf{y}_p - \hat{\mathbf{G}}(\mathbf{W}, \mathbf{x}_p + \mathbf{n}) |^2 \tag{11}$$

Minimizing eq.(11) is equivalent to minimizing both eq.(3) and eq.(7). It shows that the random noise in the training samples may improve the network's ability to resist random noise. On the other hand, more training samples may need to be included in the training to achieve this statistical property.

5. Tests with Synthetic Seismic Data

The synthetic model is designed to simulate a thin lens-shaped reservoir on top of a formation boundary. The velocity of the lens-shaped reservoir is about 4% lower than the formation where the reservoir is located. The objective is to predict the lateral extent of the lens-shaped reservoir.

The model with a target layer of 10 m ($\approx 1/8$ wavelength) thickness is shown in Figure 1. The corresponding synthetic seismic section using a 35-Hz Ricker wavelet is shown in Figure 2. Four traces located at ***S1***, ***S2***, ***S3*** and ***S4*** in Figure 2 were used to train the neural network. After training, all seismic traces were presented to the network to generate a prediction. Figure 3 shows the results. The lateral extent of the target formation was accurately predicted. The trained neural network exhibited very good generalization at the boundary.

To further test the adaptability of the neural network when strong coherent noise exists, such as stable strong reflections near the reservoir or multiples, sine

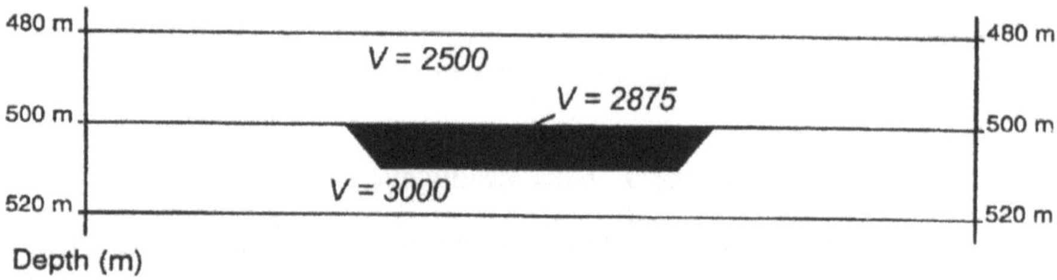

Figure 1. Lens-shaped reservoir model. Target thickness is about 1/8 wavelength.

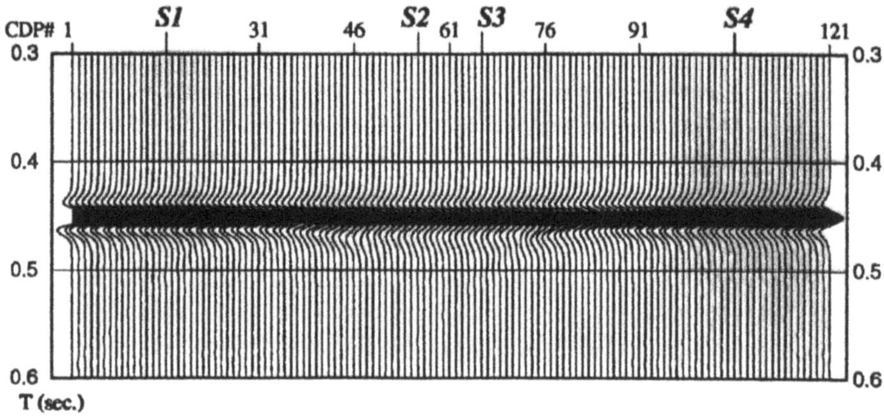

Figure 2. Synthetic section for the model in Figure 1.

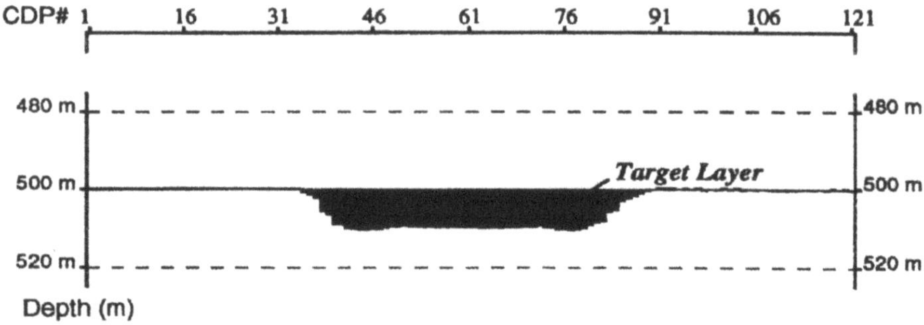

Figure 3. Neural network prediction of the lens-shaped model (shown in Figure 1).

wave noise was added to the synthetic data. This was of the same frequency as the Ricker wavelet used to generate the synthetic section, and of an amplitude twice the maximum amplitude of the synthetic seismic data.

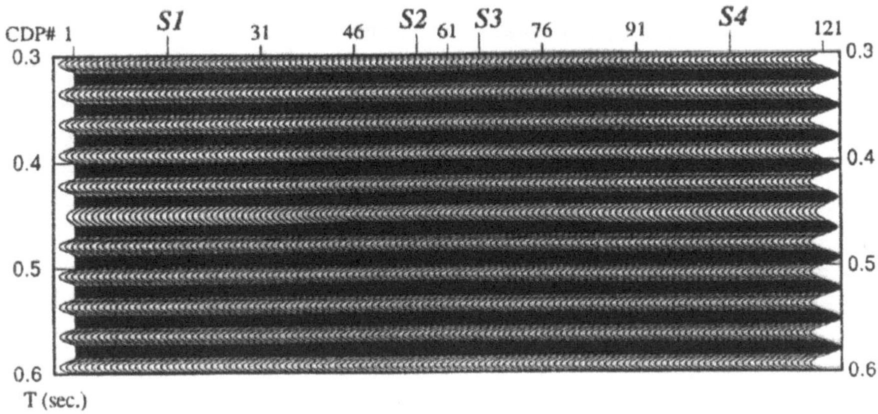

Figure 4. Strong sine signal added to simulate consistent coherent noise.

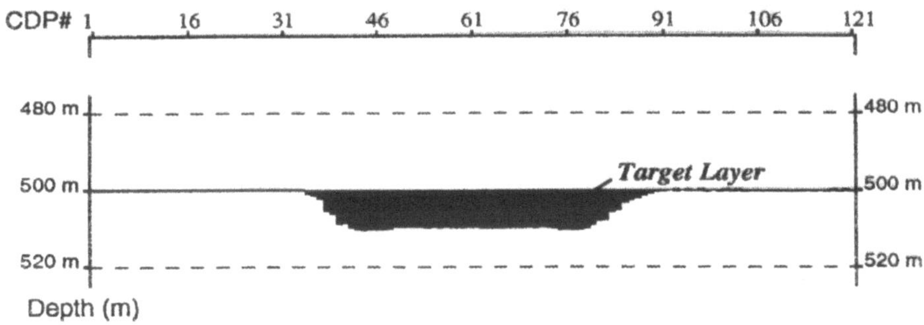

Figure 5. Neural network prediction from Figure 4.

Although the sine noise is not similar to the reflection or multiples, the effect on training and testing of the neural network should be the same, mathematically. The resulting section is shown in Figure 4. The neural network was trained using traces marked ***S1***, ***S2***, ***S3*** and ***S4*** in Figure 4, then all traces were presented to the trained neural network, which again accurately predicted the boundary (Figure 5).

Comparing Figure 5 with Figure 3, it can be seen that the two predictions, one from the data contaminated with strong sine wave noise (Figure 5) and the other from the data without the noise (Figure 3), are identical. This result is consistent with the theoretical analysis, presented earlier.

The synthetic seismic data (Figure 2) was next imposed on a random noise background (Figure 6). The ratio of maximum amplitude of synthetic signal to maximum amplitude of random noise is about 1.0. Four traces at ***S1***, ***S2***, ***S3***, ***S4*** from the synthetic data (Figure 2) were selected, and combined with random noise of a similar amplitude. The neural network was then trained using these noisy traces. All traces in Figure 6 were then presented to the trained neural network. The result in

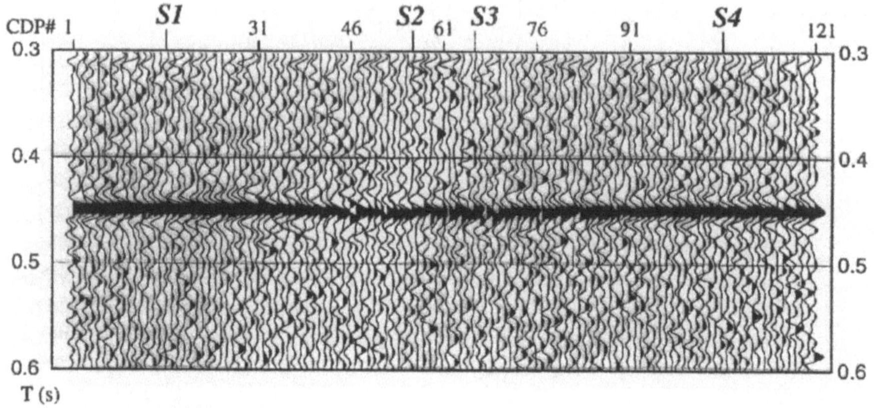

Figure 6. Synthetic data (Figure 2) added to strong random noise. Maximum amplitude of noise is equal to the maximum amplitude of the signal.

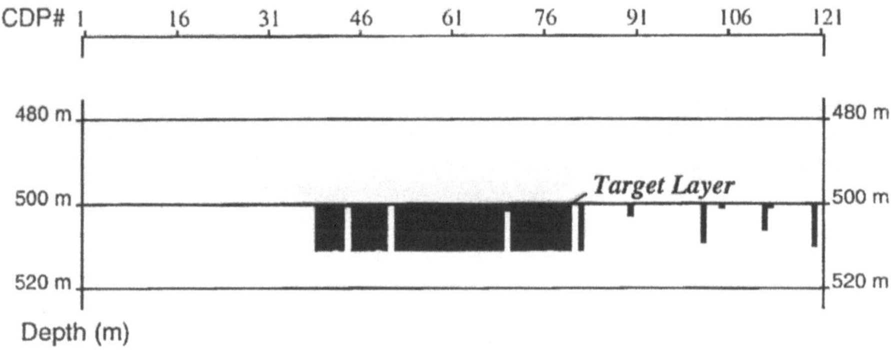

Figure 7. Neural network prediction from Figure 6.

Figure 7 shows that the boundaries of the target layer have been accurately predicted, although the smooth change of thickness at the boundaries has been lost.

Examples of further testing using synthetic seismic data, can be found in An and Moon (1993) and An (1994). Similar models with a target layer as thin as 3 m ($\approx \frac{1}{25}$ wavelength) were tested using strong additive sine noise, and the target boundaries were again accurately predicted. More complicated models with simultaneous changes in thickness and average velocity were also tested, and the trained neural networks again predicted both changes simultaneously.

6. Devonian Reef Porosity Mapping

The above approach, using a 3-D seismic waveform, was used to predict the product of porosity and net pay thickness (ΦH) in the Swan Hills Unit 1 oil field in Northern Alberta, Canada (Kalantzis and An, 1996). The net pay thickness is defined using 5% porosity cut-off and no water saturation. This is a mature oil field with 1,413 million bbl original oil in place; 550 million bbl of cumulative production; 22,100 B/D oil production; 220,000 B/D water production, average field water cut of 91.5% and 15% of production decline per year (production data as of 1996).

The oil in this field is mainly produced from a Devonian stromatoporoid limestone reef complex at a depth of about 2500 m. The reef complex is well developed and is composed of lagoonal, reef margin and foreslope facies. The reef is subdivided from its base to full buildup, into eight layers or stages with porosity as high as 20% in some layers ("hot streaks"). The reef complex is developed on top of an areal extensive and stable carbonate platform. There is a thin porous coral zone within the lower platform that produces oil and which is spatially stable in the study area. Above the reef complex is the Beaverhill Lake shale, which overlies directly the carbonate platform where the reef vanishes.

In 1992, Home Oil Company Limited in Calgary, Canada, conducted a 3-D seismic survey on the northeast edge of the Swan Hills Unit 1 oil field (Uffen, 1995). The objective of the survey was to delineate the reef edge, identify un-drained parts of the reservoir (e.g. up-dip clean reef edge oil), enhance recovery schemes (e.g. injection pattern design and monitoring) and mitigate drilling risk (e.g. drilling location and horizontal trajectory design). In this project, feedforward neural networks were used to estimate the distribution of ΦH of the Devonian reef using the 3-D seismic data. The results were used in the subsequent drilling location planning and trajectory design for horizontal drilling.

The 3-D seismic survey covers an area about 23 km^2, with 20m × 20m bin size and 37 wells. Input training samples were collected from the relevant interval of the seismic waveform from the well-tie traces. Output training samples were the product of reef porosity and net pay thickness (ΦH), derived from the corresponding well-log data. The well locations and their ΦH values are shown in Figure 8. Neural networks were used to model the relationship between the seismic waveform and the ΦH values.

The main result is the ΦH map from the neural network prediction (Figure 9). The validation tests show that the average prediction error is about 20%. The ΦH map outlines the reservoir boundary and predicted possible tight streaks inside the reservoir. It also shows possible porous reefs either connected to or isolated from the major reservoir. Three wells, one horizontal and two vertical, were drilled and were very successful.

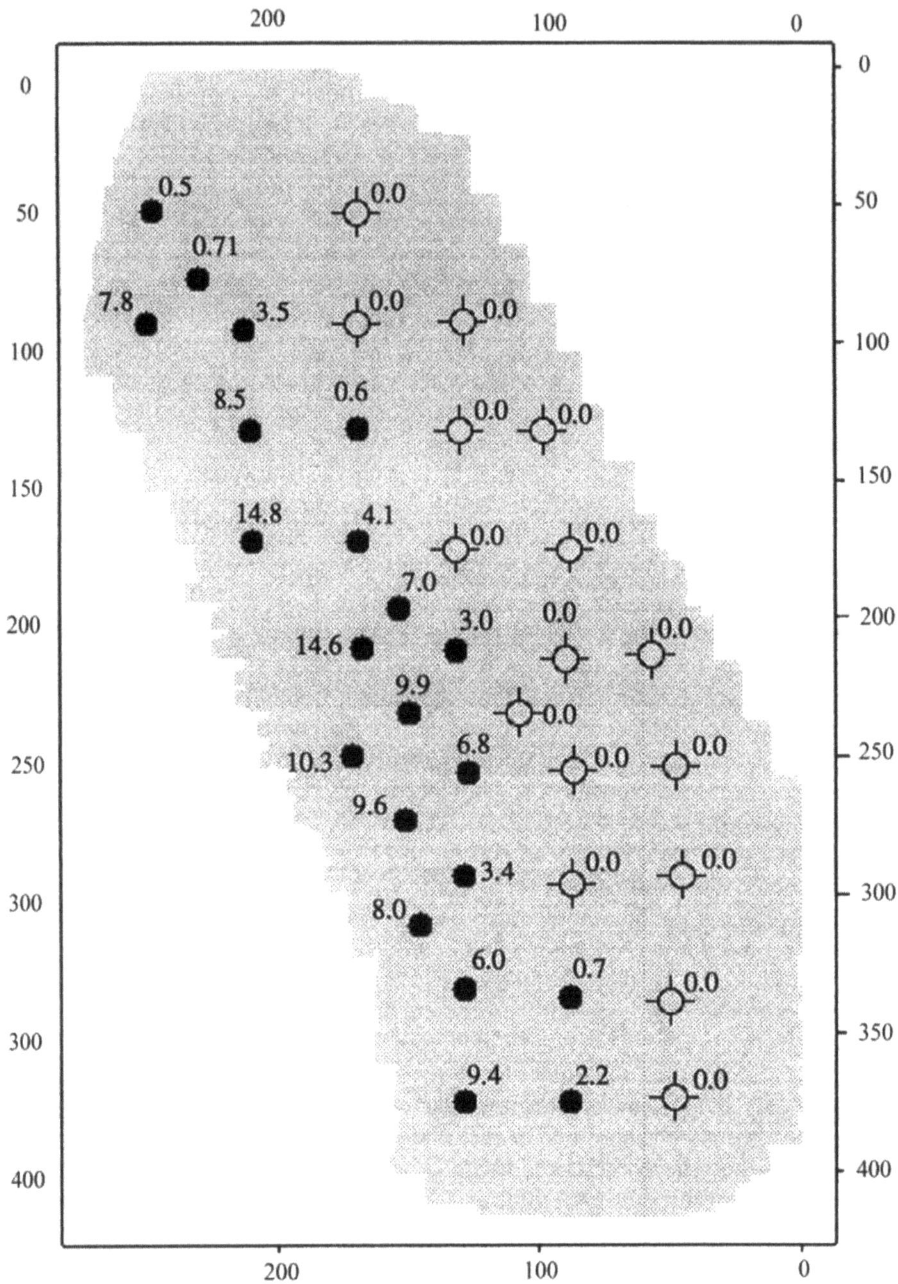

Figure 8. 3-D seismic survey bin map and well locations. The numbers at the well locations show the ΦH value from the well-log data.

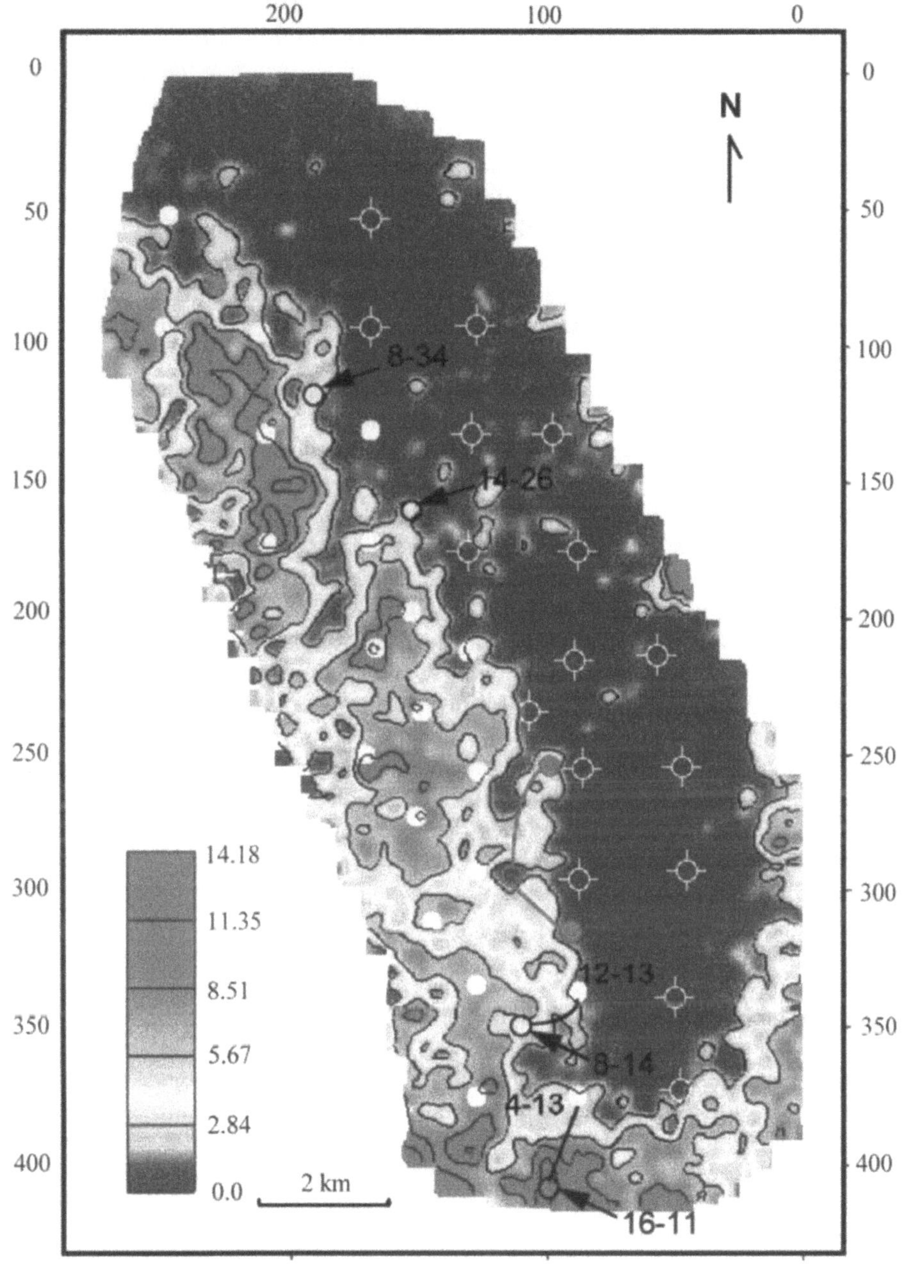

Figure 9. ΦH prediction map. The red well locations were drilled before the ΦH map was generated but were not used in training the neural network. Well 16-11 was drilled based on another 3-D survey, which overlaps the southern end of the study area.

Vertical well 8-34 drilled a north-to-south trend ΦH anomaly (Figure 9). It penetrated 16m of reef margin facies and was drilled into the platform. The predicted ΦH was approximately 4.5 around the well. The actual ΦH from the drilling results was approximately 3.2 in the reef, and 1.0 in the platform. The total encountered ΦH was 4.2.

Vertical well 14-26 drilled an isolated anomaly with a predicted ΦH of 3 around the well. The well penetrated 17m of fore-reef facies with a net pay thickness of 10m and was drilled into the platform with an additional 14m of net pay. The ΦH from the reef was 1.5, and 1.2 in the platform. The total encountered ΦH was 2.7. This well enabled the production of 750 B/D of clean oil, including the oil from the platform.

Vertical well 12-13 was virtually a dry hole, with a ΦH = 0.7. The inferred ΦH map (Figure 9) shows good reef porosity not far from the well. The 12-13 well-bore was re-entered and the 8-14 horizontal well was drilled along the indicated trajectory. As predicted by the ΦH mapping, a porous reef was encountered 95m away from the 12-13 vertical well. Drilling then penetrated a tight saddle and entered another porous zone. The well drilled 555m horizontally of fore-reef facies with 240m of net pay producing 450 B/D of clean oil.

Another 3-D seismic survey which overlaps the south part of the study area, was also processed using the above approach, and again generated a ΦH prediction map (Kalantzis and An, 1996). Based on the ΦH map, horizontal well 16-11 was successfully drilled by re-entering the 4-13 vertical well-bore. The well encountered a porous channel 134m away from the vertical well and penetrated 644m of channel, fore-slope, and reef margin facies. It drilled 444m of net pay and produced at 96% water cut and 80 B/D of clean oil.

The resulting ΦH maps provided valuable and accurate information for planning drilling locations and designing trajectories for horizontal drilling. The maps were also used to estimate volumetric reserves for economics, and identified previously undetected and undrained parts of the reservoir along the reef edge. The successful drilling of the above four wells added reserves and increased daily production of the field by 10%.

7. Advantages, Pitfalls, and Discussion of the Approach

The approach described above, using neural networks for reservoir property estimation, is, in general, model-free, nonlinear, data driven and localized to a specific area and play. The model-free capability is essential because there is realistically no known model between seismic data and reservoir properties, although it is known that seismic data may contain important information regarding reservoir properties. Hence, the nonlinear property is very useful for establishing a model between the input seismic data and output reservoir properties. The term "data-driven" is a counterpart of knowledge-driven. The knowledge-driven approach relies mainly on known theories and expertise to solve a problem; in this case, predict the reservoir properties. The knowledge may be from textbooks, publications or experience from other old fields.

The data-driven approach, on the other hand, uses the given data to generate a prediction, in general. There is no crisp boundary between data-driven and knowledge-driven. Interpreters inspect the data when trying to make a decision, and knowledge is required to evaluate and interpret the prediction from a data-driven approach. The neural network methodology described above is a data-driven approach. Neural networks are trained by sample data from a specific play in a project area. The trained neural network may be applicable only to the play in that area.

Neural networks can also be applied using seismic attributes or features as input, to estimate reservoir properties (Schultz et al., 1994). Seismic attributes are computed from the interval of seismic data relevant to the reservoir. They are representative of the seismic waveform. Each attribute should represent in a specific way, information pertaining to certain characteristics of the seismic waveform. Many different attributes can be computed from seismic data. Some of these are robust and some may be biased by noise. It normally takes significant effort to identify the subset of the attributes that are useful variables as input to a neural network.

Using the seismic waveform as input, all information is provided to the neural network. During training, the neural networks learn from the sample data, identify the relevant information, and establish a model (or mapping) between the input seismic waveform and the output reservoir properties. In this case, there is no concern about the possibility of missing useful information during the process of computing attributes or features.

The number of training samples (control wells) required for a statistically stable prediction depends on the quality of the seismic data and complexity of the geology in the study area. Theoretically, fewer training samples may be required for the same prediction accuracy, in an area of simple geology with good quality seismic data, than in an area of complex geology with moderate quality seismic data.

As with many other neural network applications, there is not a standard neural network paradigm that can be followed. During training, neural networks with different architectures can be trained to satisfy a common standard or accuracy with the same training sample set. For prediction, neural networks with different architectures are not guaranteed to provide the same prediction. Although some tests using synthetic data show that the predicted results are very similar when different numbers of hidden nodes are used (An and Moon, 1993), a predictive validation should always be carried out to determine an optimal architecture.

Predictive validation is also recommended for choosing an optimal seismic data window size. Neural networks can be trained satisfactorily in terms of training error using different window sizes. Predictions using different window sizes can be significantly different. A larger window size may contain more information about the reservoir, since the seismic events below the reservoir have to penetrate the reservoir twice to return to the ground. A larger window size may also include possible geological change below the reservoir. The effect on the neural network is the same as increasing the complexity of the background geology. It normally increases the required number of training samples for an accurate prediction. A predictive validation test may be performed to determine an optimal window size.

8. Concluding Remarks

Theoretical analysis, synthetic data tests, and application to a Devonian reservoir reef edge oil exploitation project, have shown that feedforward neural networks and seismic waveforms are a very powerful approach to characterize reservoir properties. Tests using synthetic data have demonstrated that FNNs can identify small changes in reservoir properties even when there exists strong coherent noise. FNNs can also accurately identify reservoir boundaries when the synthetic seismic data are in a strong random noise background. This approach has been successfully applied to 3-D seismic data to delineate a Devonian reef edge, identify areas with undrained reef edge oil, and predict an areal distribution of reservoir quality. The neural network prediction of ΦH was used to estimate volumetric reserves and to plan drilling locations and horizontal well trajectories. Drilling results showed that the neural network predicted the ΦH distribution accurately.

Acknowledgments

The authors wish to acknowledge the support of Schlumberger-GeoQuest in the preparation and publication of this manuscript. The authors would like to thank Home Oil Company and its unit partners for allowing publication of the data and results. Partial research support to W. M. Moon was funded by NSERC of Canada (Operating Grant #7400). We wish to acknowledge Doug Uffen, Judi Dobin and Brend Arnell of Home Oil Company for their support and contribution to the Devonian reef case study, which was performed while authors Ping An and Fotis Kalantzis were affiliated with Signal Estimation Technology and Home Oil Company.

References

An, P., and Moon, W. M., 1993, Reservoir characterization using feedforward neural networks: 63rd Ann. Internat. Mtg., Soc. Expl. Geophys., Expanded Abstracts, 258-262.

An, P., 1994, The effect of random noise in lateral reservoir characterization using feedforward neural networks: 64th Ann. Internat. Mtg., Soc. Expl. Geophys., Expanded Abstracts, 787-790.

Kalantzis, F., and An, P., 1996, Porosity mapping in a Devonian reef using 3-D seismic waveform and neural networks: 58th Ann. Internat. Mtg., Eur. Assn. Geosci. Eng., Extended Abstracts, X004.

Hornik, K., 1991, Approximation capabilities of multilayer feedforward networks: Neural Networks, **4**, 251-257.

Matsuoka, K., 1992, Noise injection into inputs in back-propagation learning: IEEE Trans. Syst., Man, Cybern., **22**, 436-440.

Rumelhart, D. E., and Williams, R. J., 1986, Parallel distributed processing: exploration in microstructures of cognition: Vol. 1, MIT Press, Cambridge, NJ.

Schultz, P. S., Ronen, S., Mantran, P., Hattori, M., and Corbett, C., 1994, Seismic-guided estimation of log properties, Part 3: a controlled study: The Leading Edge, **13**, No. 7, 770-776.

Uffen, J. D., 1995, Swan Hills Unit #1: adding value with seismic through reservoir delineation and characterization: 65th Ann. Internat. Mtg., Soc. Expl. Geophys., Expanded Abstracts, 115-117.

AN INFORMATION INTEGRATED APPROACH FOR RESERVOIR CHARACTERIZATION

Joint Lithologic Inversion

LI-YUN FU

CSIRO Petroleum,
PO Box 1130, Technology Park,
Bentley, Perth,
W.AUSTRALIA
E-Mail: lfu@per.dem.csiro.au

Formerly Institute of Tectonics,
University of California at Santa Cruz,
Santa Cruz, CA 95064,
USA
E-Mail: lfu@es.ucsc.edu

Abstract

Ambiguous dependence of observed data related to lithologic parameters suggests that practical lithologic inversion problems are characterized by both deterministic mechanism and statistical behavior. The Caianiello neural network method is presented in this paper, including neural wavelet estimation, input signal reconstruction, and nonlinear factor optimization. A joint inversion scheme for porosity and clay-content estimations is established based on the combination of the Caianiello neural network with some deterministic petrophysical models. First, inverse neural wavelets are extracted using known solutions, and then the inverse-operator-based inversion is used to estimate an initial parameter model. Second, forward neural wavelets are estimated likewise, and then the forward operator-based reconstruction can improve the initial parameter model. The scheme has been applied in a complex continental deposit in western China and significantly improves the spatial description of reservoirs.

1. Introduction

A reservoir system is generally composed of both the minimal set of parameters describing the system and the physical relationships relating the parameters to the results of measurements on the system. On the other hand, some parameter properties of the system are statistically described. This results in a probabilistic description of the

W. Sandham and M. Leggett (eds.),
Geophysical Applications of Artificial Neural Networks and Fuzzy Logic, 157-178.

measured results. Therefore, the reservoir system is a typical physical system that is characterized by both deterministic mechanism and statistical behavior. The goal of reservoir characterization is to provide a comprehensive and detailed picture of the reservoir distribution by using both deterministic and geostatistical approaches to integrate and analyze vast amounts of data generated from different associated disciplines. A comprehensive integrated information system is needed to provide not only an extensive database to store information, but more importantly, a capacity for efficient and comprehensive data analysis in the reservoir characterization problem-solving environment. An artificial neural network system for space- and time-varying signal processing, as an integrated approach proposed in this paper, will provide significant potential for constructing this integrated systematic platform oriented to reservoir detecting, characterizing, and monitoring.

Reservoir characterization problems are strongly related to inexact data (e.g., information-incomplete, information-overlapping, and noise-contaminated),ambiguous physical relationships, and the needs for adaptive fast history matching. In principle, the neural network-based scheme is an optimal integrated method due to its well-known advantages. In contrast to conventional pattern recognition and statistical techniques, the advantages of neural networks arise from their powerful pattern matching and nonlinear mapping capabilities with adaptive learning, tolerance to component failure, and robustness to noises. The advantages of neural networks are expected to make them most suitable for comprehensively analysing large volumes of data to identify relationships, recognise patterns, learn associations and automatically make predictions for reservoir characterization. During the last decades, artificial neural networks have been extensively used in exploration geophysics, for example, automatic first-break picking and seismic data trace editing (McCormack et al., 1993; Hart, 1996), seismic deconvolution (Wang and Mendel, 1992; Essenreiter and Karrenbach, 1996), mineral and lithologic identification from well logs (Baldwin et al., 1990; Rogers et al., 1992; Huang et al., 1996), seismic horizon picking (Huang, 1990; Leggett et al., 1994; Clark et al., 1996), location of subsurface targets (Poulton et al., 1992), reservoir clustering analysis (Zhang et al., 1995), velocity picking (Fish and Kasuma, 1994) and automatic NMO correction (Calderon-Macias et al., 1996), mineral exploration (Pan, 1996; Bolt et al., 1997), and a number of applications to reservoir parameter prediction (An and Moon, 1993; Ecoublet et al., 1997; Himmer and Link, 1997; Wang et al., 1997).

The algorithms used for the above tasks are generally based on conventional neural networks. For instance, the Hopfield network is used to solve a problem through relating the network cost function to the objective function of the problem. This may not fully employ the statistical population codes of numerous neurons to increase the robustness of the algorithm. Conversely, the multilayer perceptron (MLP) is often used to perform inversion mappings based on a black-box training. However in some cases, the rules obtained by neural networks through learning from known solutions cannot account for the underlying physical relationship between datasets. In addition, most traditional neural networks have found limited use due to the lack of fast network operation and convergence algorithms for vast amounts of data processing. A new neural network method for information integration is developed in this paper with an attempt to overcome the disadvantages of conventional methods. It consists mainly of the following contents.

In most cases, the data available for reservoir characterization are signals with different scales of spatiotemporal distributions. However,most existing neural networks are based on the MP neuron model (McCulloch and Pitts, 1943) and are unsuited to process spatiotemporal information. The neural network presented in this paper is based on the well-known Caianiello neuron model (Caianiello, 1961). The corresponding neural wavelets, acting as the weight functions of the network, can be constructed with a particular amplitude-phase spectrum in terms of temporal frequencies, and hence provide the Caianiello neural network with an information processing ability for time-varying patterns. The Caianiello neuron has been extended into a 4D filtering neuron to include spatial frequencies for both space- and time-varying signal processing (Fu, 1999a).

Regularly, one of the major ways to use neural networks for reservoir characterization is implicitly through network training with several different examples of solutions to a problem. The examples selected as the solutions become very important to the problem, even more than the neural network method itself. This black-box-based mapping requires that the examples be selected with enough information representation to cover the underlying physical relationship inside the problem. This is a comparably difficult task. However, if some deterministic physical mechanisms about the reservoir system under study are introduced into the Caianiello neural network via the activation functions of neurons, the black-box mapping may become a grey-box problem with inexact input data and approximate physical relationships. It will reduce the harsh requirements for the examples used to train the network and meantime take advantage of the statistical population codes of neurons.

The widely used model-based seismic lithologic inversion is to adjust an initial parameter model in an iterative manner until a good fit between the modeled response of the physical system and the actual observations is achieved. Perturbing the inputs and observing the response of a neural network with the hope of achieving a better fit between the real and desired outputs will lead to a model-based inversion method using neural networks. I will show in this paper that the Caianiello neural network can not only update its neural wavelets so that it can adapt to the current information environment, but also adjust the input information environment to adapt to the trained network. The combination of neural wavelet estimation and input signal reconstruction of the Caianiello neural network will provide a potentially significant application to reservoir parameter inversion and reservoir numerical modeling.

The Caianiello neural network has been successfully incorporated with the Robinson seismic convolutional model (Robinson, 1957) to construct a joint inversion scheme for single-well-controlled seismic impedance inversion (Fu, 1995; Fu et al., 1997) and multi-well-controlled seismic impedance inversion (Fu, 1997; Fu, 1999b). In this paper, the neural network will be further modified for reservoir parameters estimates. Following a brief introduction to recently developed reservoir parameter estimation techniques, a joint lithologic inversion scheme is established based on the combination of the Caianiello neural network with deterministic petrophysical models that relate acoustic properties to reservoir parameters (e.g., porosity and clay content).

The deterministic petrophysical models are used to provide a physical relationship for the Caianiello neural network. Finally, the scheme is applied in a complex continental deposit in western China to estimate porosity and clay content by integrating well data, seismic impedance data, and geological knowledge. The results show that the joint lithologic inversion can significantly improve the spatial description of reservoirs.

2. Lithologic Inversion from Seismic Data

Traditionally, the petrophysical properties are distributed laterally by using an interpolated contour map according to the detailed zonation of wells so that well-to-well correlation is performed on a fence diagram through several wells. Seismic data can be incorporated into porosity estimation (Maureau and Van Wijhe, 1979; Angeleri and Carpi, 1982; de Buyl et al., 1986) by means of the porosity-velocity relationship expressed mainly by Wyllie's time-average equation (Wyllie et al., 1958). These results may capture large-scale lateral variations, but contain less detail. Much effort has been made to combine seismic data, well data, and other information into the lithologic inversion for modeling detailed geological heterogeneities. These approaches can be grouped below into two categories of deterministic and statistical.

The former assumes a deterministic forward relation for the lithologic inverse problem. A generalized linear inversion method (Lines and Treitel, 1984) is commonly used to construct algorithms for porosity estimation from seismic data (Gelfand and Larner, 1984; Martinez, 1985; He and Reynolds, 1995). An initial guess of the porosity model becomes important to assure algorithm convergence. To obtain a reasonable initial parameter model as well as alleviate the nonuniqueness problem inherent in the inversion to some degree, more exact treatments are presented with the initial parameter model determined by the petrophysical-based forward modelings (Neff, 1990). These synthetic seismograms systematically demonstrate the effects of changes in petrophysical properties (porosity, shale volume, and saturation) on seismic waveform responses. As a result, the Robinson seismic convolutional model and a petrophysical model that is obtained from the modification of the Raymer's equation (Raymer et al., 1980) to include shaly sand and gaseous hydrocarbons are combined to construct a model-based inversion for porosity and pay-thickness estimates (Burge and Neff, 1998). It is clear that the deterministic lithologic inversion methods rely on an accurate petrophysical model transforming petrophysical parameters into acoustic properties. The accurate model can narrow the deviations occurring in the transformation and improve algorithm convergence.

To overcome the disadvantages of the deterministic methods, statistical techniques based mostly on Bayesian and kriging (including cokriging) methods are becoming increasingly popular for lithologic parameter prediction (Tarantola and Valette, 1982; Jackson and Matsu'ura, 1985; Thadani et al., 1987; Doyen, 1988; Duijndam, 1988; Lortzer and Berkhout, 1992). These methods demonstrate the ability to integrate different sources of information, to incorporate spatial patterns of the variations of reservoir properties, and to provide measures of uncertainty. A probabilistic forward model is assumed in the inversion to relate seismic properties to lithologic parameters.

Furthermore, statistical relationships between log properties and seismic attributes can be built using geostatistical techniques from well data and adjacent seismic traces, and then applied to seismic-guided statistical interpretation of log properties between wells (Schultz et al., 1994; Fournier and Derain, 1995; Russell et al., 1997). Obviously, the successful applications of these methods require that the statistical relationship be constructed to cover such a complex physical system as the reservoir which is primarily described by deterministic theories with the minimal set of parameters.

In general, the main difficulties arising in the solution of lithologic inverse problems are instability, nonuniqueness, and uncertainty. These problems are strongly related to inexact observed data and ambiguous physical relationships. The deterministic inversion approaches are less robust and suffer from uncertain factors; conversely, if the physical relationship for the system is assumed to be statistical only, spurious results obtained by the statistical inversion are sometimes unavoidable, particularly in the cases of sparse well control and low signal-to-noise ratio data applications. The optimal inversion method is the one with the ability to aptly merge certain deterministic physical mechanisms into a statistical algorithm.

3. Methodology

3.1. CAIANIELLO NEURON-BASED MULTILAYER NETWORK

The well known Caianiello neuron equation (Caianiello, 1961) is defined as

$$o_i(t) = f\left(\sum_{j=1}^{N} \int_0^t w_{ij}(\tau) o_j(t-\tau) d\tau - \theta_i(t)\right), \tag{1}$$

where the neuron's input, output, and threshold are represented by $o_j(t)$, $o_i(t)$, and $\theta_i(t)$, respectively, $f()$ is the neuron activation function, and $w_{ij}(t)$ is the time-varying connection weight (neural wavelet) with which the firing of the *j*th neuron affects the *i*th neuron after the t time-units. The neuron eq.(1) represents a neuron with its spatial integration of inputs being a dot-product operation similar to the MP model, but with its temporal integration of inputs being a convolution. The neuron's filtering mechanism, intrinsically, is that its neural wavelets crosscorrelate with the inputs from other neurons, and large correlation coefficients denote a good match between the input information and the neuron's filtering property. The neurons with similar temporal spectra gather to complete the same task using what are known as statistical population codes. The adaptive changes of the neural wavelets in eq.(1) can provide the neuron with immense learning power.

The architecture of a multilayer network based on the Caianiello neuron is similar to the conventional MLP neural network (Rumelhart et al., 1986), except that each parameter becomes a time sequence instead of a constant value. Figure 1 illustrates the Caianiello neural network architecture with three layers. Each connection line has an associated neural wavelet (weight) $w_{ij}(t)$ that scales signals travelling between the *i*th neuron and the *j*th neuron in different layers. Each neuron receives a number of time signals from other neurons and produces a single signal output that can fan out to other neurons. If the dataset used to train the neural network consists of an input matrix $o_i(t)$ ($i = 1,2,\ldots,I$, where I is the number of input time signals) and the desired output matrix $o_k(t)$ ($k = 1,2,\ldots,K$, where K is the number of output time signals), one can select an appropriate network architecture with I neurons in the input layer and K neurons in the output layer. For a general problem, one hidden layer between the input and output layers is enough.

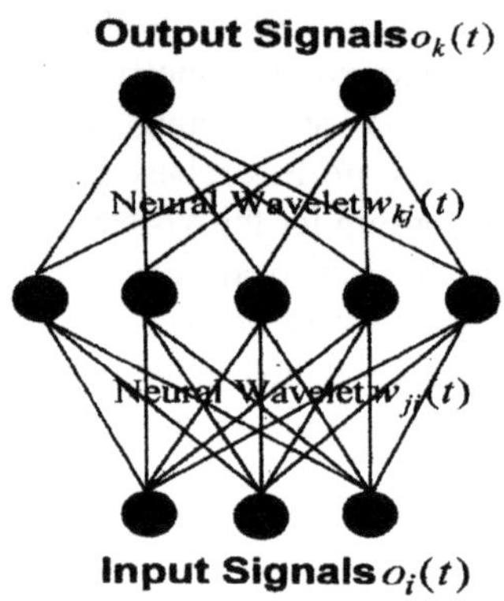

Figure 1. Three-layer Caianiello neural network architecture.

The advantages of the Caianiello neural network over the traditional multilayer network are temporal pattern processing, fast operation with FFTs, and fast convergence by adaptive changes of learning rate. These features enable the network to process volumes of seismic data rapidly. In addition, lithologic inversion using zero-offset seismic data often treats individual seismic traces as spatially independent observations and neglects the spatial correlation in the variations of reservoir properties. The spatial summation in eq.(1) offers an ability to simultaneously consider a number of seismic traces in the inversion. The mapping ability of the Caianiello neural network results mainly from the nonlinear activation function in eq.(1). In general, the sigmoid nonlinearity of neurons is used. In the joint inversion scheme discussed in this study, some geophysically meaningful transforms can be used instead of the conventional sigmoid activation function.

3.2. DETERMINATION OF NONLINEAR PETROPHYSICAL MODEL

The nonlinear transform function in the neuron model of eq.(1) provides a possible means to incorporate some deterministic petrophysical relationships into the Caianiello neural network for the joint lithologic inversion. It has been recognized that the variations in reservoir acoustic properties integrate the effects of a number of different geologic variables such as lithology, porosity, clay content, fluid saturation, pore pressure, temperature, etc. Although the joint lithologic inversion scheme presented in this paper can perform multiparameter estimation, it has been a formidable task to build a deterministic petrophysical model that can take into account all the variables simultaneously. If the joint lithologic inversion is aimed at using amplitude-preserved seismic data to achieve a lateral lithology variation that can basically reflect the relative changes in the real model of the subsurface reservoir, we can only consider the effects of such important rock parameters as porosity and clay content, and neglect other parameters.

During the last two decades, much effort has been made to reach a detailed understanding of the relationship between reservoir acoustic properties and porosity and/or clay content. The Wyllie's time-average equation ($1/v_p = (1-\phi)/v_m + \phi/v_f$, where v_m is the P-wave velocity of the rock matrix and v_f is the velocity of the pore fluid), due to its simplicity, has been widely used to perform a velocity-to-porosity transform valid in the middle range of porosity. The Raymer's equation ($v_p = (1-\phi)^2 v_m + \phi v_f$) was proposed as a modification of the time-average equation by suggesting different laws for different porosity ranges (Nur, 1998). The two models appear adequate for clay-free sandstones, but fail to describe shaly sandstones. Numerous advances later are involved with the combined effects of porosity and clay content on acoustic properties (Tosaya and Nur, 1982; Kowallis et al., 1984; Han et al., 1986). It is noteworthy that the Han's linear relation ($v_p = A_0 - A_1\phi - A_2 c$, where c is the clay content, and the coefficients A_0, A_1, and A_2 can be obtained by least-square regression) fits laboratory data quite well for a variety of lithologies in a relatively wide range of porosity. This suggests that empirically-derived multivariate linear regression equations can be constructed by relating the velocities to porosity and clay content.

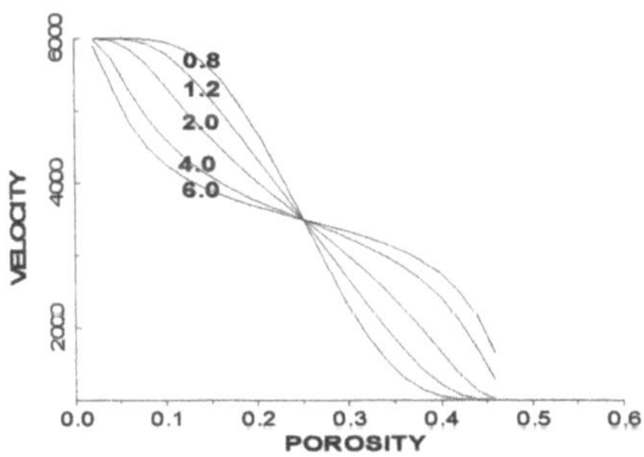

Figure 2. A schematic diagram from eq.(2) showing the acoustic velocity versus porosity under different nonlinear factors.

The above deterministic models can be used as the neuron's activation function $f()$ in the Caianiello neural network to estimate porosity and clay content. Considering the lithologic inversion in complex geological environments, an empirically-derived, relatively flexible model is presented here with the intention to fit well log data for unconsolidated sandstone reservoirs in the continental sediments in western China,

$$\frac{\phi_m(\phi_m - 2\phi)}{\phi(\phi_m - \phi)} = \lambda \ln\left(\frac{v_p - v_f}{v_m - v_p}\right), \tag{2}$$

where ϕ_m is the maximum porosity of sandstones in the reservoir under study and λ is a nonlinear factor that can nonlinearly adjust the functional form of the equation to an appropriate shape for practical data points. Figure 2 shows the performance of the model with different nonlinear factors. We see that for a fixed value of λ, different porosity ranges have distinct gradients which reflect different velocity-porosity relationships. On the other hand, different λ values corresponding to different curves can give different models applied to different types of rocks. Moreover, we can "fudge" the v_m, v_f, and ϕ_m values for various lithologies and fluids to match practically any dataset in complex deposits. This simple equation can also model laboratory data by the appropriate selections of λ for different types of rocks. Several aspects are considered below for its construction and applications to lithologic inversion.

Regularly, for a deterministic lithologic inversion, first a reservoir zonation of wells is performed, then an appropriate petrophysical model relating well properties to seismic attributes (e.g., seismic transit times and impedances) is chosen for each reservoir zone according to the lithology and fluid content of the reservoir. These seismic attributes are inverted away from wells along a short time window defined by the picks of the top and bottom of the reservoir interval. The joint lithologic inversion presented in this study, however, is performed over the entire seismic trace or the range matched to the valid segment of a well, which contains a wide range of lithologic composition, even covering distinctly different depositional environments. A flexible expression to include various lithologies in the same model can be obtained from eq.(2),

$$\frac{\phi_m(t)(\phi_m(t)-2\phi(t))}{\phi(t)(\phi_m(t)-\phi(t))}=\lambda(t)\ln\left(\frac{v_p(t)-v_f(t)}{v_m(t)-v_p(t)}\right), \tag{3}$$

where $\phi(t)$ is the porosity curve in vertical time and $v_p(t)$ is the velocity curve. The accurate estimation of the time-varying nonlinear factor $\lambda(t)$ for different lithologies at different depths is a crucial point in the applications of the model to the joint lithologic inversion. Fortunately, the Caianiello neural network provides an optimization algorithm to iteratively adjust $\lambda(t)$ in adaptive response to lithologic variations vertically along a well log. I will discuss this algorithm in the next section.

Recent research (Neff, 1990, 1993; Vernik and Nur, 1992; Vernik, 1994) indicates that because the petrophysical properties of reservoir units are highly variable vertically and horizontally, accurate porosity estimate or lithology prediction from acoustic velocities requires the determination of petrophysical relationships to be based on the detailed petrophysical classification, particularly in unconsolidated sediments. Figure 3 is one of the cases proposed by Burge and Neff (1998), which illustrates the distinct variation in the impedance versus porosity relationship due to the lithologic variation and the change in fluid type of gas condensate versus water within the siliciclastic unit, each distinct lithologic unit having a unique set of petrophysical constants and equations. From the figure, we see that the rule from eq.(2) can also describe the impedance-porosity relationships for different lithologic units. This indicates that the eq.(3) may provide a possible means to facilitate implementation of this petrophysical classification scheme for practical lithologic inversion.

A class of functions similar to eq.(2) and their evolving versions have been widely applied to describe a physical process with stronger state variations occurring in the early and late stages than in the middle. This physical phenomenon widely exists in the natural world. This implies a local sudden change occurring in the process. In fact, numerous experimental data from rock physics laboratories suggest that there exists a critical porosity that separates the entire porosity range (from 0 ~ 100%) into different porosity domains with different velocity-porosity behavior (Nur, 1992; Nur et al., 1998). This critical porosity becomes a key to relating acoustic properties to porosity for the reservoir interval with a remarkably wide range of porosity distribution. The nonlinear transform of eq.(2) is constructed with an attempt to apply the critical porosity concept to the joint lithologic inversion.

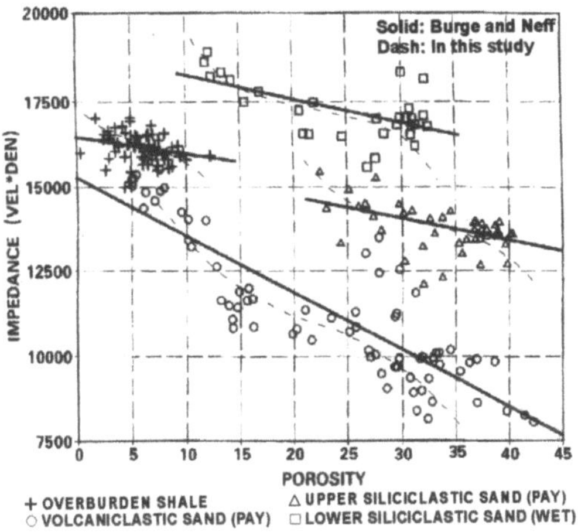

Figure 3. Acoustic impedance versus porosity for a well in a Gulf of Mexico field. Note that the different lithologies have distinct relationships of impedance to porosity (trend curves) and different regions of scatter distributions of data points in the impedance-porosity space. This figure from Burge and Neff (1998), used with the permission of *The Leading Edge*.

The last aspect considered for the application of eq.(2) is explicitly illustrated in Figure 3. That is, a certain value of acoustic impedance corresponds to a wide range of porosity units. An experimental data-based crossplot of compressional velocity against porosity for 104 samples of carbonate-poor siliciclastics at 40 MPa effective stress (compiled from Han et al., 1986; Klimentos and McCann, 1990; Domenico, 1977; and Vernik and Nur, 1992) shows that the scatter of the data point distribution reaches up to about 15 porosity units at a certain value of compressional velocity. This implies that the scale of such a parameter as velocity is not matched to that of porosity in rocks; the statistical behavior of the reservoir system should be considered by a probabilistic description of the measured results; or the effects of other parameters should also be incorporated to narrow the ambiguity. In this case, even if the velocities of both solid and fluid phases are estimated correctly, no optimal petrophysical models can yield accurate porosity estimates. To alleviate this problem to some degree in lithology or porosity prediction from seismic velocities, more exact treatments are proposed based on the detailed petrophysical classification of siliciclastics (Vernik and Nur, 1992;

Vernik, 1994) and the use of shear-wave velocities (Han et al., 1986).

The joint lithologic inversion scheme presented in this study takes this problem into account by crosscorrelating a scanning operator with porosity curves. Neural wavelets in the Caianiello neural network provide an effective means to facilitate the implementation of this strategy. The use of neural wavelets cannot narrow the deviations of data points from the trend given by the deterministic velocity-porosity eq.(3) unless other seismic attributes are incorporated, but can cover the deviations with a boundary of arbitrary shape to make a distinction between different lithologies. This is actually an integration of the neural network-based pattern classification with deterministic velocity-porosity equations, which can provide an areal approximation to velocity-porosity datasets. Especially in the case of shale shown in Figure 3, the pattern classification will be dominant in the procedure of lithologic inversion. The aperture of the neural wavelet depends on the range of scatter distribution of data points. Sandstones containing different amounts of volume clay have different regions of scatter distributions of data points in the velocity-porosity space as well as different deviation levels from the trends, which correspond to different apertures and spectral contents of neural wavelets.

Similarly, an empirically-based, simple deterministic relationship between acoustic velocities and clay content for clay-rich sandstones can be given as follows:

$$\frac{c_m(c_m - 2c)}{c(c_m - c)} = \lambda \ln\left(\frac{c_m(v_p - v_c)}{c_m(v_m - v_p) - c(v_m - v_c)}\right), \tag{4}$$

where v_c is the velocity of the pure clay and c_m is the maximum value of the clay content in the reservoir interval under study. The discussions about eq.(2) can also be applicable to this model. This model will be used in the joint lithologic inversion for clay-content estimate. In general, the effect of clay content on acoustic velocities for clay-bearing sandstones is next to that of porosity. If appropriate petrophysical models are available, the incorporation of seismic waveform information into the joint lithologic inversion due to its stronger sensitivity to the changes in lithologic parameters than velocities (Klimentos and McCann, 1990; Neff, 1990) will allow for more accurate clay-content estimates than only using velocity information.

3.3. NEURAL WAVELET ESTIMATION

The neural wavelet of each neuron in the Caianiello network can be adjusted iteratively to match the input and desired output signals. The cost function for this problem is the following mean-square error performance function

$$E = \frac{1}{2}\sum_k \sum_t e_k{}^2(t) = \frac{1}{2}\sum_k \sum_t [d_k(t) - o_k(t)]^2 , \tag{5}$$

where $d_k(t)$ is the desired output signal. The changes in the neural wavelet for any neuron of all the layers are in the same form. For the jth layer, we have

$$\Delta w_{ji}(t) = \eta(t)\delta_j(t) \otimes o_i(t), \tag{6}$$

where $\otimes$ is the crosscorrelation operation symbol and $\eta(t)$ is the learn rate, which can be determined by automatic searching. For the output layer, the backprop error $\delta_k(t)$

through the kth layer is expressed as

$$\delta_k(t) = e_k(t) f'(net_k(t) - \theta_k(t)), \tag{7}$$

with $net_k(t) = \sum_j w_{kj}(t) * o_j(t)$. For any hidden layer, $\delta_j(t)$ is obtained by the chain rule

$$\delta_j(t) = f'(net_j(t) - \theta_j(t)) \sum_k \delta_k(t) \otimes w_{kj}(t). \tag{8}$$

Compared with the information forward propagation formulated by temporal convolution operations, the error back propagation and neural wavelet update take crosscorrelation operations. A block frequency-domain implementation with FFTs for the forward and back propagation can be used in the Caianiello network.

3.4. INPUT SIGNAL RECONSTRUCTION

In general, artificial neural networks are used first through learning (weight adjustments) in an information environment known both for the inputs and the desired outputs. Once trained, they can be applied to any new input dataset in a new information environment with known inputs but unknown outputs. However, the new information environment, in many cases, is present with known or incomplete outputs, but with unknown or inexact inputs. Therefore, the new information environment also needs to be changed to adapt to the trained neural network.

Relating the adjustments in the inputs to the known neural wavelets and desired outputs leads to a model-based algorithm for input signal reconstruction. The cost function for this case is similar to eq.(5). For instance, the input signal modification in the hidden layer of the network shown in Figure 1 can be formulated as

$$\Delta o_j(t) = \mu(t) \delta_k(t) \otimes w_{kj}(t), \tag{9}$$

where $\mu(t)$ is the gain constant vector and the backprop error $\delta_k(t)$ through the kth layer is determined by eq.(7). Likewise, defining the backprop error $\delta_j(t)$ through the jth layer as eq.(8) leads to the update equation for $o_i(t)$ in the input layer:

$$\Delta o_i(t) = \mu(t) \delta_j(t) \otimes w_{ji}(t). \tag{10}$$

3.5. NONLINEAR FACTOR OPTIMIZATION

As mentioned in the last section, the adjustment of the time-varying nonlinear factor $\lambda(t)$ in eq.(3) is needed for obtaining an optimal petrophysical model to fit well data with varieties of lithologic composition. The application of the error-back-propagation technique to neurons of each layer yields an update equation for the nonlinear factors in this layer.

Define the cost function for this problem as eq.(5). The update equation for $\lambda(t)$ has a general recursion form for any neuron in any layer. For instance, the nonlinear factor modification for $\lambda_i(t)$ in the input layer of Figure 1 can be expressed as

$$\Delta \lambda_i(t) = \beta(t) r_i(t) f'(\lambda_i(t)), \tag{11}$$

where $\beta(t)$ is the gain vector and the correlation function $r_i(t) = \sum_j \delta_j(t) \otimes w_{ji}(t)$ with $\delta_j(t)$ being eq.(8).

4. Joint Lithologic Inversion

In this section, the Caianiello neural network methods (including neural wavelet estimation, input signal reconstruction, and nonlinear factor optimization) are incorporated with the deterministic petrophysical models of eqs.(3) and (4) into a joint lithologic inversion scheme for porosity and clay-content estimations from a broad-band seismic impedance obtained by a joint impedance inversion scheme (Fu, 1997; Fu, 1999b). The well-derived porosity and clay-content logs are used as the particular solutions of the joint inversion at the wells, which are combined with the areally dense seismic impedance to estimate porosity and clay content away from the wells.

The joint lithologic inversion scheme includes two studies. First, to achieve appropriate deterministic petrophysical models, a large number of well data-based numerical modelings on the relationships of acoustic impedance and reservoir parameters (porosity and clay content) are needed to determine cutoff parameters (e.g., minimum velocity, maximum velocity, maximum porosity, maximum clay content, etc.) and initial nonlinear factors for different reservoir rocks and lithologies in the area under study. Second, even if the petrophysical model is determined optimally, it provides only a trend to fit data points. To quantify the deviations of data points from the trend for each lithology, neural wavelets are used as scanning operators to discern the scatter distributions and separate different lithologies in the impedance-porosity space. The joint lithologic inversion consists of two processes. First, inverse neural wavelets are extracted at the wells, and then the inverse-operator-based inversion is used to estimate initial porosity and clay-content models away from the wells. Second, forward neural wavelets are estimated at the wells, and then the forward-operator-based reconstruction is employed to improve the porosity and clay-content models.

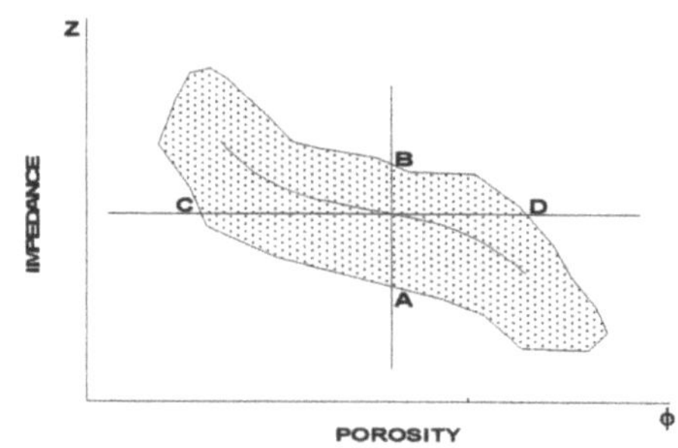

Figure 4. Schematic description of the joint lithologic inversion. First a deterministic petrophysical model defines the overall trend across the cloud of data points. Next neural wavelets determine the scatter distribution of data points around the trend curve along both the ϕ -axis (e.g., Line **CD**) and the z -axis (e.g., Line **AB**).

4.1. JOINT LITHOLOGIC INVERSION STEP 1: INITIAL MODEL ESTIMATION

As mentioned above, impedance dependence on porosity in sandstones is ambiguous probably due to the effects of clay content and textural position of clay. From Figure 4, we see that the optimum deterministic petrophysical model can only capture the overall trend across the cloud of data points in the impedance (z)-porosity (ϕ) space. Taking the scatter of data points in the vicinity of the trend into account for different rock/lithology types is a key to transforming the impedance into porosity. Neural wavelets are used in this study as scanning operators to determine the receptive field of the trend curve in two ways with one along the lines parallel to the z -axis (e.g., Line **AB** in Figure 4) and another along the lines parallel to the ϕ -axis (e.g., Line **CD**). The former can be expressed as: $\phi(t) = f(z(t), w_z(t), \lambda(t))$ where the petrophysical model $f()$ together with its nonlinear factor $\lambda(t)$ and cutoff parameters can define the trend curve, and the crosscorrelation of the impedance $z(t)$ with the inverse neural wavelets $w_z(t)$ can determine the data-point scatters around the trend curve in the direction of the z -axis. It should be mentioned that the statistical population codes of numerous neurons in the Caianiello network are used in this procedure.

The purpose of this joint inversion step is to combine the Caianiello neural network with the deterministic petrophysical model of eq.(3) to extract the inverse neural wavelets and then perform the inverse-operator-based initial porosity estimation. A number of seismic impedance traces at each well are used as the input to the neural network, and the well-derived porosity log of the well acts as the desired output. The estimation for the inverse neural wavelets demonstrates a hierarchical approximation from seismic impedance to porosity. Specifically, the total training set consists of an input matrix $z_{il}(t)$ ($l = 1,2,\ldots,L$, where L is the number of wells in the interesting area; $i = 1,2,\ldots,I$, where I is the number of input seismic impedance traces at the lth well, also denoting the number of neurons in the input layer), and a desired output matrix $\phi_{kl}(t)$ ($l = 1,2,\ldots,L$; $k = 1,2,\ldots,K$, where K is the number of porosity logs and/or the other relevant signals associated with the lth well, also representing the number of neurons in the output layer). In general, the parameter I is chosen large enough to take advantage of spatial correlation of properties among adjacent traces. The network training is actually a procedure of optimization. The main difference from regular applications of neural networks is that the direction and size of the weight adjustment that is made during each back propagation cycle are controlled by eq.(3) as an underlying physical relationship.

Once trained for all wells, one can have a neural network system for a seismic section or an interesting region, which, to some degree, gives a representation (i.e., $\phi(t) = f(z(t), w_z(t), \lambda(t))$) of the relationship of seismic impedance data (as inputs) and the subsurface porosity (as outputs). The neural network with numerous Caianiello neurons implicitly allows for different suites of inverse neural wavelets and nonlinear factors for different wells. Obviously, the information representation can be sufficient and reliable if more wells are available. For laterally stable depositional units, the neural wavelets are less laterally variable and a sparse well control is also applicable. Case studies indicated that a single-well controlled porosity inversion can also give a comparably good result in a laterally heterogeneous sediment. This may be a main advantage, benefited from eq.(3), of the method over traditional geostatistical techniques that have been also used to establish mapping relationships between well properties and seismic parameters. Nevertheless, the neural network system can be gradually improved by feeding new well data into it during the lifetime of an oil field.

The above trained neural network then can be used to estimate the initial porosity model away from the wells. In this processing, the network remains unchanged, and seismic impedance traces are fed successively into the network and inverted at the outputs. This method is a direct inversion method that attempts to estimate the porosity directly from seismic data. Although the network remains intact during the extrapolation phase, a set of new inverse neural wavelets and nonlinear factors can be autonomously produced by means of automatic interpolation of the network for the inversion of individual seismic impedance trace between wells.

4.2. JOINT LITHOLOGIC INVERSION STEP 2: MODEL-BASED INVERSION

From Figure 4, we see that in contrast with the joint inversion step 1, another manner for neural wavelets to determine the scatter range deviated from the trend can be expressed as $z(t) = f(\phi(t), w_\phi(t), \lambda(t))$. The crosscorrelation of the porosity $\phi(t)$ with the forward neural wavelets $w_\phi(t)$ can evaluate the deviations from the trend along the lines parallel to the ϕ -axis.

In this step, I first extract the forward neural wavelets by inputting the porosity log of each well to the Caianiello neural network and using the seismic impedance traces at the well as the desired output. The extracted forward neural wavelets for all wells in an interesting area are stored in one Caianiello neural network which can be used to model seismic impedance data directly from any known porosity model. The information representation of the network can be easily refined by updating the existing network to honor new wells. Rather than setting initial neural wavelets and training the network all over again, it is often faster to update the forward neural wavelets from a previously trained network.

The neural network trained above then can be used for the model-based inversion away from the wells to produce a final porosity profile. Here, seismic impedance traces are used as the desired output of the network, and the initial porosity model obtained in step 1 is used as the input. Similarly, for each trace to be inverted, a number of seismic impedance traces around this trace can be employed to compose its desired output matrix. In this processing, the forward neural wavelets and nonlinear factors remain intact and the algorithm scheme for the input signal reconstruction in the Caianiello network is applied to perform the model-based inversion with the intention to improve the initial porosity model.

It should be stressed that the lateral variations of the inverse and forward neural wavelets are assumed to be gradual in each large depositional unit associated with the blocky nature of impedance distribution. Each such distinct zone of deposition has a family of the inverse and forward neural wavelets (including the nonlinear factors) to represent its petrophysical properties. In the areas of complex geologic structures, such as faults with large throws, pinchouts, and sharp dips, a specified stratal geometry to control main reflectors must be used as a stratigraphic constraint during the extrapolation in the joint lithologic inversion. This constraint ensures that the applications of the neural wavelets and nonlinear factors are laterally restricted within the same seismic impedance sequences that often correspond to a large-scale geological unit. That is, during the extrapolation phase the different sets of neural wavelets and nonlinear factors should be confined to different depositional units from which they are extracted at the well. The stratal geometry is determined as follows: First a geological interpretation of the seismic section studied is conducted under well data control, determining the spatial distributions of some main geological units. Next a polynomial fitting technique is used to track main events and build a reliable stratal geometry.

5. Application to a Field Dataset

The above described methodology for lithologic inversion has been applied to a clastic field in western China. The interesting reservoir interval in this area corresponds to an areal continental deltaic sedimentary facies where the delta front facies deposited with sandstone-mudstone sequence is the interesting zone with a number of reservoir distributions. I process 9 main seismic lines using the joint lithologic inversion scheme with the aid of 11 wells in the interesting area. Figure 5 shows the location map of these seismic lines and wells. The seismic data in Figure 6 correspond to the seismic line marked with a bold line in Figure 5, which crosses three wells. Inversions of the data, under the control of these three wells, are demonstrated respectively in Figure 7 for porosity and in Figure 8 for clay content. The well-derived porosity and clay-content logs of these three wells are partly inserted at the wells on the corresponding sections respectively so that one can track and correlate layers on the logs away from the wells. These results significantly improve the spatial description of reservoirs in this area. The wells designed based on the results of the study are under drilling and will be addressed in another paper.

6. Discussion and Conclusions

The Caianiello neuron model is used to construct a new multilayer neural network for time-varying signal processing, including neural wavelet estimation, input signal reconstruction, and nonlinear factor optimization. Some deterministic petrophysical relationships, relating reservoir parameters to measured results of a reservoir system, can be introduced into the Caianiello neural network via nonlinear activation functions of neurons. This leads to an information integrated approach for reservoir characterization. As a result, a new joint lithologic inversion scheme for porosity and clay-content estimations has been built by integrating broad-band seismic impedance data, well data, and geological knowledge. The main conclusions can be summarized as follows:

1) An empirically-derived, relatively flexible deterministic petrophysical model relating acoustic velocities to porosity is given for clay-bearing sandstone reservoirs. This is based on the facts that first the different porosity ranges have different gradients of trends, and secondly different rock/lithology types have distinct petrophysical relationships and different regions of scatter distributions of data points in the velocity- porosity space.

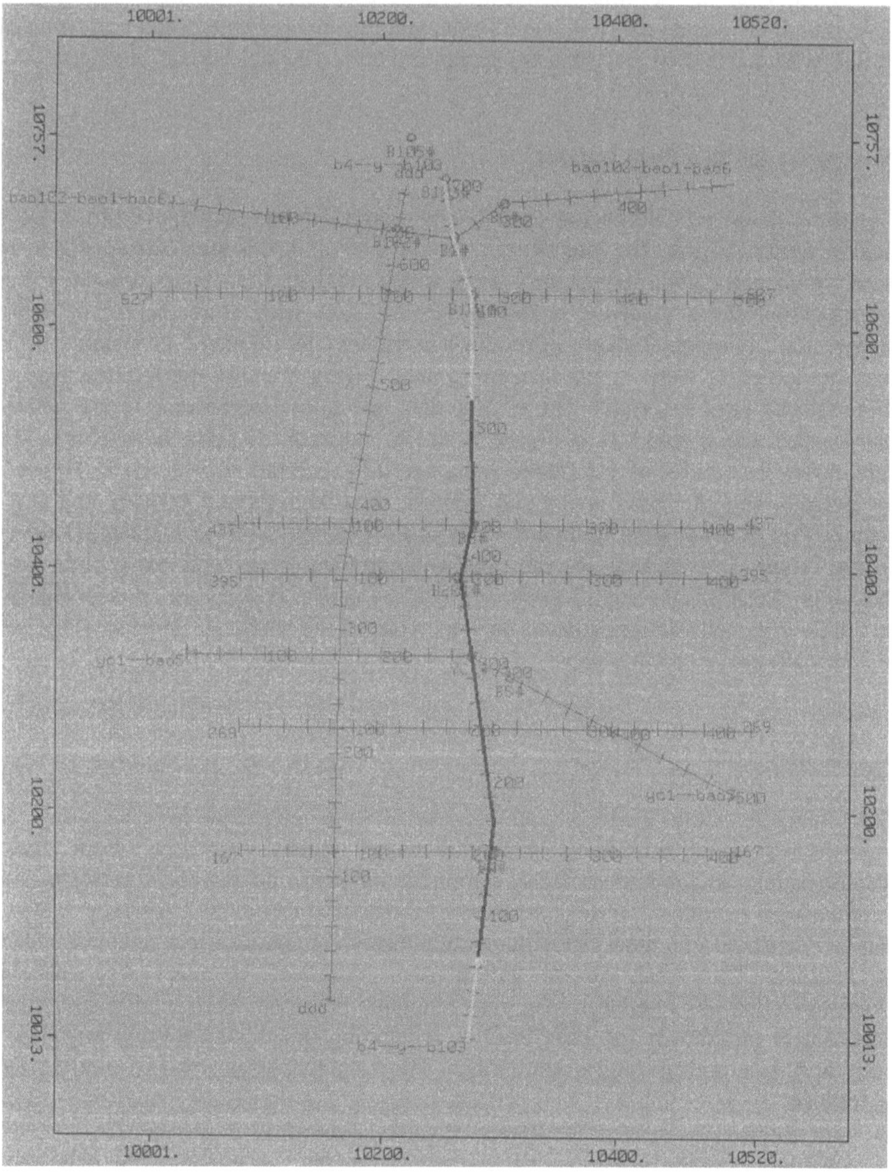

Figure 5. Location map of seismic lines and wells in a clastic field with complex continental deltaic sediments in western China. The borehole is indicated by the symbol #.

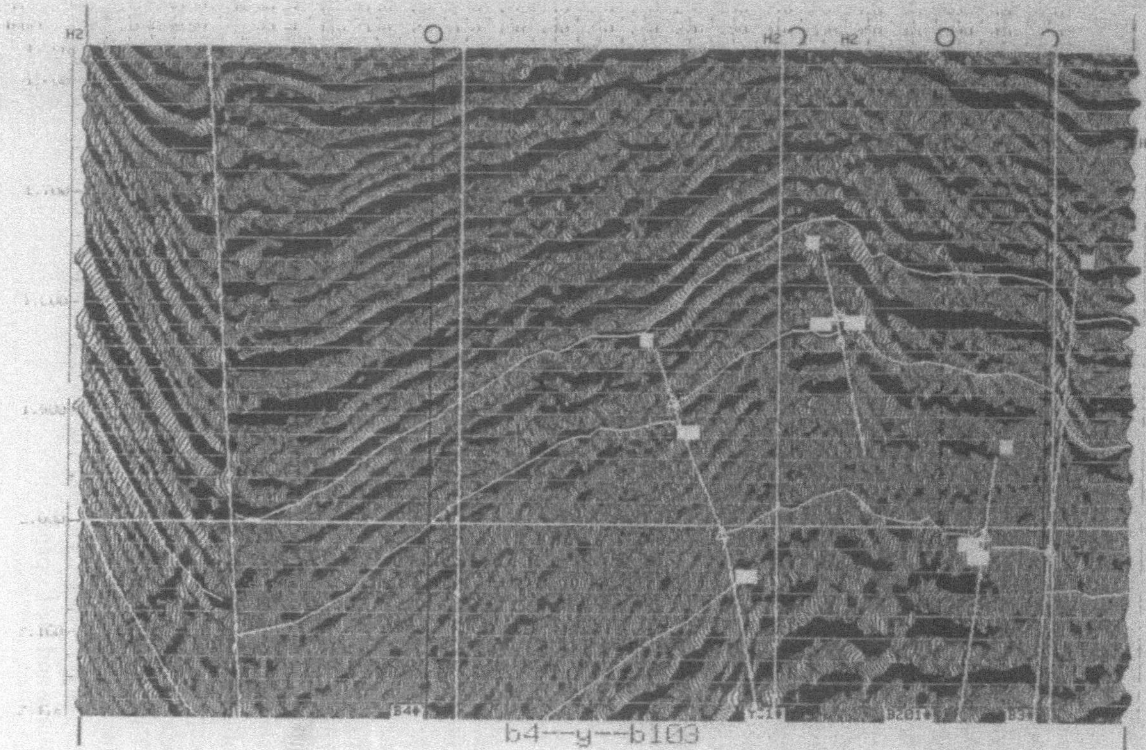

Figure 6. Seismic data from the seismic line that is marked with a bold line in Figure 5 and crosses three wells.

2) The joint inversion scheme tends to combine the concepts and algorithms of both statistical and deterministic approaches to use their advantages and avoid their disadvantages. First the appropriate deterministic petrophysical models can define optimal overall trends to fit data points for different rock/lithology types in the velocity-porosity space. Next the neural wavelets are used as scanning operators to determine the scatter distributions of data points that deviate from the trends so that the different rocks and lithologies can be separated in the space.
3) Sandstones containing different amounts of volume clay have different regions of scatter distributions of data points in the velocity-porosity space as well as different deviation levels from the trends, which correspond to different apertures and spectral contents of neural wavelets.
4) The joint lithologic inversion scheme consists of two processes. First, inverse neural wavelets are extracted at the wells, and then the inverse-operator-based inversion is used to estimate initial porosity and clay-content models away from the wells. Second, forward neural wavelets are estimated at the wells, and then the forward-operator-based reconstruction is implemented to improve the porosity and clay-content models.
5) The use of neural wavelets cannot narrow the deviations of data points from the trends. If appropriate petrophysical models are available, the incorporation of seismic waveform information into the joint lithologic inversion will allow for more accurate porosity and clay-content estimates than only using velocity information.

Acknowledgements

I would like to express my deep gratitude to Yongguang Mu who supervised the Ph.D. thesis on which this work is based, and to Caifang Fu and Wenbo Liu for providing all necessary assistance related to this research during my work at China Offshore Oil Exploration & Development Research Center (EDRC). I would also like to thank Huey-Ju Yang for her help in many aspects related to this study during my postdoctorial research in Tsinghua University in China. The helpful discussions and many insights into this study from Yanda Li are greatly appreciated. This material is based on the research supported partly by the Doctorial Foundation of the China National Education Committee and partly by the China National Postdoctorial Foundation. I thank EDRC for computing support. Thanks are extended to Liaohe, Shengli, Dagang, and Tuha Petroleum Administration Bureaus of China for providing field data. I thank the W.M. Keck Foundation at Institute of Tectonics, University of California, Santa Cruz for allowing me to use its facility to complete this paper. Finally, I am grateful to David Warren Burge, Dennis Neff and *The Leading Edge* for allowing me to use the figure from Burge and Neff's paper.

References

An, P., and Moon, W.M., 1993, Reservoir characterization using feedforward neural networks: 63th Ann.

Internat. Mtg., Soc. Expl. Geophys., Expanded Abstracts, 258-262.
Angeleri, G.P., and Carpi, R., 1982, Porosity prediction from seismic data: Geophys. Prosp., **30**, 580-607.
Baldwin, J.L., Bateman, R.M., and Wheatley, C.L., 1990, Application of a neural network to the problem of mineral identification from well logs: The Log Analyst, **31**, 279-293.
Bolt, D.G., Daneel, G., and Clare, A., 1997, Prospect Explorer: an exploration neural analysis tool: 67th Ann. Internat. Mtg., Soc. Expl. Geophys., Expanded Abstracts, 2041-2044.
Burge, D.W., and Neff, D.B., 1998, Well-based seismic lithology inversion for porosity and pay-thickness mapping: The Leading Edge, **17**, 166-171.
Caianniello, E.R., 1961, Outline of a theory of thought-processes and thinking machines: J. Theoret. Biol., **2**, 204-235.
Calderon-Macias, C., Sen, M.K., and Stoffa, P.L., 1996, A neural network optimization approach for automatic NMO correction and velocity estimation: 66th Ann. Internat. Mtg., Soc. Expl. Geophys., Expanded Abstracts, 1979-1982.
Clark, G.A., and Glinsky, M.E., Sandhya Devi, K.R., Robinson, J.H., Cheng, P.K.Z., and Ford, G.E., 1996, Automatic event picking in pre-stack migrated gathers using a probabilistic neural network: 66th Ann. Internat. Mtg., Soc. Expl. Geophys., Expanded Abstracts, 735-738.
de Buyl, M., Guidish, T., and Bell, F., 1986, Reservoir description from seismic lithologic parameter estimation: J. Petr. Tech., **40**, 475-482.
Doyen, P.M., 1988, Porosity from seismic data: A geostatistical approach: Geophysics, **53**, 1263-1275.
Duijndam, A.J.W., 1988a, Bayesian estimation in seismic inversion. Part I: Principles: Geophys. Prosp., **36**, 878-898.
Duijndam, A.J.W., 1988b, Bayesian estimation in seismic inversion. Part II: Uncertainty analysis: Geophys. Prosp., **36**, 899-918.
Ecoublet, P., Symes, W.W., and Levin, S., 1997, Seismic inversion for porosity using a backpropagation neural network: TRIP Annual Report 97, Rice University.
Essenreiter, R., and Karrenbach, M., 1996, Deconvolution using neural networks: 66th Ann. Internat. Mtg., Soc. Expl. Geophys., Expanded Abstracts, 1611-1614.
Fish, B., and Kasuma, T., 1994, A neural network approach to automate velocity picking: 64th Ann. Internat. Mtg., Soc. Expl. Geophys., Expanded Abstracts, 185-188.
Fournier, F., and Derain, J.F., 1995, A statistical methodology for deriving reservoir properties from seismic data: Geophysics, **60**, 1437-1450.
Fu, L.Y., 1995, An artificial neural network theory and its application to seismic data processing: PhD thesis, University of Petroleum, Beijing, PRC.
Fu, L.Y., Chen, S., and Duan, Y., 1997, ANNLOG technique for seismic wave impedance inversion and its application effect: Oil Geophysical Prospecting: **32**, 34-44.
Fu, L.Y., 1997, Application of the Caianiello neuron-based network to joint inversion: 67th Ann. Internat. Mtg., Soc. Expl. Geophys., Expanded Abstracts, 1624-1627.
Fu, L.Y., 1999a, A neuron filtering model and its neural network for space- and time-varying signal processing: Third International Conference on Cognitive and Neural systems, Boston University, Paper Vision B03.
Fu L.Y., 1999b, Joint inversion for acoustic impedance: *Geophysics*, accepted.
Gelfand, V.A., and Larner, K.L., 1984, Seismic lithologic modeling: The leading Edge, **3**, no.11, 30-34.
Han, D.H., Nur, A., and Morgan, D., 1986, Effects of porosity and clay content on wave velocities in sandstones: Geophysics, **51**, 2093-2107.
Hart, D.I., 1996, Improving the reliability of first-break picking with neural networks: 66th Ann. Internat. Mtg., Soc. Expl. Geophys., Expanded Abstracts, 1662-1665.
He, N.Q., and Reynolds, A.C., 1995, Estimation of porosity in thin-layered reservoirs by seismic inversion: 65th Ann. Internat. Mtg., Soc. Expl. Geophys., Expanded Abstracts, 1022-1024.
Himmer, P., and Link, C., 1997, Reservoir porosity prediction from 3-D seismic data using neural networks: 67th Ann. Internat. Mtg., Soc. Expl. Geophys., Expanded Abstracts, 842-845.
Huang, K. -Y., Chang, W.R.I., and Yen, H.T., 1990, Self-organising neural network for picking seismic horizons: 60th Ann. Internat. Mtg., Soc. Expl. Geophys., Expanded Abstracts, 313-315.
Huang, Z., Shimeld, J., Williamson, M., and Katsube, J., 1996, Permeability prediction with artificial neural network modeling in the Venture gas field, offshore eastern Canada: Geophysics, **61**, 422-436.
Jackson, D.D., and Matsu'ura, M., 1985, A Bayesian approach to nonlinear inversion: J. Geophys. Res., **90**, 581-591.
Klimentos, T., and McCann, C., 1990, Relationships among compressional wave attenuation, porosity, clay content, and permeability in sandstones: Geophysics, **55**, 998-1014.
Kowallis, B., Jones, L.E., and Wang, H.F., 1984, Velocity-porosity-clay content; systematics of poorly

consolidated sandstones: J. Geophys. Res., **89**, 10355-10364.
Leggett, M., Sandham, W.A., and Durrani, T.S., 1994, 3-D Seismic horizon tracking using an artificial neural network: 56th Meeting, Eur. Assoc. Expl. Geophys., Paper B049.
Lines, L.R. and Treitel, S., 1984, A Review of least square inversion and its application to geophysical problems: Geophys. Prosp., **32**, 159-186.
Lortzer, G.J.M., and Berkhout, A.J., 1992, An integrated approach to lithologic inversion. Part I: Theory: Geophysics, **57**, 233-244.
Martinez, R. D., 1985, Deterministic estimation of porosity and formation pressure from seismic data: 55th Ann. Internat. Mtg., Soc. Expl. Geophys., Expanded Abstracts, 461-464.
Maureau, G.T.F.R., and Van Wijhe, D.H., 1979, The prediction of porosity in the Permian carbonate of eastern Netherlands using seismic data: Geophysics, **44**, 1502-1517.
McCormack, M.D., Zuacha, D.E., and Dushek, D.W., 1993, First-break refraction event picking and seismic data trace editing using neural networks: Geophysics, **58**, 67-78.
McCulloch, W.S., and Pitts, W., 1943, A logical calculus of the ideas immanent in nervousactivity: Bull. of Math. Bio., **5**, 115-133.
Neff, D.B., 1990a, Incremental pay thickness modeling of hydrocarbon reservoirs: Geophysics, **55**, 558-566.
Neff, D.B., 1990b, Estimated pay mapping using three-dimensional seismic data and incremental pay thickness modeling: **55**, 567-575..
Neff, D.B., 1993, Amplitude map analysis using forward modeling in sandstone and carbonate reservoirs: Geophysics, **58**, 1428-1441.
Nur, A., 1992, The role of critical porosity in the physical response of rocks: EOS, Trans. AGU, **43**, 66.
Nur, A., Mavko, G., Dvorkin, J., and Galmudi, D., 1998, Critical porosity: A key to relating physical properties to porosity in rocks: The Leading Edge, **17**, 357-362.
Pan, G.C., 1996, Recent development for optimum data integration in mineral exploration: 66th Ann. Internat. Mtg., Soc. Expl. Geophys., Expanded Abstracts, 646-649.
Poulton, M.M., Sternberg, B.K., and Glass, C.E., 1992, Location of subsurface targets in geophysical data using neural networks: Geophysics, **57**, 1534-1544.
Raymer, D.S., Hunt, E.R., and Gardner, J.S., 1980, An improved sonic transit time-to-porosity transform: Presented at the Soc. Prof. Well Log Anal. 21st Ann. Mtg., Paper P.
Robinson, E.A., 1957, Predictive decomposition of seismic traces: Geophysics, **22**, 767-778.
Rogers, S., Fang, J.H., Karr, C.L., and Stanley, D.A., 1992, Determination of lithology from well logs using a neural network: AAPG Bull., **76**, 731-739.
Russell, B., Hampson, D., Schuelke, J., and Quirein, J., 1997, Multiattribute seismic analysis: The Leading Edge, **16**, 1439-1443.
Schultz, P.S., Ronen, S., Hattori, M., and Corbett, C., 1994, Seismic guided estimation of log properties, Part 1: The Leading Edge, **13**, part 1, 305-310, part 2, 674-678, part 3, 770-776.
Tarantola, A., and Valette, B., 1982, Inverse problems: Quest for information: J. Geophys., **50**, 159-170.
Thadani, S.G., Alabert, F., and Journel, A.G., 1987, An integrated geostatistical/pattern recognition technique for characterization of reservoir spatial variability: 57th Ann. Internat. Mtg., Soc. Expl. Geophys., Expanded Abstracts, 372-375.
Tosaya, C., and Nur, A., 1982, Effects of diagenesis and clays on compressional velocities in rocks: Geophys. Res. Letts., **9**, 5-8.
Vernik, L., 1994, Predicting lithology and transport properties from acoustic velocities based on petrophysical classification of siliciclastics: Geophysics, **63**, 420-427.
Vernik, L., and Nur, A., 1992, Petrophysical classification of siliciclastics for lithology and porosity prediction from seismic velocities: AAPG Bull., **76**, 1295-1309.
Wang, B., Pann, K., Schuelke, J., Shirley, T., and Ferguson, B., 1997, View of neural network training as constrained optimization and applications to rock porosity prediction: 67th Ann. Internat. Mtg., Soc. Expl. Geophys., Expanded Abstracts, 838-841.
Wang, L.X., and Mendel, J.M., 1992, Adaptive minimum prediction-error deconvolution and source wavelet estimation using Hopfield neural networks: Geophysics, **57**, 670-679.
Wyllie, M.R.J., Gregory, A.R., and Gardner, G.H.F., 1958, An experimental investigation of factors affecting elastic wave velocities in porous media: Geophysics, **23**, 459-493.
Zhang, X., Li, Y., Hu, Q., and Feng, D., 1995, Early-stage reservoir analysis with SOMA: A neural network approach: 65th Ann. Internat. Mtg., Soc. Expl. Geophys., Expanded Abstracts, 138-141.

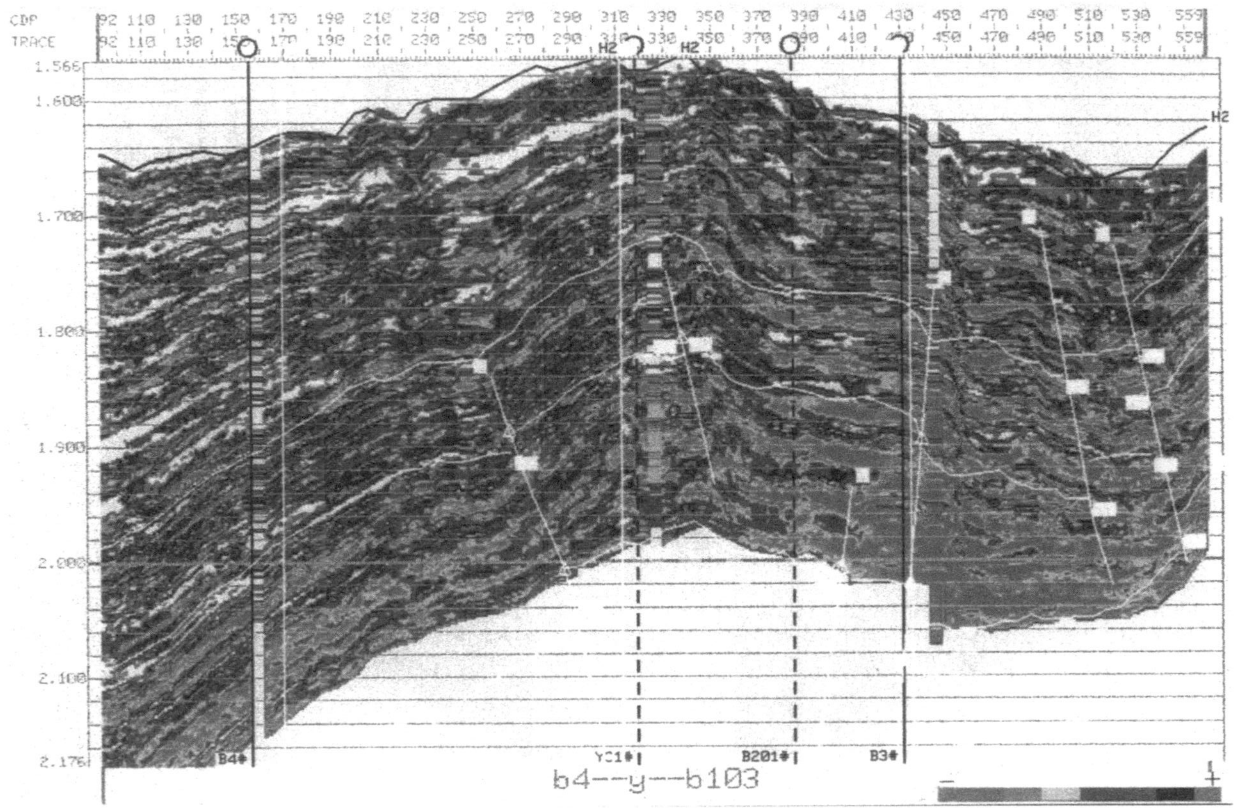

Figure 7. Porosity estimate under control of these three wells. The well-derived porosity logs of these wells are partly inserted at the wells, respectively.

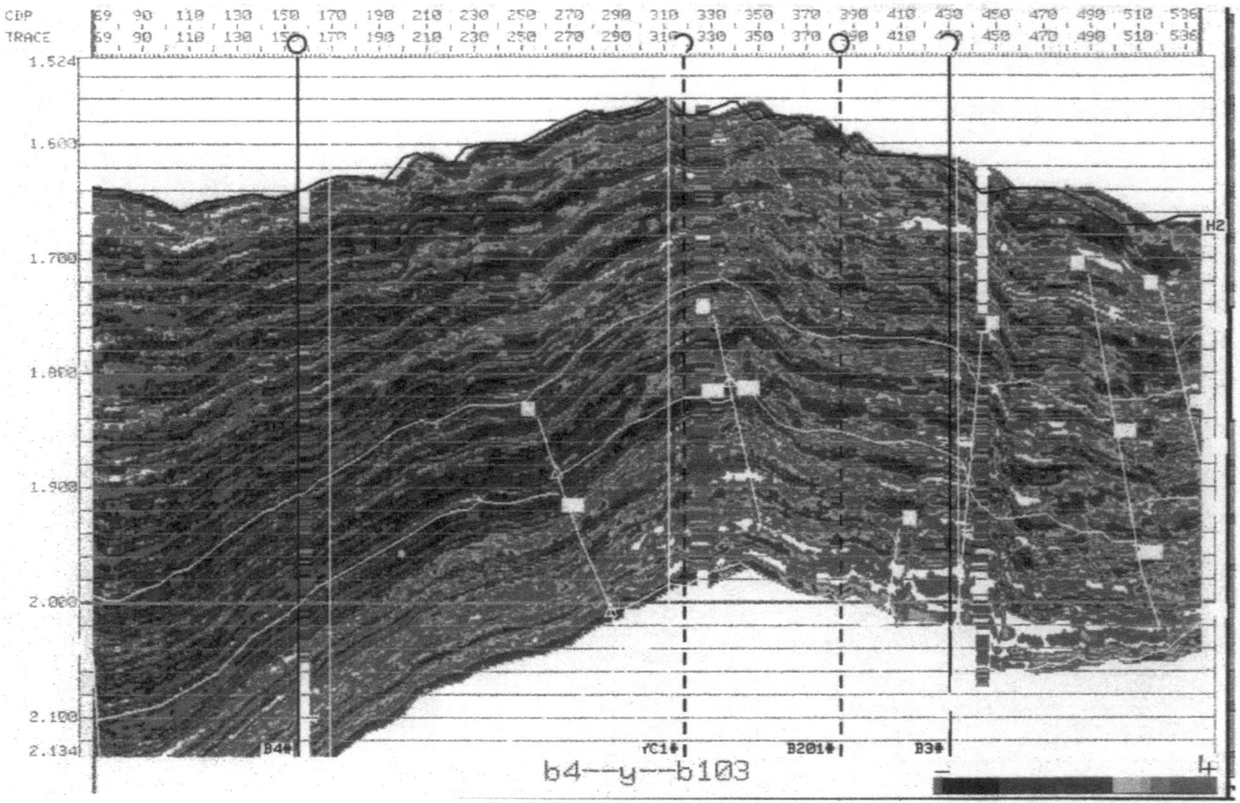

Figure 8. Clay-content estimate under control of these three wells. The well-derived clay-content logs of these wells are partly inserted at the wells, respectively.

AN ARTIFICIAL NEURAL NETWORK METHOD FOR MINERAL PROSPECTIVITY MAPPING: A COMPARISON WITH FUZZY LOGIC AND BAYESIAN PROBABILITY METHODS

WARICK M. BROWN[†], DAVID I. GROVES[†] and
TAMÁS D. GEDEON[‡]

† *Centre for Global Metallogeny,*
Department of Geology and Geophysics,
University of Western Australia,
Perth, WA 6907,
AUSTRALIA
E-mail: wbrown@geol.uwa.edu.au
E-mail: dgroves@geol.uwa.edu.au

‡ *School of Information Technology,*
Murdoch University,
Perth, WA 6150,
AUSTRALIA
E-mail: tgedeon@murdoch.edu.au

Abstract

A multilayer perceptron (MLP) neural network is used to combine multi-source exploration data in a Geographic Information System (GIS) database and produce a mineral prospectivity map for gold deposits in the Tenterfield 1:100,000 sheet area, NSW, Australia. Statistical and probability measures of map quality indicate that the neural network method performs better than the empirical weights-of-evidence (Bayesian probability) and conceptual fuzzy logic methods in the generation of the mineral prospectivity map. The neural network has several important advantages over existing methods, including the ability to respond to critical combinations of parameters, rather than automatically increasing the prospectivity estimate in response to every favourable parameter.

1. Introduction

The volume and complexity of regional data available to the exploration geologist are increasing rapidly, from a variety of sources such as remote sensing, airborne geophysics and large commercially-available geochemical databases compiled from

W. Sandham and M. Leggett (eds.),
Geophysical Applications of Artificial Neural Networks and Fuzzy Logic, 179-196.

open-file exploration reports. As the amount of data increases, so too will the need for an effective means of integration and analysis (Cucuzza and Goode, 1998).

A mineral prospectivity map, showing areas in different colours according to their potential to host mineral deposits of a particular type, represents one method of combining and summarizing regional data sets within a GIS in a way that can help geologists make decisions about ground acquisition and target selection. The two most commonly used methods of producing prospectivity maps are the empirical weights-of-evidence method (Bonham-Carter et al., 1988) and the fuzzy logic method, although both these methods have important limitations.

The weights-of-evidence method is based on Bayesian probability and requires a statistically significant number of mineral occurrences to estimate the relative importance of geological evidence associated with the deposits. Consequently, the method can not be used in poorly explored areas containing few known deposits. The conceptual fuzzy logic method overcomes this problem, but relies on a deposit expert to make subjective assessments. Conceptual methods also depend on the application of a deposit model to make decisions about which geological parameters might be significant. However, new deposit types are still being discovered (e.g. Olympic Dam at the time of discovery: Woodall, 1994), and models for well-investigated deposit styles are continually being updated in the light of new discoveries (e.g. Century: Legge, 1995).

Artificial neural networks offer a new approach to the problem of mineral prospectivity mapping. Neural networks are particularly suitable for pattern recognition and classification tasks. The ability of neural networks to function without pre-existing knowledge (e.g. a deposit model), and to operate at acceptable accuracy when the data are noisy or contain outliers, suggest that neural networks could be well suited to the integration of mineral exploration data. If digital geological, geophysical and geochemical maps can be superimposed in a GIS system to produce an image from which the signatures of ore deposits can be extracted, then neural networks could be trained to recognise these signatures in order to predict the location of new deposits. To test this concept, a neural network was used to integrate regional exploration data from a small GIS database into a single prospectivity map, and the performance of the neural network approach compared with Bayesian probability and fuzzy logic methods.

2. Previous Work

The potential of combining neural network and GIS technologies was recognised by Jeffrey and Rossner (1986) and Ritter et al. (1988), and much of the subsequent research focussed on automated pattern recognition and the classification of remotely-sensed imagery (Decateur, 1989; Ritter and Hepner, 1990). Many of these studies showed that neural networks could be used to combine large numbers of data sets to perform classification (Benediktsson et al., 1993; Bruzzone et al., 1997). Although neural networks have been applied to a wide range of remote-sensing and geophysical problems, few studies describe their application to mineral exploration or mineral prospectivity mapping.

Bonham-Carter (1994) has described previous work in mineral prospectivity mapping as either empirical (data-driven) or conceptual (knowledge-driven). An empirical approach generally involves a statistical analysis of the spatial relationships of map features (e.g. faults and lithological units) to mineral deposit locations. Empirical techniques include; multiple regression (Sinclair and Woodsmith, 1970), logistic regression (Agterberg, 1974; Reddy et al., 1991), canonical correlation (Pan and Harris, 1992), a Bayesian-spatial method (Singer and Kouda, 1988) and weights-of-evidence (Bonham-Carter et al., 1988).

Conceptual methods include evidential-belief theory (Moon, 1990), fuzzy logic (An et al., 1991), expert systems (Katz, 1991), Boolean algebra and index-overlay method (Bonham-Carter, 1994), and a mineral-systems method (Wyborn et al., 1995).

Singer and Kouda (1996) applied a multilayer feed-forward network to Kuroko-type deposits to produce a map depicting the distance to the nearest deposit, and the use of a proprietary neural network system for the classification of anomalies in multi-source data has been described by Clare et al. (1997). The use of a multilayer perceptron (MLP) to produce a prospectivity map was compared with fuzzy logic and weights-of-evidence methods by Brown et al. (1998, 1999). Harris and Pan (1999) have compared probabilistic neural network, logistic regression and discriminant analysis methods of mineral prospectivity mapping. Although the approach has been an empirical one in these studies, the neural network approach could equally well be applied to the processing of input map layers based on geological knowledge and subjective judgment.

3. Artificial Neural Networks

Although many different neural-network architectures exist, an MLP neural network was chosen for this study, because it performs very well for a large variety of problem types and has powerful function-approximation capabilities (Hornik, 1991; Masters, 1993). Learning in MLP neural networks is achieved by modifying the connection weights while a set of training examples is repeatedly processed. This training method is referred to as supervised because each example specifies all the input parameters together with the desired outputs. The set of inputs to the network constitutes a real-valued vector, $[x_1, \ldots, x_n]$. As the neural network used in this study contains only a single output unit, the training data consist of a list of input vectors together with the corresponding scalar output-value. The difference between the desired output and the output produced by the network represents the error.

Learning proceeds by applying the error back-propagation algorithm (Werbos, 1974; Rummelhart et al., 1986) to iteratively adjust the weight values in order to minimize the network error. After training is completed, the network should be able to generalize: that is, produce appropriate responses for inputs that have not been used during training.

4. GIS Database

The GIS database for the trial study corresponds to the area of the Tenterfield 1:100,000 sheet (map 9339), and was compiled by the Geological Survey of New South Wales (Barnes, 1993; Barnes et al., 1995). The database comprises the following thematic layers; solid geology, regional-scale faults, airborne total magnetic-intensity, airborne gamma-ray spectrometry (U, Th, K and total count) and deposit locations. Magnetic and radiometric data were supplied as gridded, grey-scale images. The mineral deposit layer includes mineral occurrences, and small and medium sized mines. Each cell in the grid data represents a 200 metre square on the ground. There are a total of 248 cells in the x-direction and 280 cells in the y-direction, giving a total of 69,440 cells.

5. Geology

The Tenterfield 1:100,000 sheet is located in the southern portion of the New England Orogen (Figure 1). The study area consists of three main lithostratigraphic units: 1) a Silurian to Carboniferous metasedimentary and volcanic basement, 2) the Mid to Late Permian Wandsworth Volcanic Group, comprising intermediate to felsic volcanic

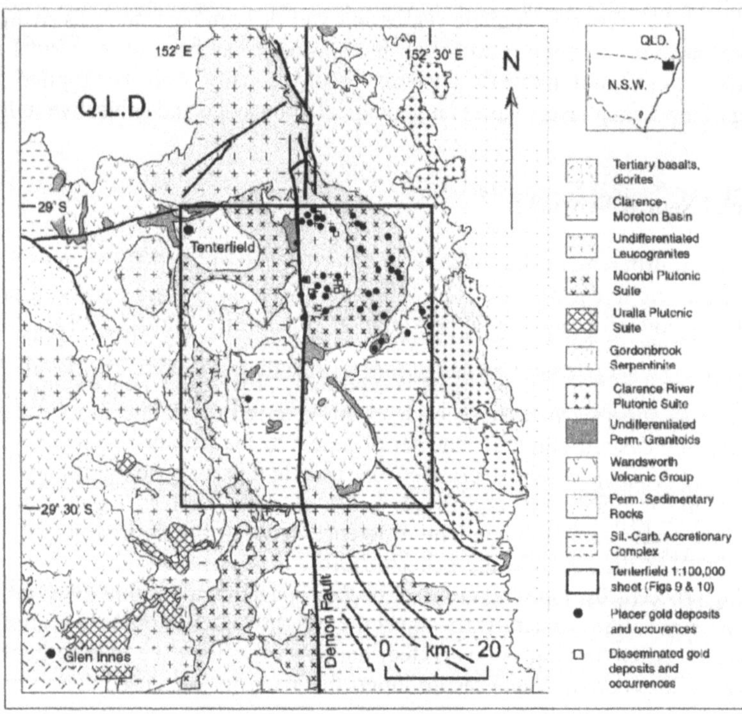

Figure 1. Location and simplified geology of the Tenterfield 1:100,000 sheet area adapted from an unpublished regional digital map supplied by the Geological Survey of New South Wales.

rocks and comagmatic felsic intrusions, and 3) Late Permian to Early Triassic leucogranitoids. The main granitoids in the subject area are those of the Clarence River and Moonbi Suites and an undifferentiated group of leucogranitoids.

The highly fractionated leucogranites are the most important hosts to gold mineralization in this area. The mid-Triassic Stanthorpe Adamellite, appearing as a semi-circular body in the north of the study area, east of the Demon Fault, hosts gold mineralization in the form of large low-grade primary disseminations in poorly-defined irregular lenses near the upper surface of the granite body (e.g. the Timbarra-Poverty Point deposits) and placer deposits. Numerous alluvial gold occurrences are scattered across the Bungulla Porphyritic Adamellite, which forms an arcuate body on the eastern margin of the Stanthorpe Adamellite (Barnes et al., 1995). Forty-seven secondary, thirteen primary and three spatially-associated granitoid-hosted gold deposits of unknown form were selected as the basis for the prospectivity analysis in this study.

Placer gold deposits are included in the analysis, because the placer deposits are likely to be favourable indicators for the presence of primary gold deposits, since most of the placer deposits are derived from the primary deposits in this area and have not been transported far. This conclusion is supported by; 1) gold grains, which have a faceted morphology, reflecting the shape of crystals on which they are deposited (Bampton, 1988), 2) the fineness of both the alluvial and primary gold (Barnes et al., 1995), 3) the flat and poorly-drained top of the plateau formed by the Stanthorpe Adamellite, and 4) the fact that much of the historical production from the Timbarra-Poverty Point goldfield came from eluvial deposits.

6. Neural Network Method

6.1. PRE-PROCESSING

All thematic layers in the GIS database were converted to raster format prior to further processing. The original solid-geology layer contained 41 rock units, many of which contain zero or very few deposits. The reliability of probability estimates used in the weights-of-evidence method depends on the existence of a statistically significant number of known deposits. Therefore, units belonging to the same rock suite or stratigraphic group were combined, resulting in a simplified geology layer consisting of 12 rock units. Values in the magnetic and radiometric grey-scale images were subdivided into eight classes. The GIS database layer containing regional faults was converted to a grid in which the cell values indicate the distance to the nearest fault. The thematic layers in the Tenterfield GIS database were combined into a set of input feature vectors. At each cell location in the Tenterfield map grid, a value from each thematic layer was combined to form a vector (Figure 2). These vectors, known as input feature vectors, formed the input to the neural network.

Linear scaling, based on the maximum and minimum values contained in each thematic layer, was applied to all input feature vectors. Target values used in the training data were scaled to the range [0.1,0.9]. This is a result of applying a simple linear mapping of the binary output values, zero and one, to the practical limits of the network output. For a feedforward network with an asymptotic logistic activation

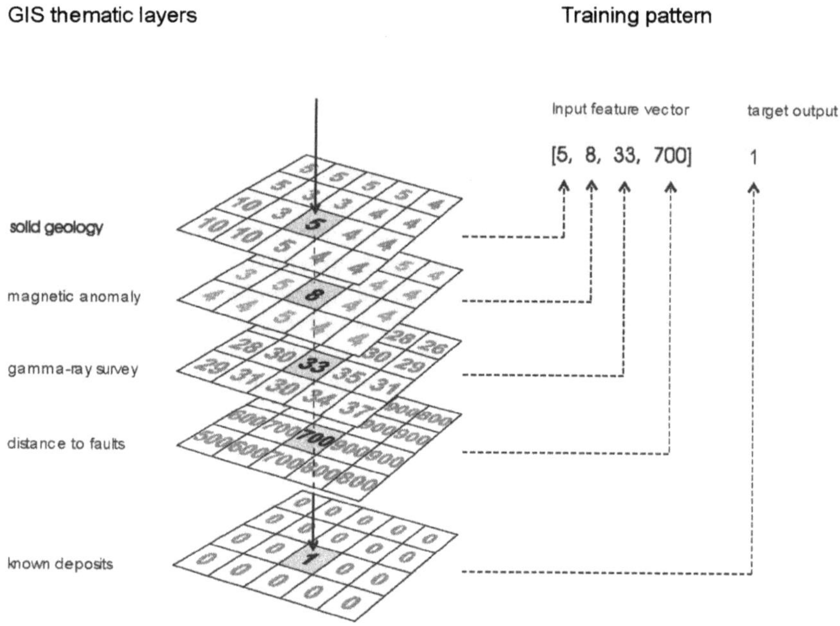

Figure 2. Relationship between GIS thematic layers and feature vectors used as input to the neural network. At each location on the map grid, the cell values for each thematic layer are combined to form an input feature vector. Patterns used to train the network consist of an input feature vector paired together with the desired output, represented by the value of the binary layer showing the locations of known deposits.

function, the output activation values are typically limited to 0.1 and 0.9 (Masters, 1993).

Apart from the preparation of the map layers, all processing was performed outside the GIS using a series of utility programs written in C++ and a neural network back-propagation simulation program (McClelland and Rumelhart, 1988). Data processing in the neural network method consisted of three main stages: 1) pre-processing of data to a format suitable as inputs to the neural network, 2) neural network training and processing the entire data set after training, and 3) post-processing required to convert the output values to a map. Processing in the weights-of-evidence and fuzzy logic methods was performed using the GEODIPS GIS prospectivity-analysis package (Jusmady and Taylor, 1997; Taylor, 1997).

6.2. INPUT CODING

A single input unit was assigned to each of the magnetic, four radiometric and fault-proximity layers. A one-of-n coding scheme was applied to the solid geology layer, so that each of the 12 rock types in the simplified geology layer was assigned a separate

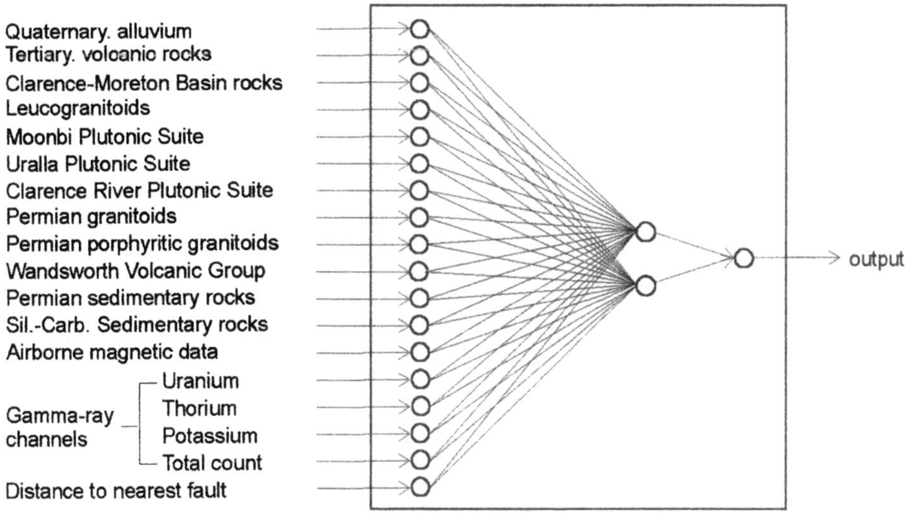

Figure 3. Architecture of the MLP neural network used in this study. The network has an 18-2-1 topology (where the numbers refer to input, hidden and output units, respectively). Each map layer in the GIS database was assigned an input unit except the geology layer, for which 1-of-n encoding was used.

input layer unit in the neural network. For a particular input pattern, only one of the input units is set to one and all the rest are set to zero. The network topology and inputs used are shown in Figure 3. Binary coding was used for the single output unit, with one and zero representing the presence and absence of deposits, respectively.

6.3. TRAINING DATA

In each training run, the total sum of squares (TSS) error follows a decreasing trend and asymptotically approaches zero as training proceeds. However, excessive learning can lead to what is known as overfitting, associated with poor generalization performance. This is due to the presence of noise in the data set. If the number of modifiable weights in an MLP network is high in relation to the number of patterns in the training data set, and training is continued beyond the stage at which the general trends in the data are learnt, the network begins to learn the noise in the data set. This is harmful to the ability of the network to generalize with new data sets.

To avoid overtraining, a procedure known as early stopping is applied (Wang et al., 1994). The data available for training are divided into three parts referred to as training, test and validation sets, respectively. The concept is to improve generalization by stopping learning before the global minimum of the training data set TSS error; i.e. before the idiosyncrasies of the data set are learnt. The training data set is used to adapt the weights in the network and the test data set is used to determine when to stop training. The validation data set plays no role in training and is used to assess the generalization performance of the trained network.

Although the ratio of deposit to non-deposit cells in the Tenterfield grid is approximately 1:1000, the ratio of deposit to non-deposit vectors used in the training set was close to 1:1. If deposit patterns are represented in the training set in the same proportion as they appear in the total data population, the network would not learn to recognize the rare deposit patterns at all, or would perform very poorly for this class of patterns (Masters, 1993; Zaknich, 1999). Consequently, the training, test and validation data sets were randomly assigned one third of the available deposit vectors (21 each), and two non-deposit vectors were randomly selected for each of the 12 main rock units, giving a total of 24 non-deposit vectors in each data set.

Since there are only 63 gold deposits of the required type in the Tenterfield area, compared with 69,377 non-deposit cells, the size of the data sets used for training were limited by the number of known deposits.

6.4. NETWORK TOPOLOGY AND NETWORK PARAMETERS

The number of units in the hidden layer of the network was determined experimentally, resulting in an 18-2-1 network topology. The number of hidden units was kept to a minimum, in order to limit the number of modifiable weights in the network and thereby reduce the ability of the network to learn the noise in the training data.

A learning rate of 0.5 was used for all networks. Weights were randomly initialized in the range [-0.5,+0.5]. The value of the momentum parameter (used to add a fraction of the previous weight change to the current weight change) was set to zero. Training patterns were presented to the network in a randomly permutated order, and the weights were updated after the presentation of each training pattern (incremental update).

6.5. NETWORK TRAINING

A series of networks were trained using ten different random sets of initial weights. In each case, the initial weights were saved. The total sum of squares (TSS) error was monitored for both the training and test data sets. Although the training error decreases during training, the test error decreases to a minimum and then begins to increase as training continues. Training is stopped at the point at which the test TSS error reaches a minimum value, and the weights corresponding to the optimum number of cycles are used for all subsequent processing. From the ten neural networks, the one giving the best classification performance for the independent validation data set was used to process the input vectors for the entire Tenterfield grid. The output values resulting from this step were re-scaled and classified into integer values to produce an nine-class prospectivity map.

7. Fuzzy Logic Method

Fuzzy systems represent a conceptual or knowledge-based approach to mineral prospectivity mapping. In contrast to neural networks, which can learn from data, fuzzy systems require explicit statements about uncertain knowledge (Kasabov, 1996). The method applied in this study has been described in detail by Bonham-Carter (1994) and only a short description is given here. Each cell value in the input map layers is assigned a fuzzy membership value in the range [0,1], which expresses the degree to which the value belongs to the set "favourable for mineralization". Fuzzy membership values are assigned according to subjective judgement. For the Tenterfield area, parameter values and the corresponding fuzzy membership values are shown in Table 1. Using this table, each cell value in GIS map layers was replaced by a fuzzy membership value. The fuzzy membership values from the different GIS layers were then combined using fuzzy operators to produce a single output membership value for each grid cell location representing the overall favourability for mineralization.

TABLE 1. **Map class values and fuzzy membership values for maps used in the fuzzy logic method.**

Lithostratigraphic Units		*Distance to faults (metres)*		*U count (classed data)*		*Th count (classed data)*	
Quaternary Alluvium	0.1	<200	0.1	1	0.1	1	0.2
Tertiary Volcanic Rocks	0.0	200-400	0.1	2	0.2	2	0.2
Clarence-Moreton Basin	0.0	400-600	0.1	3	0.2	3	0.25
Leucogranitoids	0.95	600-800	0.1	4	0.4	4	0.4
Moonbi Plutonic Suite	0.95	800-1000	0.1	5	0.7	5	0.6
Uralla Plutonic Suite	0.0	1000-1200	0.1	6	0.8	6	0.8
Clarence River Plutonic. Suite	0.1	1200-1400	0.1	7	0.9	7	0.8
Permian. Granitoids	0.4	1400-1600	0.1	8	0.9	8	0.9
Permian. Porphyritic. Granitoids	0.0	1600-1800	0.1				
Wandsworth Volcanic. Group	0.1	1800-2000	0.1				
Permian Sedimentary. Rocks	0.1	>2000	0.75				
Sil.-Carb. Sedimentary. Rocks.	0.1						

A variety of different fuzzy logic inference nets using different fuzzy operators (in the case of the gamma fuzzy operator, different values of the gamma parameter) were tested in order to combine membership. The prospectivity map that best accounts for the known deposit points, while minimizing the area of the highest prospectivity classes, was chosen as the final fuzzy logic map. The relationship between deposit points and magnetic intensity was ambiguous, and therefore magnetic data were not used as an input in the fuzzy logic and weights-of-evidence methods. The fuzzy gamma operator with $\gamma = 0.95$ was used to combine membership values after the values for the U and Th values were summed. The fuzzy gamma operator is given by eq.(1);

$$\mu_{comb} = \left[1 - \prod_{i=1}^{N}(1-\mu_i)\right]^{\gamma}\left(\prod_{i=1}^{N}\mu_i\right)^{1-\gamma} \tag{1}$$

where γ is a number in the range [0,1], μ_{comb} is the fuzzy membership value for the combined data sets, and μ_i is the fuzzy membership value for the ith data set. Descriptions of fuzzy operators are given by Zimmermann (1985), An et al. (1991), and Bonham-Carter (1994).

8. Weights-of-Evidence Method

Originally developed for medical diagnosis (Lusted, 1968), the weights-of-evidence method was adapted to spatial data and mineral exploration by Bonham-Carter et al. (1988). A full description of the weights-of-evidence method is given by Bonham-Carter (1994) and only a brief description is given here. The method is based on the concept of prior and posterior probabilities. If each deposit is assumed to occupy a single grid cell (where the map layer is in raster format), then the prior probability of any cell chosen at random containing a deposit is the ratio of the number of deposit cells to the total number of cells, that is,

$$P(D) = n(D)/n(T) \qquad (2)$$

This prior probability can be updated using binary maps that show whether a parameter favourable for mineralization is present or absent. The updated probability estimate is the conditional probability of a deposit occurring in a grid cell, given the presence of a binary map pattern, *B*. Bayes' Rule and the Theorem of Total Probability can be used to write;

$$P(D/B) = P(D) \; * \; P(B/D) \,/\, P(B) \qquad (3)$$

Here, the prior probability *P*(*D*) is updated with the factor *P*(*B*/*D*)/*P*(*B*) to give the posterior conditional probability, *P*(*D*/*B*). The log of odds formulation of eq.(3) allows the effect of each map layer, which depicts parameters important for mineralization, to be calculated independently and then combined (Bonham-Carter et al., 1989);

$$\ln\left[Odds\left(D / B_1 \cap B_2 \cap \ldots \cap B_n\right)\right] = \ln\left[Odds(D)\right] + \sum_{i=1}^{N} W_i^{\pm} \qquad (4)$$

where
$$W_i^{\pm} = \begin{cases} W_i^{+} = \ln\left[\dfrac{P(B_i / D)}{P(B_i / \overline{D})}\right] & \text{if pattern B present} \\ W_i^{-} = \ln\left[\dfrac{P(\overline{B_i} / D)}{P(\overline{B_i} / \overline{D})}\right] & \text{if pattern B is absent} \\ 0 & \text{if data are missing} \end{cases}$$

and B_n means a favourable evidence pattern is present in a binary predictor map and *D* means a deposit is present. A bar over the letter indicates that the favourable feature is absent. The positive weight (W_i^+) is used in eq.(4), when the favourable pattern is present, and increases the prior probability, and the negative weight (W_i^-) is used when the pattern is absent and decreases the prior probability.

A binary predictor map is produced by reclassifying multi-class maps, so that the spatial association between deposit points and the pattern is maximized. A pair of weights is then calculated for each binary predictor map, based on the conditional probability ratios estimated from the numbers of grid cells. The final prospectivity map is presented in terms of probability values calculated from the posterior log-odds (shown on the left-hand side of eq.(4)).

In order to apply eq.(4), an assumption that parameters in the binary evidence maps are conditionally independent with respect to deposits must be made; i.e., the effects of the interactions between parameters, used as evidence to modify the prior probability, must be ignored. This assumption is rarely satisfied completely in geological data sets (Bonham-Carter, 1994). Where a significant correlation between binary map parameters exists, one of the dependent maps must be omitted or dependent maps must be combined using Boolean operators (AND or OR), multiple regression or principal component analysis (Bonham-Carter, 1994). Weights-of-evidence maps were prepared from input map layers exported from a GIS using the GEODIPS GIS prospectivity analysis package (Jusmady and Taylor, 1997).

9. Measuring Prospectivity Map Quality

The quality of maps produced using different methods was compared quantitatively by measuring the extent to which high prospectivity areas of the maps include the locations of known mineral deposits and occurrences. A variety of statistical and probability indices of map quality were used.

The Chi-square statistic measures the extent to which observed and expected numbers of deposits differ. Although statistical tests of significance are not valid in the case of spatial associations, the Chi-square statistic can provide explorative and descriptive measures of spatial correlation (Bonham-Carter, 1994). The observed number of deposits in the map classes with the highest favourability should be much higher than the number that would be expected if the map class areas were distributed randomly with respect to deposits. A problem with the Chi-square statistic is that it is only sensitive to the total difference between observed and expected numbers of deposits for all the map classes, and not to whether it is the higher prospectivity-map classes that contain higher than expected numbers of the known deposits.

Spearman's and Kendall's rank correlation coefficients were used to assess the degree to which the probability of occurrence of known mineral deposits increases with prospectivity map class. For both these statistics, a value of zero indicates that no correlation exists and a value of one indicates a perfect correlation.

Conditional probabilities of the occurrence of a known mineral deposit, given the grid cell is located within a particular prospectivity class, were calculated for each prospectivity map class using the numbers of grid cells. Each deposit was assumed to occupy only a single cell. A disadvantage of conditional probabilities is that they are very small numbers and depend on the total number of known deposits in a region.

The ratio of the conditional probability of a deposit given a particular prospectivity class area, $P(D/A)$, to the probability of a deposit for the map area as a whole, $P(D)$, results in numbers that are easier to compare. This ratio is equivalent to an intuitive notion of a predictive prospectivity map; i.e., a high prospectivity map class should define an area in which a high proportion of the known deposits are predicted in a small proportion of the total map area. This can be expressed as the following ratio:

$$\frac{n(D_A)/n(D_{total})}{n(A)/n(T)} = \frac{P(D/A)}{P(D)} \quad (5)$$

where $n(D_A)$ represents the number of deposits in map class A, $n(D_{total})$ is the total number of deposits, $n(A)$ is the area of map class A, and $n(T)$ is the total area.

If a prospectivity map is to be useful, then the probability of discovering a deposit in the highest group of prospectivity map classes should be significantly upgraded compared to selecting a grid cell at random. Consequently, the ratio in eq.(5) should be greater than one. Conversely, map classes representing the lowest prospectivity classes should have ratios significantly less than one, indicating that the probability of finding a deposit is much lower than average. In the middle group of map classes, the ratio should be approximately equal to one, indicating that cells in these areas are no more likely to contain a deposit than cells chosen at random.

The fuzzy logic and neural network maps were reclassified from nine to four classes of favourability prior to calculating the probability-based statistics, to allow a more direct comparison with the weights-of-evidence map, which, due to problems of conditional independence, only contains four prospectivity classes.

10. Results

Mineral prospectivity maps produced using the weights-of-evidence, fuzzy logic and neural network methods are shown, together with known gold occurrence and deposit points, in Figure 4. Both the neural network and fuzzy logic favourability results are classified into nine prospectivity classes. Only three colours are displayed in the weights-of-evidence map (Figure 4a). Due to the necessity of combining several input maps to avoid problems of conditional dependence, the weights-of-evidence method results in only three discrete probability levels. Classifying these values results in a four-class prospectivity map, in which one of the classes (class 3 in Figure 4a and Table 3) does not correspond to any cells on the map. Pairwise tests of all possible binary evidence maps from the weights-of-evidence method show that the assumption of conditional independence is violated for all map pairs except those involving the map showing distance to the nearest fault. To overcome this, dependent maps (geology, U, Th, K, and total count gamma-ray channels) are combined using the Boolean AND operator.

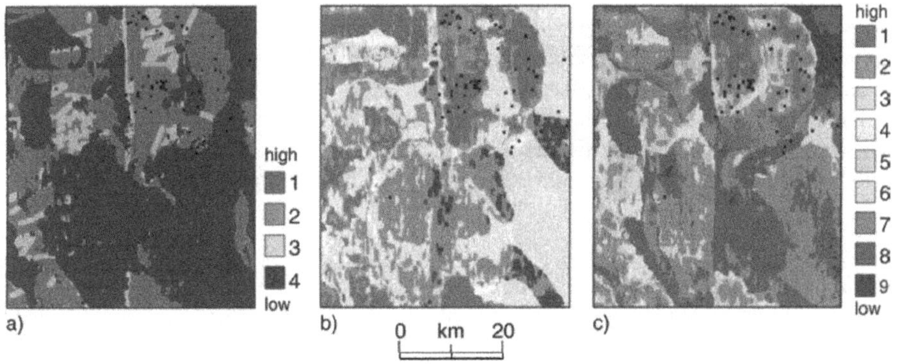

Figure 4. Prospectivity maps for the Tenterfield 1:100,000 area. a) Weights-of-evidence method. b) Fuzzy logic method. c) Neural network method. Crosses, dots and open squares represent known primary, alluvial and undifferentiated gold deposits and occurrences, respectively.

Values of the Chi-square statistic and Spearman's and Kendall's rank correlation coefficients are shown for each of the three prospectivity mapping methods in Table 2. Table 3 shows probability-based measures of map quality in which favourability values for the fuzzy logic and neural network methods are reclassified from nine to four classes. Although no cells in the weights-of-evidence map correspond to class 3, it was assumed that just one cell value corresponds to this class in order to calculate the statistics shown in Tables 2 and 3. The upper part of Table 3 shows the conditional probability that a cell contains a known deposit, given that the area of interest is restricted to a particular prospectivity class. The lower part of Table 3 shows the ratio of conditional probability for a given map-class area to prior probability for the study area as a whole.

TABLE 2. Statistical measures of map quality for prospectivity maps produced using different methods.

Quality statistic	Method		
	Weights-of -Evidence	Fuzzy logic	Neural network
Chi-square	66	63	83
Spearman's ρ	0.80	0.50	0.93
Kendall's τ	0.67	0.53	0.83

11. Discussion

The neural network method produces geologically meaningful results. The similarity of the neural network and conceptually-based fuzzy logic maps indicates that the neural network result conforms to current understanding of geological factors important for gold mineralization in the Tenterfield area. The broad similarity between the neural network and the empirically-derived weights-of-evidence maps indicates that the neural network result accounts for spatial relationships between known mineral deposits and the parameters of the GIS database.

The output of the trained neural network; i.e., values in the range [0.1,0.9], can be interpreted as the degree of similarity of a particular output to a composite of all deposit vectors presented to the network during training. Such an interpretation fits with the

TABLE 3. **Probability measures of map quality for prospectivity maps produced using different methods. The fuzzy logic and neural network maps are reclassified from nine to four classes to enable a direct comparison with the weights-of-evidence map. The upper part of the table shows the conditional probability of a gold deposit given that a cell occurs in the particular map class. The ratio of the posterior to prior probability is shown in the lower part. In both cases, prospectivity is low for class 4 and high for class 1. The weights-of-evidence classes are the same as those shown in Figure 4a.**

Map Class	Method		
	Weights-of-Evidence	Fuzzy Logic	Neural Networks
P(D/A) X 10^4			
4	2.52	2.85	2.60
3	0.00	1.65	2.15
2	8.67	6.89	11.66
1	23.45	18.91	26.74
P(D/A)/P(D)			
4	0.28	0.31	0.29
3	0.00	0.18	0.24
2	0.96	0.76	1.29
1	2.58	2.08	2.95

intuitive notion that a prospective area has characteristics that closely resemble areas known to contain mineral deposits. Therefore, the network output is also a measure of how favourable the area is for mineral deposits of the type included in the training data set.

The main difference between the prospectivity maps is that the neural network map contains much smaller high prospecitivty areas (classes 1 and 2 in Figure 4). This is interpreted to be due to the ability of the neural network to respond in a highly non-linear way to combinations of input parameters, so that a high favourability value is only assigned to areas where there is a combination of the critical favourable parameters. This contrasts with the weights-of-evidence and fuzzy logic methods, where, for a given grid location, each favourable input-parameter value automatically results in an increase in the final prospectivity estimate. The broad band of high prospectivity (coloured red and orange) in the northeastern corner of the fuzzy logic map appears to an example of this effect (Figure 4b). The high favourability estimate is the result of high radiometric responses in all channels of the gamma-ray survey data. This area is not highlighted in the neural network map, where both a favourable leucogranite host and a high radiometric response are required to produce a high favourability estimate. The high prospectivity band in the fuzzy logic map is absent in the weights-of-evidence map because gamma-ray survey map layers, which contain very similar information, violate the assumption of conditional independence and were therefore converted to a single input layer.

Both the fuzzy logic and weights-of-evidence methods are quite sensitive to data layers that are strongly correlated; i.e., contain similar information. This characteristic is common in geological exploration data sets, particularly those involving geochemistry. As explained above, the result of incorporating data sets containing similar information in the fuzzy logic method is a map containing spuriously high favourability estimates. With the weights-of-evidence method, correlated data sets violate the assumption of conditional independence. The solution requires the dependent data sets to be either discarded or combined in some way, but Boolean methods of combining data sets reduce multiple parameters to a single parameter and therefore result in a loss of information. The results of this study, in which four layers of gamma-ray survey data were used as inputs to the neural network, suggest that the neural network method is very robust with respect to data sets containing similar information.

The neural network method allows data sets to be combined without the loss of information inherent in the weights-of-evidence method and some implementations of the fuzzy logic method. The weights-of-evidence method requires that multi-class data be converted to a binary format prior to integration. Values below the threshold used to reclassify multi-class to binary maps play no role in determining prospectivity, and values above the threshold (included in the binary evidence map) are treated as though they are equal because they are assigned the same weight. The fuzzy logic method overcomes the problem of applying a fixed cut-off value. Instead of converting input map layers to binary values, the raw cell values are replaced with fuzzy membership values in the range [0, 1]. However, loss of information also occurs where single fuzzy-membership values are applied to a whole class or range of values, as in the case of studies by Eddy et al. (1995) and An et al. (1991). This simple method was also applied here. In contrast, the neural network method does not require ratio data to be reclassified prior to combining data sets.

Inconsistent or misleading data are unlikely to represent as serious a problem for neural networks as for statistical methods such as the weights-of-evidence methods and combinations of statistical and fuzzy logic methods. The GIS thematic layer containing regional faults and lineaments used in this study is an example of a possibly spurious data set. An examination of the digital elevation model (DEM) for the Tenterfield area reveals many major lineaments not present in the structural layer. Thus, the structural map represents incomplete information. The weak inverse relationship between deposit points and faults that emerged from the statistical analysis performed in the weights-of-evidence method is likely to be misleading. The effect of this input on the estimated prospectivity is clearly illustrated in the weights-of-evidence prospectivity map in the form of elongate zones of moderate prospectivity throughout the map (Figure 4a). In contrast, the neural network prospectivity map does not show any evidence that major faults have played an important role in determining prospectivity. A property of neural networks is the ability to respond to spurious input data by adjusting the weights connected to the relevant input to very low values, thereby reducing or eliminating the contribution made by that data to the output.

The statistical measures of prospectivity map quality (Table 2) are all highest for the neural network map. The Chi-square statistic indicates greater differences between the number of deposits observed in the class areas of the neural network map and those that would be expected if the class areas were located at random, compared with maps produced by the fuzzy logic and weights-of-evidence methods. The larger values of both Spearman's and Kendall's rank correlation coefficients show that there is a stronger correlation between map class and probability of a cell containing a known deposit than for the other two methods.

The probability and probability ratio measures of map quality (Table 3) indicate that the highest prospectivity map classes in the neural network map are stronger predictors of deposits than the highest prospectivity map classes in the fuzzy logic map and slightly stronger than those in the weights-of-evidence map. As noted above, in an ideal prospectivity map, the ratio $P(D/A)/P(D)$ should be greater than one in the high prospectivity class, approximately equal to one for the middle prospectivity class, and considerably less than one for the low prospectivity class.

The prospectivity map produced using the neural network method is based on about two thirds of the data that are used in the weights-of-evidence method. Approximately one third of the available data was reserved as a validation data set in the neural network method and used to check the generalization performance of the network after it had been trained. Only one third of the available data, the training data set, was used directly in training the network and a further third, the test data set, was used indirectly. The validation data set played no role in training. In contrast, all of the available deposit points were used in the weights-of-evidence method. There were an insufficient number of gold deposits and occurrences available to allow a third of the data to be reserved as a validation data set in the weights-of-evidence method.

12. Conclusions

The neural network method, using an MLP network trained with the error back-propagation algorithm, produces a geologically-plausible mineral prospectivity map similar, but superior, to the fuzzy logic and weights-of-evidence maps. The use of neural networks for the integration of large multi-source data sets used in regional mineral exploration, and for the prediction of mineral prospectivity, offers several advantages over existing methods. These include the ability to: 1) respond to critical combinations of parameters, rather than automatically increasing the prospectivity due to all favourable parameters, 2) combine data sets without the loss of information inherent in existing methods, and 3) produce results that are relatively unaffected by redundant data, spurious data and data containing multiple populations. In addition, the neural network method in this study used approximately a third less data than the weights-of-evidence method.

Acknowledgments

The authors gratefully acknowledge Robert Barnes of the Geological Survey of New South Wales, Department of Mineral Resources, for providing the GIS data used in this study. The use of the GEODIPS GIS prospectivity analysis package, written by Jusmady and Geoff Taylor of the Department of Applied Geology, UNSW, in the preparation of the weights-of-evidence and fuzzy logic prospectivity maps, is also gratefully acknowledged. WMB thanks the AusIMM for supporting this study in the form of a grant from the Bicentennial Gold 88 Endowment Fund. Josef Holzschu kindly helped in preparing the diagrams. Joe Leach is thanked for contributing the initial idea for this study.

References

Agterberg, F. P., 1974, Automatic contouring of geological maps to detect target areas for mineral exploration: J. Math. Geol., **6**, 373-395.

An, P., Moon, W. M., and Rencz, A., 1991, Application of fuzzy set theory to integrated mineral exploration: Can. J. Expl. Geophys., **27**, 1-11.

Bampton, M. D., 1988, Alteration and mineralisation of the southern part of the Stanthorpe Adamellite, near Tenterfield, New South Wales: B.Sc. thesis, Univ. of Sydney. (unpubl.).

Barnes, R. G., 1993, Use of GIS and image processing for enhancing geological investigations. Case study: Tenterfield 1:100,000 sheet area, northern NSW: Geological Survey Report No. GS1993/290. Geol. Surv. N.S.W.

Barnes, R. G., Henley, H. F., and Henley, J. E., 1995, Exploration data package for the Tenterfield and Coaldale 1:100,000 sheet areas: Geological Survey Report No. GS1995/004. Geol. Surv. N.S.W.

Benediktsson, J. A., Swain, P. H., and Ersoy, O. K., 1993, Conjugate-gradient neural networks in classification of multisource and very-high dimensional remote sensing data: Internat. J. Remote Sensing, **14**, 2883-2903.

Bonham-Carter, G. F., 1994, Geographic information systems for geoscientists: modelling with GIS, Computer Methods in the Geosciences, **13**, Pergamon Press, Inc.

Bonham-Carter, G. F., Agterberg, F. P., and Wright, D. F., 1988, Integration of geological datasets for gold exploration in Nova Scotia: Photo. Eng. & Rem. Sens., **54**, 1585-1592.

Bonham-Carter, G. F., Agterberg, F. P., and Wright, D. F., 1989, Weights of evidence: a new approach to mapping mineral potential, *in* Agterberg, F. P., and Bonham-Carter, G. F., Eds., Statistical Applications in the Earth Sciences: Geol. Surv. Can. Paper 89-9, Geol. Surv. Can., 171-183.

Brown, W. M., Taylor, G. R., Jusmady, Groves, D. I., and Knox-Robinson, C. M., 1998, Application of artificial neural networks to prospectivity analysis in a GIS environment: a comparison with statistical and fuzzy logic methods for Au and Sn deposits of the Tenterfield area, NSW, *in* 14th Australian Geological Convention, Geological Society of Australia, Abstracts, **49**, 57.

Brown, W. M., Gedeon, T. D., and Barnes, R. G., 1999, The use of a multilayer feedforward neural network for mineral prospectivity mapping, *in* Gedeon, T., Wong, P., Halgamuge, S., Kasabov, N., Nauck, D., and Fukushima, K., Eds., ICONIP'99, Proceedings of the 6th International Conference of Neural Information Processing: **1**, 160-165, IEEE, Piscataway.

Bruzzone, L., Conese, C., Maselli, F., and Roll, F., 1997, Multisource classification of complex rural areas by statistical and neural-network approaches: Photo. Eng. & Rem. Sens., **63**, 523-533.

Clare, A., Bolt, G., and Daneel, G., 1997, The use of neural analysis as an exploration tool, *in* Proceedings of the 12th International Conference on Applied Geologic Sensing, **2**, Environmental Research Institute of Michigan (ERIM), 199-205.

Cucuzza, J., and Goode, A. D. T., 1998, Australian mineral exploration research, *in* Berkman, D. A. and MacKenzie, D. H. Eds., Geology of Australian and Papua New Guinean mineral deposits: Australasian Institute of Mining and Metallurgy Mono. **22**, 53-60.

Decateur, S. E., 1989, Application of neural networks to terrain classification: IEEE INNS International Joint Conference on Neural Networks, **1**, 283-288.

Eddy, B. G., Bonham-Carter, G. F., and Jefferson, C. W., 1995, Mineral resource assessment of the Parry Islands, high Arctic, Canada: a GIS-based fuzzy logic model, *in* Proceedings Canadian Conference on GIS, CD ROM Session C3, Paper 4.

Harris, D., and Pan, G., 1999, Mineral favourability mapping: a comparison of artificial neural networks, logistic regression, and discriminant analysis: Natural Resources Research, **8**, 93-109.

Hornik, K., 1991, Approximation capabilities of multilayer feedforward networks: Neural Networks, **4**, 251-257.

Jeffrey, W., and Rossner, R., 1986, Optimization algorithms: simulated annealing and neural network processing: Astrophysical Journal, **310**, 473-481.

Jusmady, and Taylor, G. R., 1997, Geological Data Integration and Processing System (GEODIPS) software package Version 1.03: Department of Applied Geology, Univ. of New South Wales.

Kasabov, N.K., 1996, Foundations of neural networks, fuzzy logic, and knowledge engineering: M.I.T. Press.

Katz, S. S., 1991, Emulating the Prospector expert system with a raster GIS: Comput. and Geosci., **17**, 1033-1050.

Legge, P. J., 1995, Geoscience 1994 and beyond: thoughts on geology and exploration for world-class deposits: Austral. J. Earth Sciences, **42**, 1-10.

Lusted, L. B., 1968, Introduction to medical decision making: Charles Thomas.

Masters, T., 1993, Practical neural network recipes in C++: Academic Press, Inc.

McClelland, J. L., and Rumelhart, D. E., 1988, Explorations in parallel distributed processing: M.I.T. Press.

Moon, W, M. 1990, Integration of geophysical and geological data using evidential belief function: IEEE Trans. Geosci. Remote Sensing, **28**, 711-720.

Pan, G., and Harris, D. P., 1992, Estimating a favorability equation for the integration of geodata and selection of mineral exploration targets: Mathematical Geology, **24**, 177-201.

Reddy, R. K. T., Agterberg, F. P., and Bonham-Carter, G. F., 1991, Application of GIS-based logistic models to base-metal potential mapping in Snow Lake area, *in* Proceedings of Canadian Conference on GIS, 607-618.

Ritter, N. D., and Hepner, G. F., 1990, Application of an artificial neural network to land-cover classification of thematic mapper imagery: Comput. and Geosci., **16**, 873-880.

Ritter, N. D., Logan, T. L., and Bryant, N. A., 1988, Integration of neural network technologies with geographic information systems, *in* Thomas, H. F., Ed., GIS; integrating technology and geoscience applications (Denver): National Academy of Science, 102-103.

Rummelhart, D. E., Hinton, G. E., and Williams, R. J., 1986, Learning representations by back-propagating errors: Nature, **323**, 533-536.

Sinclair, A. J., and Woodsmith, G. I., 1970, Multiple regression as a method of estimating exploration potential in an area near Terrace, British Columbia: Econ. Geol., **65**, 998-1003.

Singer, D. A., and Kouda, R., 1988, Integrating spatial and frequency information in the search for Kuroko deposits of the Hokuroku District, Japan: Econ. Geol., **83**, 18-29.

Singer, D. A., and Kouda, R., 1996, Application of a feedforward neural network in the search for Kuroko deposits in the Hokuroku District, Japan: Mathematical Geology, **28**, 1017-1023.

Taylor, G. R., 1997, Geoscience GIS: Desktop and Prospectivity Mapping: Short Course at the Key Centre for Mines, July 7-11, 1997, Univ. of New South Wales.

Wang, C., Venkatesh, S. S., and Judd, J. S., 1994, Optimal stopping and effective machine complexity in learning, *in* Cowan, J. D., and Tesauro, G., and Alspector, J., Eds., Advances in neural information processing systems (Denver, 1993): Morgan Kaufmann Publ., 303-310.

Werbos, P., 1974, Beyond regression; new tools for prediction and analysis in the behavioural sciences: Ph.D. thesis, Harvard Univ.

Woodall, R., 1994, Empiricism and concept in successful mineral exploration: Austral. J. Earth Sciences, **41**, 1-10.

Wyborn, L. A. I., Gallagher, R., and Raymond, O., 1995, Using GIS for mineral potential evaluation in areas with few known mineral occurrences, *in* Second National Forum on GIS in the Geosciences - Forum Proceedings: Australian Geological Survey Organisation, AGSO Record 1995/46, 199-211.

Zaknich, A., 1999, Artificial Neural Networks: a research dominated engineering discipline: Centre for Intelligent Information Processing Systems (CIIPS), Department of Electrical and Electronic Engineering, Univ. of Western Australia, (unpublished).

Zimmermann, H. J., 1985, Fuzzy set theory and its applications: Kluwer Academic Publ.

OIL RESERVOIR POROSITY PREDICTION USING A NEURAL NETWORK ENSEMBLE APPROACH

CURTIS A. LINK† and PHILLIP A. HIMMER‡

† *Department of Geophysical Engineering,*
Montana Tech of the University of Montana,
Butte, Montana 59701,
USA
E-Mail: clink@mtech.edu

‡ *Department of Electrical Engineering,*
Montana State University,
Bozeman, Montana 59717-3780,
USA

Abstract

The problem of parameter prediction in a hydrocarbon reservoir is typically accomplished by an interpreter using sparse well information and seismic data. The resulting maps may contain varying levels of uncertainty depending on the experience of the interpreter and the availability and quality of seismic and well data.

This chapter describes a neural network ensemble approach to the interpolation problem. An interval of seismic data representing the zone of interest is extracted from a three-dimensional (3-D) data volume. Combinations of seismic data, complex trace attributes, and geometric data are used as inputs to a multilayer perceptron (MLP) neural network. A series of networks is trained, and results are used to produce final ensembles.

Our results show that realistic prediction maps can be generated using a neural network ensemble approach with input data consisting of raw seismic amplitudes, instantaneous amplitude, and coordinate geometry. Ensemble maps created for subintervals in the zone of interest show consistent features. Also, network testing errors at finer levels of resolution are not unusually large compared to errors for a single layer prediction.

1. Introduction

Development and exploitation of a hydrocarbon reservoir typically makes use of maps describing spatial distributions of relevant parameters. These maps are generated from well information that may or may not adequately sample the reservoir in an areal sense.

W. Sandham and M. Leggett (eds.),
Geophysical Applications of Artificial Neural Networks and Fuzzy Logic, 197-213.

In addition, reservoirs often exhibit a high degree of heterogeneity that introduces high levels of uncertainties into the interpolated parameters. A recent approach has been to use complex trace attributes derived from three-dimensional seismic data volumes to track the desired parameters through heterogeneous zones. This is usually a time intensive process requiring a skilled interpreter.

Oil reservoirs are, in large part, characterized by the porosity and permeability of the oil-bearing formation. Accurate estimates of these parameters are important for calculating oil reserves and developing exploitation strategies. The motivation for this study was to be able to produce accurate porosity and permeability predictions for input into a reservoir simulator.

At present, porosity and permeability maps are created by geostatistical interpolation techniques using well log data. This approach can account for large-scale variations in lithologic parameters, but is unable to provide the level of resolution required to produce accurate production estimates or determine injection locations if the reservoir is located in a zone exhibiting a high degree of heterogeneity.

Currently, 3-D seismic surveying is one of the primary tools for characterizing a hydrocarbon reservoir. Attributes extracted from seismic data are used to qualitatively describe variations in lithology and associated physical parameters across zones of interest. By correlating seismic data with well log data, it is, in principle, possible to produce quantitative predictions at each common midpoint (CMP) location.

Conventional techniques correlate seismic attributes to lithologic parameters using deterministic and stochastic methods. The use of a neural network approach enables seismic data to be related to porosity without explicitly defining the relationships for the various parameters. More importantly, neural networks naturally utilize intervals of seismic data or combinations of attributes rather than single values. This ability increases the amount of available information from which to make predictions.

This chapter describes a set of procedures that were used to produce porosity maps for a small oil field in north-central Montana, USA. The primary focus is how porosity predictions depend on first, the type of well porosity values used and second, the number of zones of porosity in the output. Well log porosity data was available in three forms: neutron porosity, density porosity, and porosity derived from laboratory core measurements. These data types were then used as target sets in a MLP neural network using combinations of seismic data and complex attributes as input. The porosity target sets were subdivided into one, two, or three zones in the producing interval to investigate the level of resolution that could be achieved by neural network predictions.

2. Geology

This study was carried out using well data and recently-acquired 3-D seismic data from the NE Rabbit Hills oil field in Blaine County, north-central Montana, USA (Figure 1). Oil production is from the Bowes member of the upper Middle Jurassic Sawtooth formation. The Bowes unconformity overlies the Firemoon member of the Sawtooth and is overlain by the Rierdon formation (Figure 2). It lies at an average depth of 4070 ft (1233 m) and forms a structurally-enhanced stratigraphic trap in the field area.

Regionally, the Bowes member was deposited in a tidal flat setting, unconformably overlying marine shelf deposits of the Firemoon. In late Bowes time, a sea-level lowstand caused subaerial exposure, and a valley was incised into the early Bowes (Figure 3). During the subsequent rise of the Rierdon Sea, quartz sandstone, ooid limestone, and bioclastic limestone were deposited in this valley across an aggradational braid plain. The incised valley lies in a southwest-northeasterly direction. Water flow was to the northeast entering the Rierdon Sea located just northeast of the reservoir area (Porter, 1996).

Oil production in the field is from this upper Bowes valley fill ranging from 9 to 26 ft (2.7 to 7.9 m) in thickness. Within the fill, the productive lithology is the bioclastic limestone associated with secondary porosity. The average porosity of the upper Bowes valley fill is 15% compared to 7% for the lower Bowes tidal flat deposits. Of the three types of valley fill, bioclastic limestone is predominant (Figure 4).

High porosity trends of approximately 15% in the porosity prediction maps parallel to the valley would suggest thick deposits of bioclastic limestone; evidence of the braided stream deposition. Porosity lows of about 7% are interpreted as areas of the lower Bowes not eroded during sea-level lowstand.

As such, the reservoir zone represents a complex target with a large degree of stratigraphic variation both horizontally and vertically. Conventional interpolation techniques would not be able to characterize the high degree of heterogeneities due to the uneven spacing of wells.

3. Geophysical Data

A recently-acquired, high-resolution 3-D seismic survey covers an area of approximately one square mile in the study area. The processed data set results in 80 east-west cross-lines and 72 north-south in-lines giving a total of 5760 common midpoint (CMP) traces. A border of seven traces was removed from the data set due to low fold creating a data subset containing 66 cross-lines and 58 in-lines. This 66 by 58 CMP grid is the data set used for this study. Within this area, there are eight wells that have a combination of well log and laboratory core data. The portion of the log data corresponding to the upper Bowes was extracted from well log charts and core data files and averaged into one, two, and three layers for use as target data sets. The training process required matching the correct CMP location to the corresponding well location. The purpose of the neural network analysis was to find and establish a relationship between the seismic data and the porosity values. Two types of network input data were extracted from the seismic data: raw seismic amplitude and instantaneous amplitude.

3.1. SEISMIC DATA

The raw seismic amplitudes were the final processed data samples of the reservoir interval. The instantaneous amplitude attribute was calculated from the raw amplitudes using techniques described in Taner et al. (1979).

Well location corrections for well drift were made using synthetic seismograms for all wells that had sonic logs. These position adjustments correct the well log data to the correct seismic trace at the producing depth. Each well's synthetic trace was compared to all of the CMPs in a 3 by 3 grid of traces centered around the surface location of the well. These corrections are critical to relating the well's porosity and permeability logs to the correct seismic trace.

In addition to amplitude data, geometric information in the form of CMP coordinates was also used as input into the neural network (Rogers et al., 1995). A grid system was constructed with (1,1) representing the southwest corner of the seismic data and (58,66) the northeast corner where each coordinate represents a CMP location in the data set. The bin spacing for the processed seismic data was 82.5 ft by 82.5 ft (25 m by 25 m).

The Bowes member is thin with respect to the seismic wavelength, so it is not possible to directly separate the upper and lower parts of the Bowes. However, since the lower Bowes is represented by a uniform depositional environment, porosity variations should be small compared to the expected variation in the upper Bowes. Using 14 seismic samples for each CMP effectively removes any information pertaining to thickness variation of the Bowes member.

Scaling is often necessary and usually useful in neural network applications. Seismic amplitudes and instantaneous amplitudes were scaled from -0.2 to 0.2 and 0 to 0.2 respectively by scaling the seismic data by five times the maximum of the absolute value in each data set. The grid coordinates were also scaled to span the range (0.01, 0.01) to (0.66, 0.58). Scaling was seen to have a small effect on testing error and scaling the inputs to the range -0.2 to 0.2 resulted in the lowest testing error.

The final form of the network input data thus consisted of scaled combinations of seismic wavelet amplitudes, instantaneous seismic amplitudes, and grid coordinates.

3.2. ROCK PROPERTY DATA

The porosity values were obtained from analog well log charts and laboratory core analysis data. The well logs were digitized at one-foot intervals with an error of approximately 0.5% porosity units. Two porosity well logs were available: compensated neutron porosity (NPHI) and density porosity (DPHI). These logs were calibrated using a limestone matrix.

Because there are three interbedded lithologies in the producing interval, variations in the neutron porosity log could be caused by lithology variations as well as porosity variations. Therefore, a combination of porosity logs was used to try to separate the sources of the variations. There is no gas present in the Rabbit Hills field, so density and neutron porosity fluctuations due to gas are not present.

There were a total of eight NPHI logs, seven DPHI logs, and five core porosity logs available for the study area. In addition, there were five permeability logs from the core data. Denoting the wells *A* through *H*, NPHI data exist for wells *A* through *H*, DPHI data exist for wells *B* through *H*, and core data exist for wells *B*, *C*, *E*, *G*, and *H*. It is not clear whether the NPHI, DPHI, or core data more accurately represents the actual porosity. Because of varying lithologies, both well log data sets will be skewed from the true porosity values. Also, porosity obtained from cores may not be accurate due to fracturing of the cores on removal from the well. For reservoir simulation, the accuracy of the porosity input data is important, and the most accurate porosity distribution was required. An important point for this study is that there are more wells with NPHI log data making more training data available. Therefore, the NPHI predictions should, in principle, be more reliable.

As with the input data, the neural net architecture requires the same number of porosity values to be predicted for each CMP. Initial attempts at predicting porosity for a 19-foot interval at one-foot increments proved unsuccessful, most likely due to the resolution limits of the seismic data (Link, 1995). Predicting an average porosity over the 19-foot interval representing the upper Bowes resulted in acceptable testing error. Prediction errors for two- and three-layer averages over the 19 foot interval were larger but still acceptable. The average interval thickness for one-, two-, and three-layer predictions were, respectively, 22 ft (6.6m), 11/11 ft (3.4/3.2 m), and 8/7/7 ft (2.4/2.2/2.1 m).

Since permeability data were also available for the five cored wells, neural networks were also constructed to attempt permeability prediction. The permeability data were digitized in one-foot increments, and values ranged over two orders of magnitude. Two approaches were used for scaling the data: linear scaling and logarithmic. In both cases, the neural network converged to the specified training error; however, testing error showed that the trained network was not generalizing to accurately predict permeability values for other locations.

4. Horizon Extraction

Extracting the Bowes member interval from the seismic data required that a depth-time position in the data be known that could serve as a datum for the Bowes section. In the case of the Rabbit Hills data, the unconformity separating the Bowes member from the underlying Firemoon was easily located on the seismic record (Figure 5). A separate MLP backpropagation neural network was then constructed to identify and extract the Bowes interval from the data cube. Subsequent steps fine tuned the location of the datum and output the portions of the traces representing the Bowes interval.

4.1. HORIZON WINDOW

Two data slabs were created from the original seismic data cube. The first consisted of trace segments of a fixed number of samples representing the Bowes chosen from the zero crossover above the Bowes-Firemoon reflector and counting upward 14 samples. The second slab resulted from a variable width window which encompassed the Rierdon-Bowes to the Bowes-Firemoon reflectors. Zeroes were added as padding to the end of the data vector effectively flattening the data on the Rierdon-Bowes interface and providing a common length for all input vectors.

The variable-width input data set was not used in any porosity predictions because poor testing results were obtained using it (possibly due to the inclusion of data from the Rierdon formation or the effect of padding). The fixed length vector data set of 14 data points was chosen in order to include only information from the Bowes interval.

5. Neural Networks

5.1. NEURAL NETWORKS APPLIED TO POROSITY PREDICTION

Two separate neural networks were used in this project. To separate the seismic data representing only the Bowes member from the entire seismic data volume, a neural network was constructed to recognize the Bowes-Firemoon interface in the data. The position of this interface in the data was then used to extract the data representing the Bowes interval. The second neural network was used to relate the extracted seismic data and calculated attributes to porosities in the producing interval.

Instantaneous amplitude is one of many seismic attributes used to identify or track lithologic variations. In principle, with a sufficient number of neurons, neural networks should be able to recognize the patterns produced by attribute calculations without having to include them specifically. In the case of the Rabbit Hills porosity analysis, this appears to be true to some extent because the test error difference between porosity predictions made using seismic data, and seismic data combined with instantaneous amplitude, was not significant.

5.2. NEURAL NETWORK ENSEMBLE APPROACH

All neural network predictions are dependent in varying degrees upon the initialization weights. These are usually chosen to be small, random numbers based on the minima and maxima contained in the input data. The dependency occurs because multilayer networks (representing nonlinear relationships) contain multiple minima in parameter space. Thus, depending on the starting point, different minima may be reached which may not be the global minimum. The number of minima is related to the complexity of the problem.

In an attempt to make the porosity predictions more reliable, ten random seed values were used to generate the initial weights for ten different porosity network predictions. The four networks with the lowest testing errors were then averaged. This ensemble approach helps alleviate the dependence of the network solution on the network initial values.

6. Methodology

6.1. TESTING PROCEDURE

Because of the lack of a well-defined analytical relationship between seismic data and porosity, prediction error cannot be calculated explicitly. Therefore, the reliability of the neural network training process is estimated by inputting a seismic trace interval that corresponds to a known porosity distribution and comparing the neural network prediction to the known value.

This "one-out" testing procedure was performed by leaving porosity data for one well out of the network training set and training the network with the remaining wells. The seismic trace segment representing the test well was then input into the trained network. The network's porosity prediction for that well was then compared to the porosity values for the test well.

Using the NPHI (neutron porosity) data, seven of the wells were used to train the network and the eighth well to test it. This process was repeated eight times, once for each well in the group. Neural network parameters, network architecture, and input data for the neural network were adjusted based on testing errors. For the final porosity predictions, all of the wells were included in the training data set and the ensemble approach was used. The error comparison criterion was the absolute value of the difference of the predicted porosity and the measured porosity values. The same procedure was also used for the DPHI (density porosity) and core porosity analyses.

6.2. NEURAL NETWORK ARCHITECTURE

Porosity prediction was accomplished using a MLP back propagation neural network. Predictions were made for one, two, and three averaged porosity layers. The well log data used were NPHI and DPHI values and laboratory core measurements. Input data consisted of various combinations of seismic amplitude (S), instantaneous amplitude (A), and CMP coordinate position in the data grid (C).

Testing was done using the NPHI data for a single layer to determine which combination of seismic attributes should be used as input data for the network. NPHI data were used because there were eight wells with NPHI data and only seven wells with DPHI data and five wells with core data.

For a given combination of input data, single-layer porosity predictions were made and the number of hidden neurons adjusted until the smallest testing error was achieved. Ten predictions for each combination of input data were made varying only the initial values of the neural network weights. Based on this testing procedure, the final network architecture consisted of a single hidden layer with three neurons. The input layer contained up to 30 inputs consisting of combinations of 14 raw seismic values, 14 instantaneous amplitude values, and two coordinate values. A hyperbolic tangent sigmoid transfer function was used for the middle-layer neurons and a log sigmoid transfer function for the output layer. The output layer varied from 1 to 3 neurons depending on whether one-, two-, or three-layer porosities were being predicted.

6.3. INPUT DATA TYPE

Ensembles, the average of the four porosity predictions with the lowest error, were constructed to calculate an average testing error. This ensemble testing error was the basis for determining which combination of input data resulted in the most accurate predictions.

All permutations of seismic amplitude, instantaneous amplitude, and coordinates were individually input into networks to predict single layer NPHI porosity. The testing results are shown in Table 1 for single-input and combinations of input data types where C represents seismic trace coordinate input, S is raw seismic amplitude input, and A is instantaneous amplitude input. (Note: fractional porosity *100 is equal to percentage porosity.)

TABLE 1. Average ensemble prediction error (fractional porosity) for single-layer neutron porosities (NPHI).

	C	*S*	*A*	*AC*	*SC*	*AS*	*ASC*
Average	0.010	0.053	0.033	0.021	0.036	0.034	0.026

Interestingly, the lowest ensemble testing error occurred using only coordinates as input data (C). Although a slight increase in error occurred with the inclusion of seismic amplitude (SC) and instantaneous amplitude (AC), the additional information produced more realistic-appearing final prediction maps than coordinate inputs alone. Based upon these ensemble testing results, seismic amplitude, instantaneous amplitude, and coordinates were used as input data for the final porosity prediction experiments. The porosity difference between the actual and predicted values, while not as small as the coordinate only predictions, was low enough that separation between the upper Bowes and lower Bowes, in principle, is possible.

6.4. POROSITY NETWORK TESTING ERROR

6.4.1. *Neutron Porosity (NPHI) Predictions*

Ensemble NPHI porosity networks were constructed to predict one-, two-, and three-layers of average porosity using the combination of A (instantaneous amplitude), S (raw seismic amplitude), and C (location coordinates) as input data types. Average and maximum testing errors for predictions using the eight available wells with NPHI porosities are shown in Table 2.

TABLE 2. One-, two-, and three-layer NPHI average and maximum ensemble prediction errors (fractional porosity).

	1-layer	2-layer	3-layer
Average	0.026	0.049	0.031
		0.040	0.042
			0.036
Maximum	0.039	0.078	0.130
		0.074	0.097
			0.101

Prediction error for layer one increases going from the single-layer case to the two-layer case. Surprisingly, the error for layer one decreases when going from the two-layer case to the three-layer case. However, the maximum error does increase with layer number as shown by the maximum error values in Table 2. This is expected, as the degrees of freedom are increased in the network with no additional data constraints.

For the one-layer case, the mean predicted porosity was 16.0% with a standard deviation of 2.9%. For the two-layer network, the mean predicted porosity for layer one was 15.1% and 15.9% for layer two with standard deviations of 3.0% and 3.8%, respectively. For the three-layer case, the mean layer porosities and standard deviations for layers one, two, and three are 11.75%, 15.78%, and 15.65% with standard deviations of 3.34%, 4.58%, and 3.43%, respectively.

6.4.2. *Density Porosity Predictions*

Density porosity predictions were made for C, AC, and ASC input data combinations. No density porosity log was available for well A leaving seven wells available for training the network. Testing errors are shown in Table 3 for the one-, two-, and three-layer cases using ASC inputs and DPHI porosities.

Results show a significant increase in error with the addition of seismic information; a similar effect is seen with the NPHI porosities. The mean porosity for the single-layer DPHI prediction using ASC as input data was 13.48%, using AC the mean was 14.1%, and with only C as input data the mean was 12.0%. The standard deviations were 2.5%, 2.6%, and 3.3%, respectively. DPHI two-layer prediction gave a mean porosity of 12.0% in the top layer and 16.5% in the bottom layer with standard deviations of 3.5% and 2.8%. Mean values for the DPHI three-layer porosity predictions are 8.1%, 18.1%, and 16.0% porosity with standard deviations of 3.0%, 5.0%, and 4.2%.

TABLE 3. One-, two-, and three-layer DPHI average and maximum ensemble prediction error (fractional porosity).

	1-layer	2-layer	3-layer
Average	0.033	0.036	0.038
		0.048	0.042
			0.047
Maximum	0.043	0.052	0.074
		0.075	0.068
			0.081

6.4.3. *Core Porosity Predictions*

Core porosities were available for five of the eight wells in the study area. The core data also included permeability values in addition to porosity values. A one-layer porosity network was generated using coordinates (C), seismic amplitude (S), and instantaneous amplitude (A) as input data. The predicted porosities are the ensemble averages of the four predictions with the lowest average difference error. The mean core porosity for the one-layer prediction is 17.9% with a standard deviation of 3.2%. A two-layer prediction was also made using A, S, and C input data. For the two-layer predictions, the means are 17.9% and 17.3% with standard deviations of 5.5% and 2.3%, respectively. The core ensemble prediction errors are summarized in Table 4. It is not clear why the error for NPHI is lower than the average core testing error.

TABLE 4. One- and two-layer core porosity ensemble prediction error and one-layer error using the same NPHI wells (fractional porosity).

	1-layer (core)	2-layer (core)	1-layer (NPHI)
Average	0.032	0.056	0.024
		0.038	
Maximum	0.044	0.068	0.030
		0.050	

To examine how the number of training wells affects prediction accuracy, a one-layer NPHI porosity network was constructed using only the five wells that also have core data. Testing the network with the three remaining NPHI wells that do not have core data, the measured porosities are compared to the values predicted by the five well NPHI network and the predictions from the eight well NPHI network. These results are shown in Table 5.

Table 5 shows that two of three wells left out of the training showed no significant difference in prediction error with the addition of three wells to the training data set. This result suggests that there is redundancy in the training data set.

TABLE 5. One-layer NPHI porosity predictions and error for three wells without core porosities.

Well	Measured Porosity	5 well Porosity	Error	8 well Porosity	Error
A	0.162	0.189	0.027	0.180	0.018
D	0.157	0.190	0.033	0.153	0.004
F	0.127	0.163	0.036	0.159	0.032

6.5. PERMEABILITY PREDICTIONS

Permeability predictions were made using A, S, and C as network input data types. Predictions were made by scaling the core permeability by a factor of 200. Network prediction errors for one- and two-layer cases are shown in Table 6. Note, these are not ensemble prediction errors but errors for *single*-network runs.

TABLE 6. Permeability predictions and errors for single network runs for one- and two-layer cases.

	Core (md)	Pred. (md)	Av. abs. error (md)	Core (md)	Pred. (md)	Av. abs. error (md)
Average	88	84	36	75	70	87
				76	97	15
Maximum	156	135	70	163	181	145
				149	122	34

Due to the large testing error, an ensemble was not constructed. It appears that the neural network is not able to recognize a relationship that corresponds to the permeability for this particular data set.

7. Porosity Map Generation

Porosity maps were then generated using the networks trained on the various porosity types. The final trained networks used input data consisting of seismic amplitude, instantaneous amplitude, and grid coordinates. Maps were generated by passing all of the seismic trace intervals through the networks to obtain porosity predictions at each

CMP location. Predictions were made for one, two, and three layers of average porosity values using core, NPHI, and, DPHI porosity values. Permeability prediction maps are not shown due to the poor testing results referred to earlier.

A seismic amplitude map of the top of the Bowes is shown in Figure 6 for comparison with the porosity prediction maps. The porosity prediction maps are shown in Figures 7-14. Figures 7-9 contain the map predictions produced by the networks trained with neutron porosity data (NPHI). Figures 10-12 show maps generated using density porosity values (DPHI). The maps generated using laboratory core porosity values are shown in Figures 13 and 14.

A comparison of the seismic amplitude map (Figure 6) with the porosity prediction maps (Figures 7 to 14) shows some interesting similarities. The high seismic amplitude zones (red) agree with zones of high porosities on the DPHI and core porosity prediction maps. This tends to support the idea that the NPHI porosities are being influenced by hydrocarbon or other fluid content. This can be critical in producing maps for the purpose of determining in-fill drilling locations and illustrates the need for a comprehensive approach to parameter determination.

The range of measured NPHI porosity averaged over a single layer for the eight NPHI well logs is 13.6% to 17.9% porosity whereas the predicted values in the ASC NPHI single-layer porosity map (Figure 7) range from a minimum porosity of 8.3% to a maximum of 32.8%. It is worth noting that data available to train the neural networks to predict porosity is limited to wells that have geophysical logs or core data. The small set of training well porosities suggests that large regions of the porosity maps are without representation in the training set.

The differences between the NPHI and DPHI porosity maps are likely due to NPHI values giving an overestimation of the porosity in the presence of hydrocarbons while DPHI values remain relatively unaffected. In addition, well *A* did not have a DPHI log, so there were only seven training wells instead of eight. Holding all other parameters constant, as the number of prediction layers increased from one to two, the error in the predictions tended to increase. This increase in error is likely due to the limit in vertical resolution of the seismic data.

Figure 8 (a) represents the porosity prediction for the top half of the upper Bowes and Figure 8 (b) for the bottom half using NPHI porosity training data. As can be seen, these maps predict that the high-porosity regions are not vertically continuous through the upper Bowes member. The DPHI and core two-layer predictions are similar in appearance also exhibiting high porosity in the vicinity of well *A* (labeled as *EW-1* in the maps) in the top half of the upper Bowes (Figures 11 a, b and 14 a, b). The NPHI two-layer prediction does not show this area as a zone of high porosity. For the three-layer predictions using NPHI and DPHI, it becomes more difficult to distinguish features in the prediction maps (Figures 9 a, b, c and Figures 12 a, b, c).

The three-layer NPHI maps (Figure 9 a, b, c) are consistent with the NPHI two-layer prediction maps (Figures 8 a, b) but show marked differences from the DPHI prediction maps (Figures 12 a, b, c, and 11 a, b). The three-layer DPHI predictions also show good agreement with the two-layer DPHI prediction map features.

Permeability predictions were made for one and two layers. In both cases, testing errors were so large that results are open to interpretation. It is worth noting that the permeability map for one layer (not shown) shows a marked resemblance to the porosity maps for any of NPHI, DPHI, or core predictions. This suggests that permeability is not being recognized with the selection of input data for the present neural network. The two-layer permeability prediction has even higher error (Table 6). An interesting observation is that although testing error shows that the network is not predicting the permeability very well, there are reasonably coherent patterns on the prediction maps that seem to show that the network appears to have predicted some other parameter related to the seismic inputs.

The small testing error achieved using only coordinate input data emphasizes that geologic control is necessary to determine which results can be considered valid. By including seismic data, testing errors increase but are still within limits that allow the high porosity bioclastic limestone to be distinguished from the non-reservoir lithology with a difference in porosity of approximately 7%.

8. Conclusions

This study shows that a neural network ensemble approach can be a valuable technique for generating detailed porosity maps from 3-D seismic reflection data. A valid geologic model of the reservoir is necessary to constrain parameter predictions to values and distributions consistent with the geologic model and to provide an interpretive filter for the predictions.

The inclusion of coordinates in the input data is shown to be essential for the neural networks to produce realistic porosity maps. Testing results show lower errors for a single-layer porosity prediction than for two- or three-layer predictions. NPHI, DPHI, and core porosity data all yield reasonable predictions. However, which prediction is more accurate will be determined, in part, by the geologic model of the reservoir and ultimately by comparison with production data.

The use of seismic data with instantaneous amplitude data and coordinates does not cause a large increase in testing error compared to instantaneous amplitude with coordinates, and, in fact, gives a smaller testing error using DPHI data. The raw seismic data also contains amplitude sign information lost in the calculation of instantaneous amplitude. Therefore, all three types of data: seismic amplitude, instantaneous amplitude, and coordinates are used as input to the prediction networks.

Acknowledgments

The authors would like to acknowledge the United States Department of Energy, Grant No. DEFC02-91ER-75681 for supporting this research. Thanks are also due to Dr. Charles J. Wideman and Dr. Karen W. Porter for analysis and support of seismic and geological data.

References

Link, C. A., 1995, Artificial neural networks for lithology prediction and reservoir characterization: Proceedings of the SPIE: Mathematical Methods in Geophysical Imaging III, **2571**, 163-174.

Porter, K. W., 1996, A middle Jurassic Piper Formation measured section, Suction Creek, Bear Paws Mountains, Montana: Montana Bureau of Mines and Geology Report RI 2.

Rogers, S. J., Chen, H. C., Kopaska-Merkel, D. C., and Fang, J. H., 1995, Predicting permeability from porosity using artificial neural networks: AAPG Bulletin, **79**, 1786-1797.

Taner, M. T., Koehler, F. and Sheriff, R. E., 1979, Complex seismic trace analysis: Geophysics, **44**, 1041-1063.

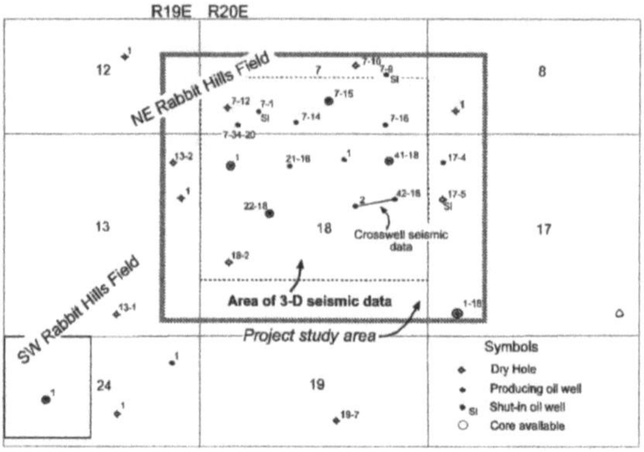

Figure 1. Map of NE Rabbit Hills oil field in northern Blaine County, Montana showing the location of the study area, wells, and seismic survey.

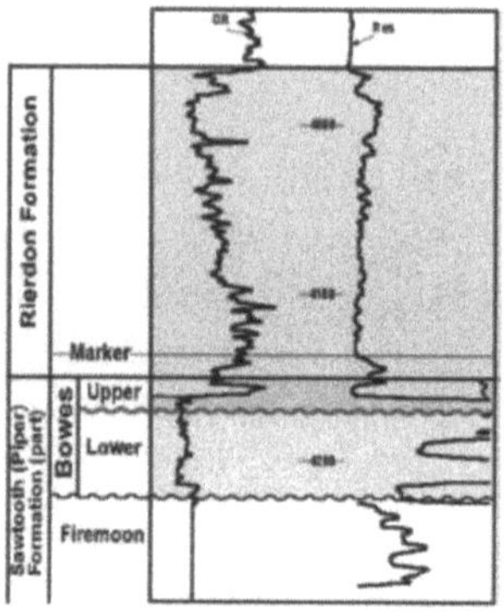

Figure 2. Typical gamma and resistivity logs showing the transition from the upper Bowes to the lower Bowes.

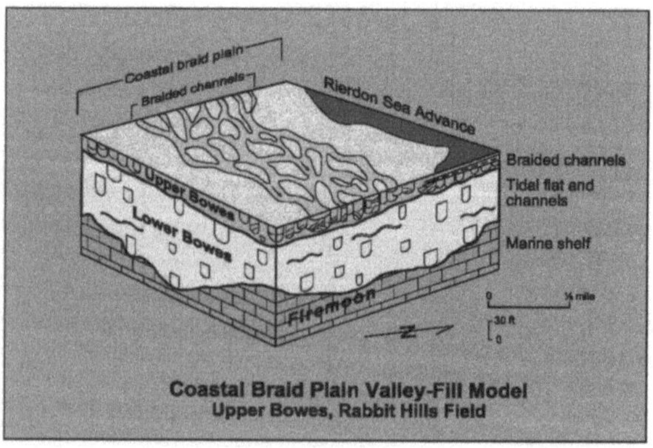

Figure 3. Three-dimensional diagram of coastal braid-plain valley-fill model for the Bowes at the NE Rabbit Hills field.

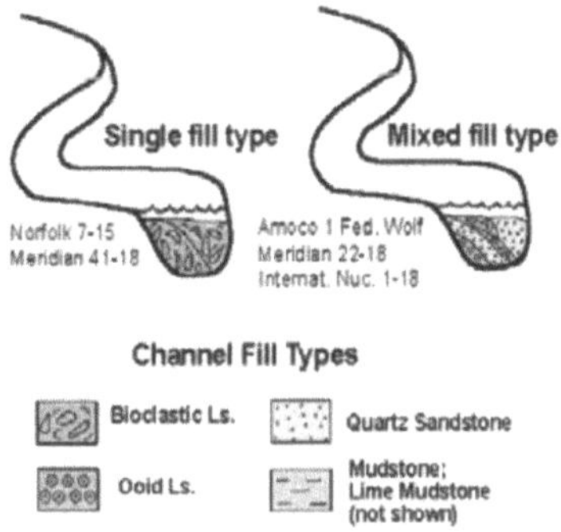

Figure 4. Geology of the various fill types for the Bowes member.

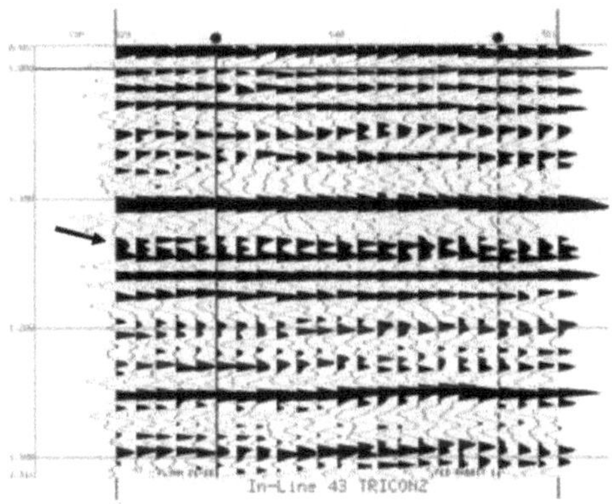

Figure 5. In-line from the 3-D data set in the center of the survey. Note the varying nature of the Bowes Member.

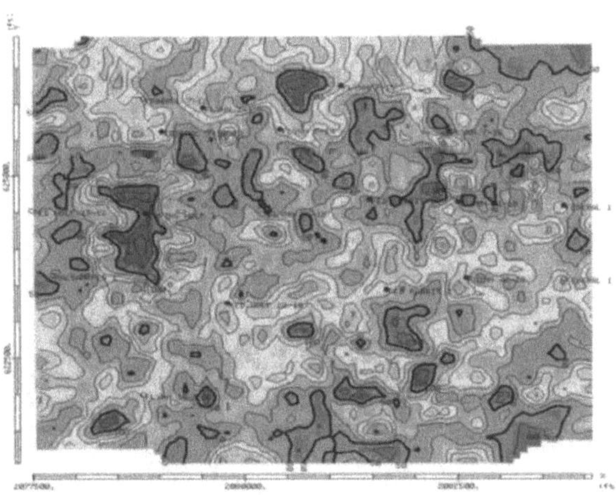

Figure 6. Seismic amplitude map from the top of the Bowes Member. Zones of higher seismic amplitude are shown in red.

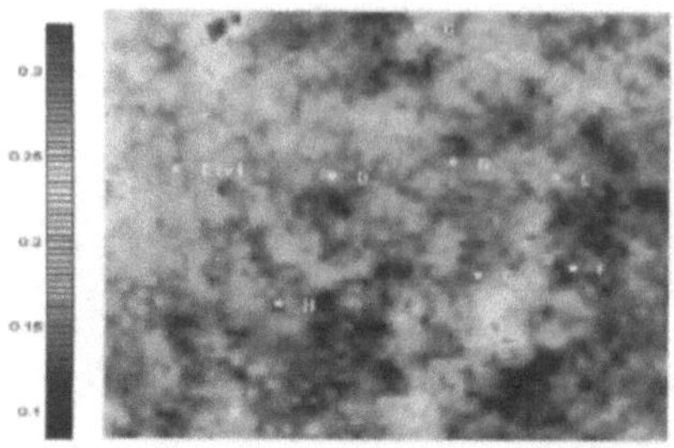

Figure 7. **NPHI porosity layer one of one.**

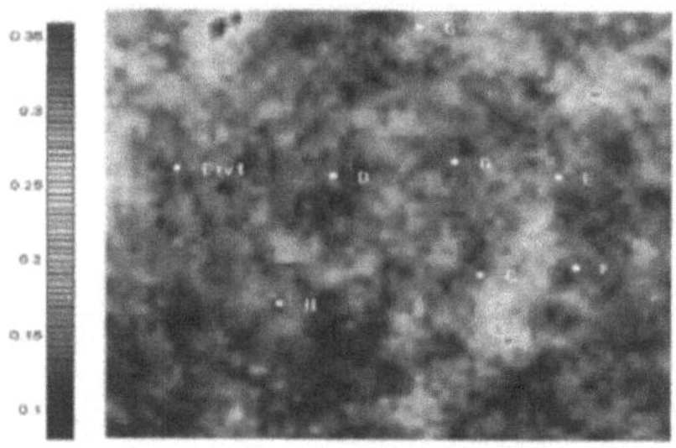

Figure 8 (a). **NPHI porosity layer one of two.**

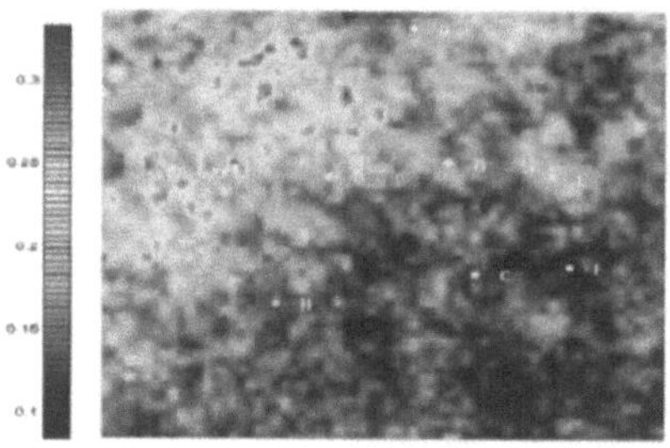

Figure 8 (b). **NPHI porosity layer two of two.**

Figure 9 (a). **NPHI porosity layer one of three.**

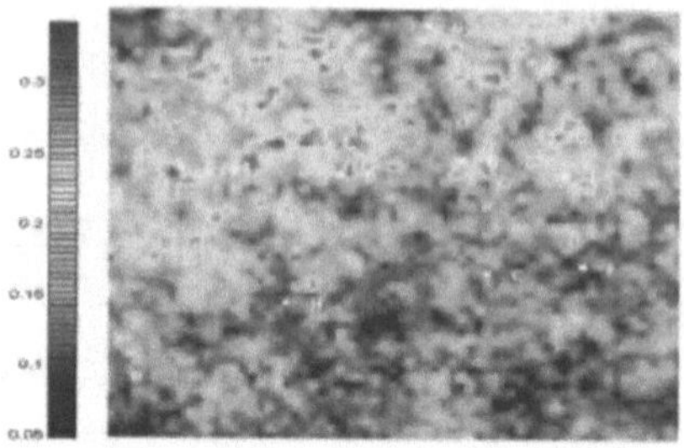

Figure 9 (b). **NPHI porosity layer two of three.**

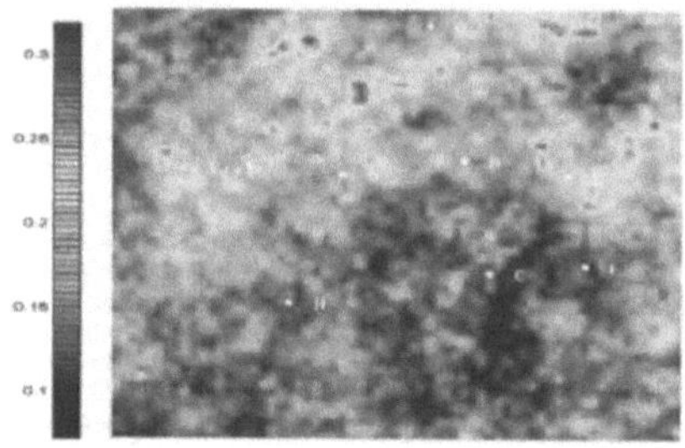

Figure 9 (c). **NPHI porosity layer three of three.**

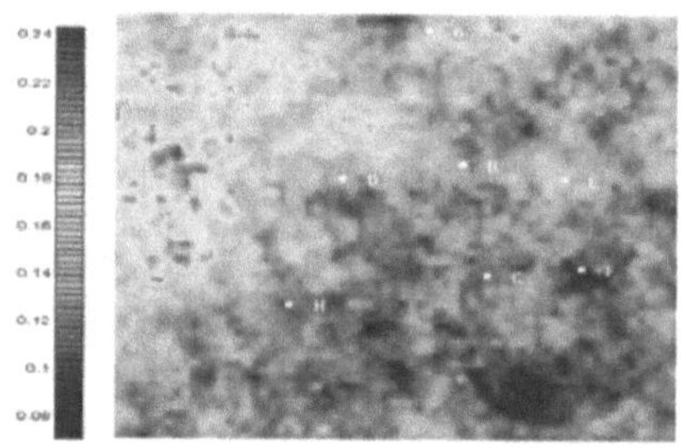

Figure 10. **DPHI porosity layer one of one.**

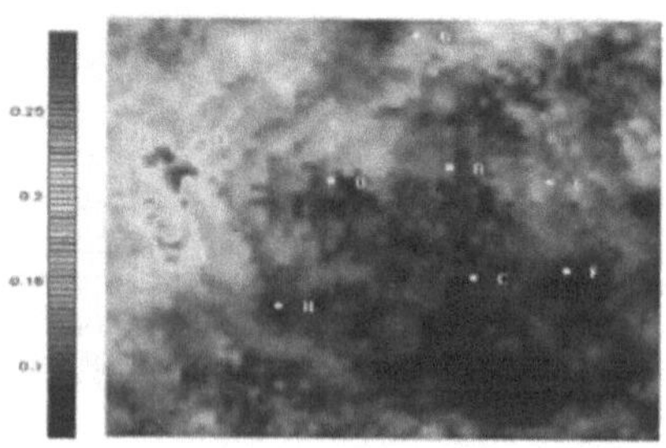

Figure 11 (a). **DPHI porosity layer one of two.**

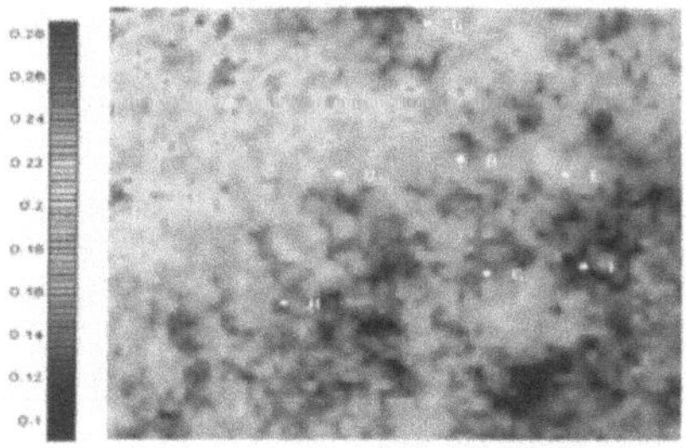

Figure 11 (b). DPHI porosity layer two of two.

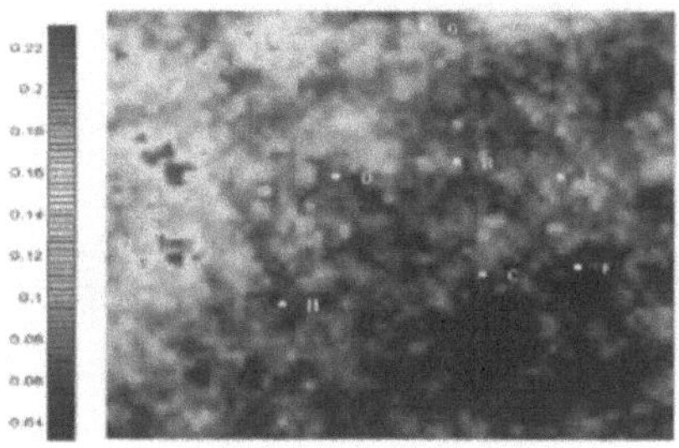

Figure 12 (a). DPHI porosity layer one of three.

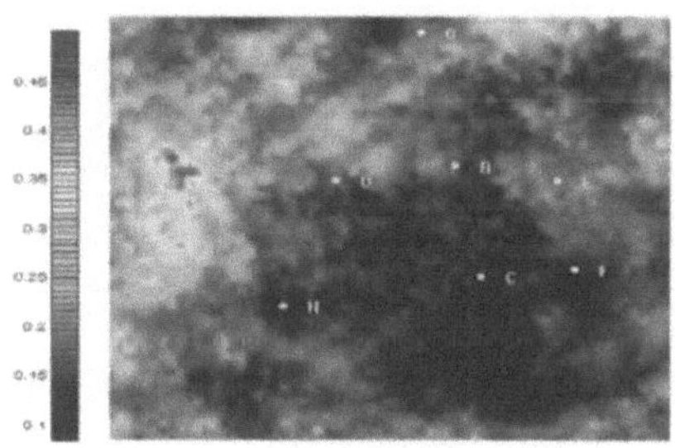

Figure 12 (b). DPHI porosity layer two of three.

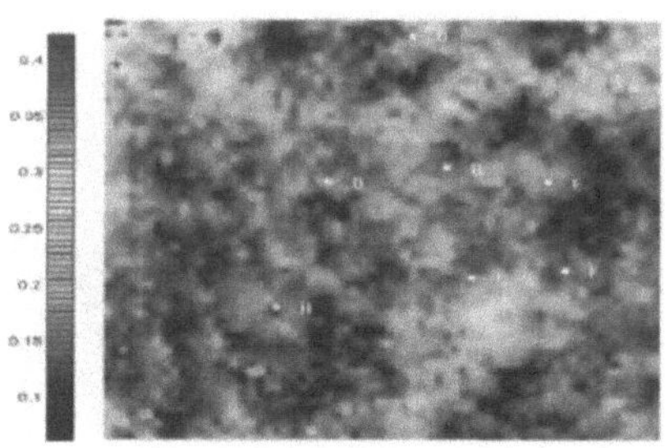

Figure 12 (c). DPHI porosity layer three of three.

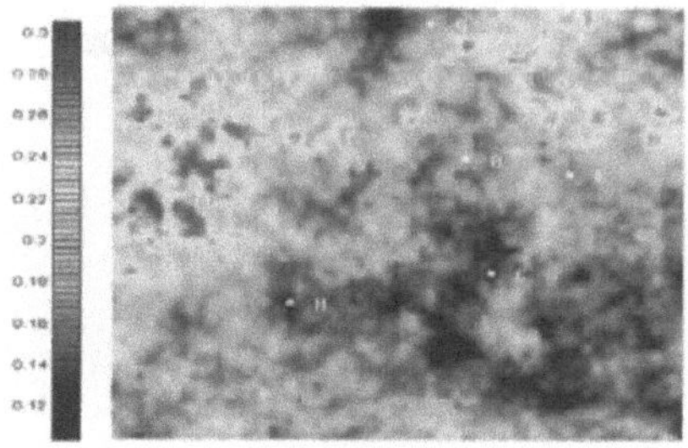

Figure 13. Core porosity layer one of one.

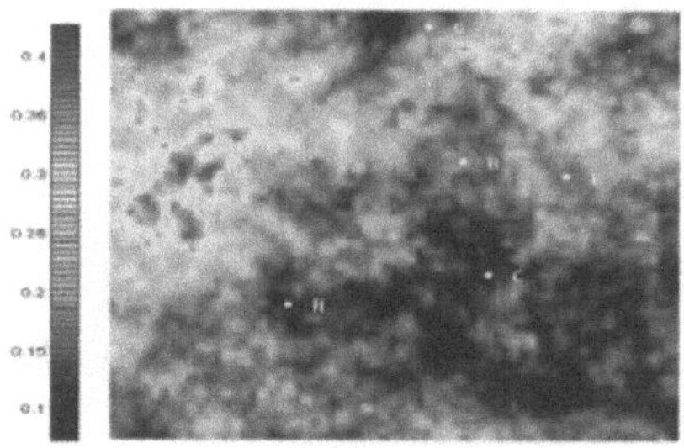

Figure 14 (a). Core porosity layer one of two.

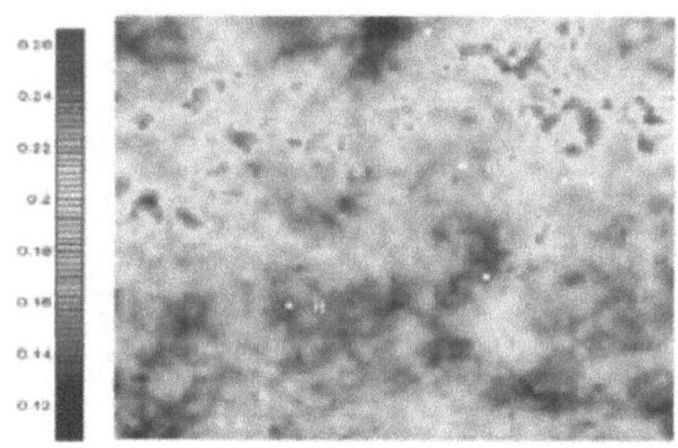

Figure 14 (b). Core porosity layer two of two.

INTERPRETATION OF SHALLOW STRATIGRAPHIC FACIES USING A SELF-ORGANIZING NEURAL NETWORK

CURTIS A. LINK† and STUART BLUNDELL‡

† *Department of Geophysical Engineering,*
Montana Tech of the University of Montana,
Butte, Montana 59701,
USA
E-Mail: clink@mtech.edu

‡ *Integrated Geoscience Inc.,*
Helena, Montana 59601,
USA
E-Mail: intgeo@MT.net

Abstract

A study was recently conducted to assess the extent of hydrocarbon impacts to groundwater and soil resources at a regional petroleum refinery. To accomplish the study, 46 groundwater-monitoring wells were installed at the site. Data collected from the wells included detailed lithologic descriptions from samples and cuttings, and suites of geophysical well logs. Because the quality of the lithologic descriptions was erratic, our approach was to produce lithofacies interpretations based on gamma ray logs, used as input to a neural network classifier system.

The type of neural network used was a self-organizing feature map (SOFM). The output classifications were plotted as *pseudo-logs,* and interpretations were performed using these pseudo-logs. Cross sections constructed with conventional well log interpretations and the neural network classifications show good, general agreement.

The main advantage of the SOFM approach is that all the well log data are analyzed simultaneously, eliminating inconsistencies that frequently occur with a conventional approach. Another benefit is the ability to vary the number of output classes to investigate how well the data fit the expected number of lithology types.

We found that the optimal number of classes is related to the level of lithologic detail that can be resolved. In addition, the network weight matrix can be used to help determine whether classifications warrant division into additional sub-classes.

W. Sandham and M. Leggett (eds.),
Geophysical Applications of Artificial Neural Networks and Fuzzy Logic, 215-230.

1. Introduction

The work described in this chapter was part of a larger project to develop an environmental information database model in a geographic information system (GIS) for a refinery site in Billings, Montana, USA. A key component of the GIS project was the reprocessing, modeling, and interpretation of the geophysical well log data in order to develop a stratigraphic facies model for the site.

The refinery is an active, crude oil refinery located on the south bank of the Yellowstone River in Billings, Montana (Figure 1). The Environmental Protection Agency (EPA) required a Remedial Investigation/Feasibility Study (RI/FS) at the facility to assess the extent of hydrocarbon impacts to groundwater and soil resources. This study was directed at that purpose.

1.1. SITE LOCATION AND GEOLOGY

The Yellowstone River is a typical, braided river characterized by coarse bedload material (gravel and cobbles), gravelly channel and braid-bar type deposits, and channel-fill sequences. The lateral and vertical distribution of these types of deposits is complex due to channel migration and abandonment processes. In the specific reach of the Yellowstone River where the refinery is located, the facies can be broadly classified into two types:

1) a lower gravel facies containing matrix material that varies from sand to silt to clay, and
2) an upper floodplain facies consisting of fine-to medium-grained sand, silt, clay, organic materials, and scattered lenses of gravel.

In December 1995, cased hole geophysical log data were collected from 46 groundwater monitoring wells to assist geologists in their interpretation of the shallow (less than 30 ft/9 m) stratigraphy at the site. Well log data consisted of gamma ray, electromagnetic conductivity, and neutron porosity logs. These borehole geophysical data acquired at the site provided the opportunity to overcome sample resolution issues, introduced by different drilling methods, as well as human inconsistencies in describing sediment samples.

In this study, the geophysical log data were first interpreted conventionally to develop a depositional model for the site in order to map three main sedimentary sequences:

1) a gravel facies deposited directly on the shaly-sandstone bedrock,
2) the lower sand of the floodplain facies deposited on the gravel facies, and
3) the upper silt/clay unit of the floodplain facies.

The well log data were then reprocessed and modeled using a self-organizing feature map (SOFM) neural network to assess the capability of a neural network approach to predict lithology in this type of depositional setting.

The vertical and lateral distribution of lithofacies associated with gravelly, braid-bar deposits is complex, which makes it difficult to correlate individual sedimentary packages from well to well. The application of an unsupervised neural network takes advantage of the ability of an *unbiased* system to recognize patterns of geophysical log curve response to individual lithologic facies.

2. Methodology

This section contains a description of the well log data and interpretations, and the neural network construction and application.

2.1. GEOPHYSICAL DATA COLLECTION AND INTERPRETATION

A total of 46 wells were logged using a downhole tool combination of gamma ray, neutron density, and electromagnetic conductivity. The geophysical interpretation consisted of two processes. First, interpretation of geophysical log response and correlation to sediment descriptions from wellsite geologists including the development of a facies model, facies isopach maps, and top structure maps, and second, analysis of the geophysical log data to assess the ability of a self-organizing neural network to classify lithology. The objective of this approach was to use an unbiased computer model to recognize sedimentary facies based on well log variations. This chapter concentrates on the analysis conducted in the second process.

2.2. APPLICATION OF NEURAL NETWORKS FOR LITHOLOGY PREDICTION

Initially, our approach was to use a supervised network with a set of 15 conventionally interpreted well logs for the target data. Well log interpretations were based on 10 Unified Soil Classification System (USCS) soil classifications that matched geologists' observations from the well samples. These interpretations were then used as training data for a MLP backpropagation network.

The goal for using the supervised ANN was to produce output similar to that of the USCS interpreted lithology from the training wells. One advantage of a supervised ANN is the format of the output. The supervised ANN can be constructed such that the output consists of sets of numbers corresponding to the original interpretation classes. The values in the sets can be specified to lie within the range 0 to 1, which can conveniently be interpreted as a *pseudo-probability* providing an indication of the reliability of the classification.

An unsupervised network was implemented after numerous unsuccessful attempts to train the supervised network. The poor performance of the supervised network was probably related to the interpretations of the well logs that were used for training data. This is discussed in more detail later.

2.3. DATA PREPARATION

The initial and most time-consuming step in neural network design is data preparation. The digitized well logs were reformatted for input into the neural network software used for this project (Matlab Neural Network Toolbox, 1993). Survey locations in the form of *xyz* coordinates were attached to the log data to place the logs at the correct sea level elevation for subsequent mapping of the well surface locations and quality control checks.

The neutron and gamma log units were converted to porosity and API units, respectively. However, the gamma ray logs were used as the primary geophysical logs for facies interpretation because of their response to rapid variations in lithology and lack of sensitivity to fluid content.

3. Unsupervised Neural Networks

The unsupervised ANN used was the self-organizing feature map (SOFM) introduced by Kohonen (1987). The SOFM is an example of a *competitive* network. Figure 2 is a schematic showing the operation of a competitive network on a simple problem. Six normalized input vectors, p_1 to p_6, are input to a competitive network that can output three classes. Weight vectors w_1 to w_3, are initially set to random values. This is represented in Step 1 in the figure. Step 2 shows the results after presenting the first input, p_2, to the system. Weight vector w_2 is closest to p_2, so it is *moved closer* to p_2. This process continues with repeated presentations of input vectors p_i until the weight vectors w_i reach the stable positions shown in Step 3. The network is now considered to be *trained*, and any new input vectors that are presented to the network will be classified into one of the three classes delineated by the boundaries shown in Step 4.

The procedure used for this study was to input a data set of well log samples and specify the number of classifications desired. The network was built to allow classifications on individual or multiple logs, using single input samples or a user-specified window of samples.

When using a data sample window size greater than one, the network classification is output at the center of the window. This procedure allows the network to *look* at data samples before and after the analysis point. This is similar to techniques used by human interpreters. The sliding window is then moved in increments equal to the depth increment of the recorded geophysical logs.

The geology of the analysis site dictates the approximate number of lithologic classifications expected. In the case of the refinery site, a minimum of four classes and possibly a maximum of twelve classes are required to describe the geology adequately. SOFM runs were made at all of these numbers of classifications. Any subsequent supervised ANN training is strongly dependent on the number of classifications and how much differentiation exists among them. SOFM neural networks can also be used to verify that the number of expected classes is warranted. This can be done by using a large number of output classes and analyzing the output for groups and subgroups.

All of the final classifications used in the geologic cross-sections were made using only gamma logs for input data with a sliding window size of 9 samples, 6 output classes, and a 21-point smoothing filter applied to the final classifications.

If smoothing is not applied to the final classifications, the SOFM results clearly show that classifications are responding to high frequency variations in the gamma logs and interpretation is almost as difficult as for the original logs. Smoothing of results is an important process in the application of SOFM to shallow environmental problems where fine scale lithology variation is the norm. It should also be noted that when larger numbers of output classes are used, the results often show a smooth gradation from class to class. This is what typically occurs in the subsurface and should be expected in a correct classification. The SOFM accomplishes this by employing a *neighborhood* parameter, which is specified in the network design.

A typical example of a well log suite and classification is shown in Figure 3. Figure

3a is the conductivity log and Figure 3b is the neutron porosity log. Both of these logs are typical of log response to fluid content in the matrix. Figure 3c shows the typical high-frequency response of the gamma ray log to varying clay content. Figure 3d shows the 6 class, smoothed SOFM output along with the scaled gamma log. The output represents 6 classes resulting from a 9-sample input window, 21-point smoothing filter, and 20,000 iterations of the SOFM.

4. Supervised Neural Networks

As discussed earlier, prior to SOFM analysis, interpretations were made on a set of 15 wells using the geologic reports and the gamma ray well logs. These interpretations were used as target values for a supervised network classification. The supervised network, a MLP back-propagation (BP) network, was designed to compare inputs of log values with their corresponding interpretations. The error between the BP output and the interpretation is used to adjust the network parameters to minimize the error in an iterative fashion. When the error decreases to a predetermined level, the process stops and the network is trained.

As noted previously, the MLP BP network analysis was not successful. The main reason for this appears to be inconsistent interpretations for the gamma logs. Similar to human interpreters, when neural networks are fed conflicting information, unpredictable or unusable results occur. There are two approaches to solve this problem. First, reinterpretations can be made from the geologists' descriptions and the geophysical logs. This reinterpretation is a time-consuming process subject not only to the bias of the particular interpreter but also to the vagaries of lithological descriptions from more than one geologist.

The second approach is to use the SOFM classifications in building revised interpretations. This approach removes a large amount of the bias that enters into the interpretation process and can be done in a fraction of the time. The process uses the SOFM classifications (consistent from well to well) and determines interpretations for them. This involves only coming up with interpretations for the particular number of classes that have been output rather than for the entire set of well logs. Thus the final interpretations are based on a non-biased classification derived only from the raw data. Once suitable interpretations have been obtained, a BP network can be constructed. (Due to client deadlines, this approach was not attempted for this study.)

It should be pointed out, that even the failure of the BP network to train is an indication that something is amiss in the data. In fact, it was this very failure to train for the monitoring well interpretations that suggested a more detailed look at the values. Inspection of the conventional USCS interpretations shows that it is easy to pick out conflicting zones that prevent the BP network from converging. For example, the interpretations GM and SW often represent very similar gamma log values (Figure 5b, left panel). Likewise, ambiguous interpretations occur for ML and SM, among others (Figure 4b and Figure 5a, left panels).

5. Geophysical Log Character of Facies

The hand interpretations of the well logs were done using the gamma ray log. Although the neutron porosity and conductivity logs were reviewed, they did not play a major role in the stratigraphic interpretation since their response was primarily a reflection of pore fluid characteristics. For conventional interpretation, our approach to log correlation was to break out packages of sediments that were reflective of coarsening-upward or fining-upward sequences. The predominant matrix material of the subsurface facies at the refinery site is silt. The gamma ray silt line was interpreted to be at 100 API units. Table 1 summarizes the gamma ray log response for silt and other sedimentary units. Note that the overlap of gamma log values requires careful judgment on the part of an interpreter.

TABLE 1. Gamma ray log response values for representative facies.

Floodplain Facies	Gamma Ray Log Response (API Units)
Fine-Medium Grained Sand (Clean)	50-75
Silty-Sand	75-100
Silt	100
Silty-Clay	100-125
Sandy-Clay	125-150
Clay	150-200
Gravel Facies	
Gravel - Sand Matrix	75-100
Gravel - Silt Matrix	100-125
Gravel - Clay Matrix	125-175

6. Results

This section describes the results obtained using the SOFM classifier.

6.1. CONVENTIONAL INTERPRETATION Vs SOFM CLASSIFICATION

Although absolute API values are used to interpret matrix material, it is the gamma ray log *signature* that is the key identifier to recognition of sedimentary facies. In other words, the *change* in the gamma log values is critical in lithofacies identification.

6.1.1. *Gravel Facies*

Figures 4a and 4b, left panels show the interpreted gamma log responses for monitoring wells W-15b (a) and W-19b (b). These wells represent sedimentary sequences of well-graded and poorly-graded gravels. Overall, the gamma ray signature is blocky and marked by pulses of fining-upward and coarsening-upward sequences. Well W-19b is interpreted to have a sand matrix based on the median value of the gamma ray of between 75-100 API units, whereas well W-15b shows a baseline shift of the median value of the gamma ray to between 125-150 API, units which indicates a silt-to-clay matrix.

The SOFM interpretation (Figures 4a and 4b, right panels) shows output class 3 correlating mainly to GM (silty gravel) and class 5 correlating with SM (silty sands). It is evident that a one-to-one correlation does not exist between the SOFM classes and the soil classifications. The SOFM classes don't seem to indicate the amount of gravel richness that is interpreted by the geologists.

The soil classifications were made by interpreters who had to look at many log curve segments to determine soil types and the difficulty of maintaining consistency becomes apparent.

Note that the left and right panel pairs in Figures 4-6 are offset slightly and are scaled differently. The right panels contain not only the SOFM classifications but also a scaled version of the gamma log for reference. The solid lines between panels connect the same points on the original and scaled versions of the gamma logs.

6.1.2. *Floodplain Sands*

Figures 5a and 5b, left panels show the log curves for monitoring wells GM3-41 (a) and GM3-34 (b), which are representative of floodplain sand deposits. The gamma ray log signature for well GM3-41 (a) shows a series of stacked sands, silty sand, and silts which probably also contain some gravel. Well GM3-34 (b) shows a cleaning-upward profile of well-graded sands.

The classes determined by the SOFM show homogeneous sequences not interpreted by the geologists (Figure 5a and 5b, right panels). Class 5 correlates well with SM (silty sands), while class 6 correlates with the more clay-rich intervals such as ML (silty or clayey sands).

6.1.3. *Clay and Silt Sequences*

Examples of clay and silt-rich floodplain sequences are shown in Figures 6a and 6b, left panels. Well W-17b (a) shows a fining-upward sequence at elevations 3085 - 3100 ft. Well GM3-36 (b) shows a similar fining-upward appearance with baseline gamma ray values between 150-175 API units.

The SOFM results represent clay-rich sequences as classes 5, 6, and 1 (Figures 6a and 6b, right panels). Figure 6 shows good agreement of the SOFM classifications with the CL (clays) and SC (clayey sands).

All of these SOFM classifications are from the same network using an output of 6 classes, 9-sample input window size, 21-point smoothing filter, and 20,000 iterations (presentations of the complete set of input vectors).

6.2. EFFECT OF SMOOTHING AND NUMBER OF CLASSES

Figures 7 and 8 show the effects of varying the amount of smoothing and the number of output classes for a single monitoring well, W-17b.

Figures 7a, 7b, and 7c show a 3-neuron classification with the number of smoothing filter points set to 5, 9 and 15. Figure 8a shows the same 3-neuron classification with a 21-point filter applied. The smoothing filter was a simple boxcar filter applied to the final classifications. The effect of smoothing is minimal except when intervals are thin or close together. Increasing the degree of smoothing would be most beneficial for aiding interpretation of logs with multiple, stacked sequences of gradational lithologies.

Figures 8a, 8b, and 8c show the result of using a 21-point smoothing filter and changing the number of output classes from 3 to 6 to 9. Note that the actual class numbers in Figure 8a do not correspond with the class numbers in Figure 8b and 8c. The SOFM does not associate a particular class number with a certain range of input values. Thus, different classification runs can have different class output numbers. It is not the value of the class number that is important, but the membership of the class itself. However, the class members in Figure 8b and 8c do appear to be consistent.

The advantage of increasing the number of classes is evident in the gradation of classes. A smooth gradation from large class numbers to smaller, for example, can indicate that a smaller number of classes may be sufficient to characterize the data.

6.3. SOFM WEIGHTS

Another benefit of SOFM classifications is the information in the weights of the final trained network. Figures 9-11 show final SOFM network weights for networks using 4 and 6 output classes and 5 and 9 input samples.

Figure 9 shows six weights of nine samples each that represent the prototype vectors representative of the set of 12,000 input vectors. Recall that the input vectors are formed from 42 digitized, gamma ray logs that have been formatted for input such that each input vector is a data window of chosen length (e.g. 9) centered on the desired output sample depth.

These final weight vectors span the range of gamma values seen on all of the logs and display a distribution sufficient to separate input vectors into the chosen number of classes. The weight vectors are closer together at the lower API values where input vector distribution is probably more dense. Wider separation is evident at higher values showing that clays distinguish themselves more readily than the silty gravels or sandy gravels at lower API values.

Figure 10 and Figure 11 show the final weight vectors for 4 output classes with 9 and 5 input samples, respectively. The distinct separation of the clay values still occurs in both cases. However, it now appears that there might be three main classes: a lower gravel-rich class, a middle class represented by two weight vectors at about 130 API units, and a clay-rich upper class. It is also interesting to note that the weight vectors display increasing magnitude for the more clay-rich values. The prototype weight vectors of higher magnitude appear to be representing fining-upward profiles. This indicates that the SOFM is recognizing trends in the input vectors. This is especially clear in looking at weight 1 (W1) for the 9 input case (Figure 10) and the 5 input case (Figure 11) and weight 4 (W4) for the 6 class case (Figure 9).

The clustering that appears in the final weights can serve as an additional analysis tool for determining how many class representations are appropriate for a particular data set.

6.4. SOFM CLASSIFICATIONS AND GEOLOGIC CROSS-SECTIONS

Two examples of cross-sections are shown with conventional, major facies interpretations and SOFM classification interpretations using six output classes (Figures 12 and 13). Comparisons of the cross-sections show generally good agreement; however, the additional classes in the SOFM results define additional stratigraphic units that are not represented in the conventional interpretations.

Cross-section B-B' (Figures 12a and 12b) is a north-south cross-section through wells GM-11, GM-18, GM3-45, and GM3-33 (location map, Figure 1). Note, the conventional interpretations were made prior to the neural network results. There is good agreement in stratigraphy for the topmost clay zone, both in presence and thickness. The large gravel unit in the lower section of GM-11 is classified as gravel and sand in the SOFM results (classes 3 and 4). The gravel and sand agreement is generally good from north to south (left to right) across the section.

Cross section D-D' (Figures 13a and 13b) is a north-south (left to right) cross-section through wells GM-5, GM3-38, GM-19, and W-17b. The conventionally interpreted section shows the silts and clays of the floodplain facies thickening at well W-17b, whereas the SOFM results show an overlying gravel unit somewhat higher on the log. The blocky sand and gravel units between GM3-38 and GM-6 are also present in the conventionally interpreted cross-section. Note that the SOFM results are able to show interfingered sand and gravel units in the lower part of W-17b.

7. Conclusions

Self-organizing neural networks can be a useful tool in the interpretation of complex stratigraphic sequences. Although a supervised neural network was unsuccessful in determining lithologic classifications for the refinery study site, an unsupervised network proved to be capable of providing realistic, lithologic classifications. The advantage of using an unsupervised network is that minimal user bias is introduced into the classifications, and additionally, very little time is required to produce classifications for every well.

The failure of the supervised network was indicative of inconsistencies in the interpretations required for the target data. Further work will include the use of unsupervised classifications as a training data set for a supervised network.

Neural network run times are relatively short for a problem of this scale. On a mid-level UNIX workstation, Matlab neural network software was used to process 12,000 input vectors, each containing 9 samples, constructed from 42 gamma ray well logs for 20,000 iterations, in about 2 minutes CPU time.

References

Hagan M. T., Demuth, H. B., and Beale, M., 1995, Neural network design: PWS Publishing Co.
Kohonen, T., 1987, Self-organization and associative memory, 2nd Ed.: Springer-Verlag.
Matlab Neural Network Toolbox User's Guide, 1993: The MathWorks, Inc.

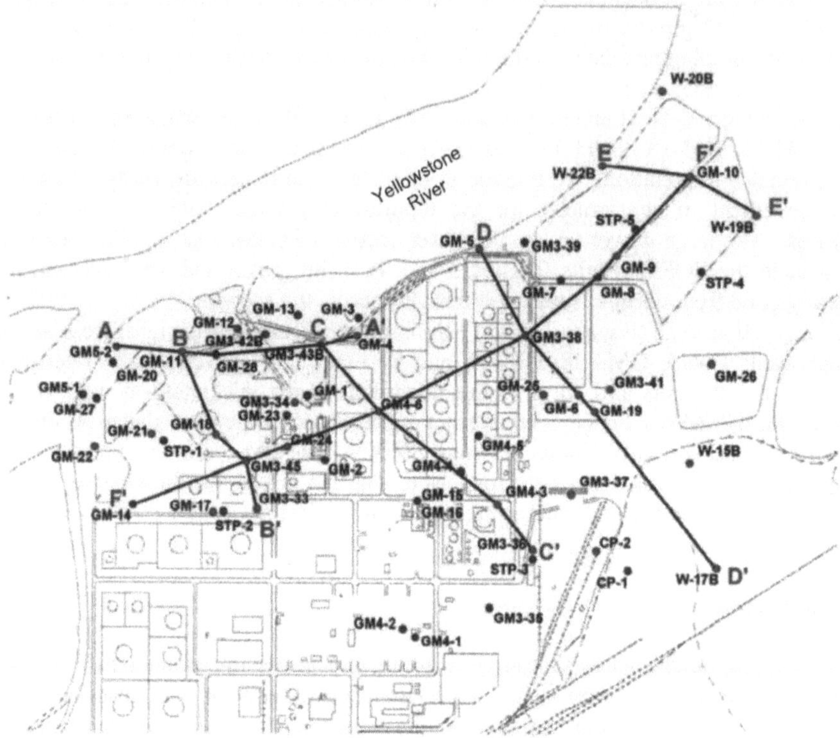

Figure 1. Map of refinery site in Billings, Montana, showing locations of monitoring wells, cross-sections, and proximity to the Yellowstone River.

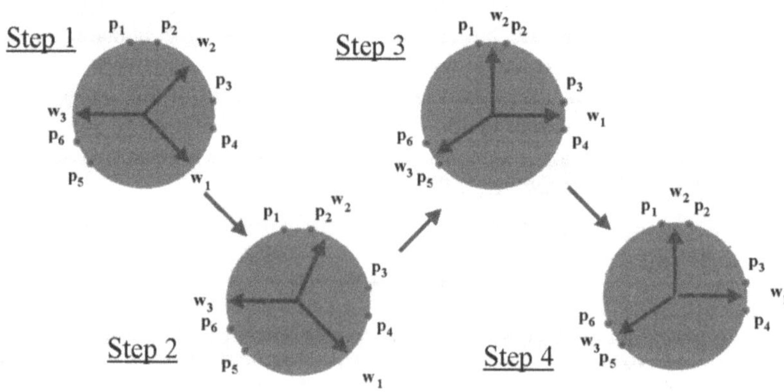

Figure 2. Sequence of steps in a simple SOFM classification problem. Step 1 - network starting point with random initial weights. Step 2 - adjustment of w_2 after presentation of p_2. Step 3 - final positions of weight vectors after repeated presentations. Step 4 - overlain boundaries showing how subsequent input vectors would be classified (after Hagan et al., 1995).

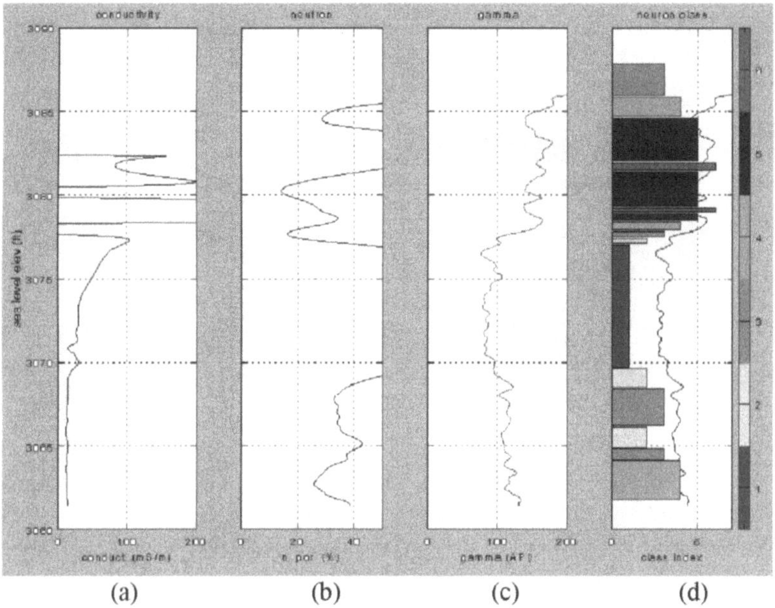

Figure 3. Typical well log suite and SOFM classification. Logs shown are conductivity (a), neutron porosity (b), gamma ray (c), and SOFM classification overlain with the scaled gamma log (d). The SOFM output represents 6 classes using a 9-sample input window, 21-point smoothing filter, and 20,000 SOFM iterations.

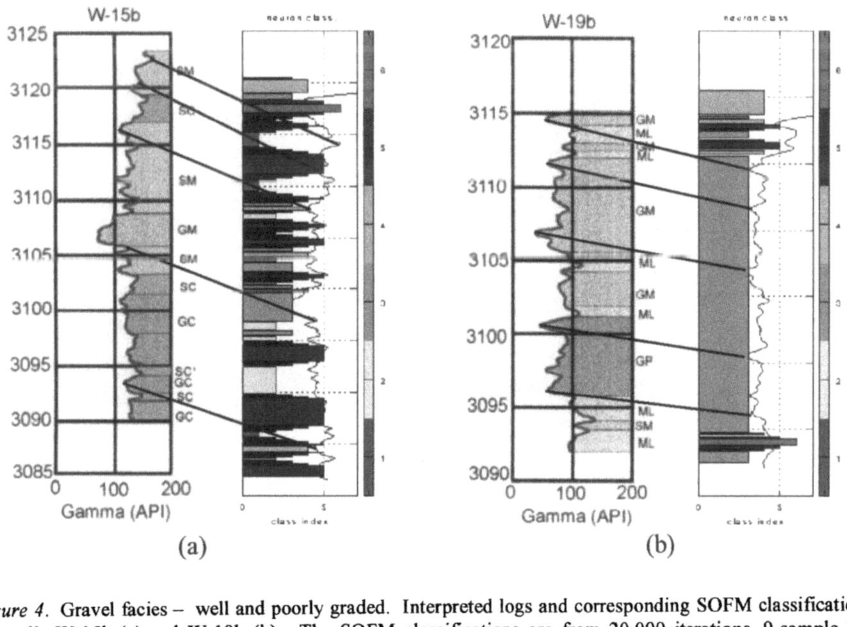

Figure 4. Gravel facies – well and poorly graded. Interpreted logs and corresponding SOFM classifications for wells W-15b (a) and W-19b (b). The SOFM classifications are from 20,000 iterations, 9-sample log window size, 7-point smoothing filter, and 6 output classes.

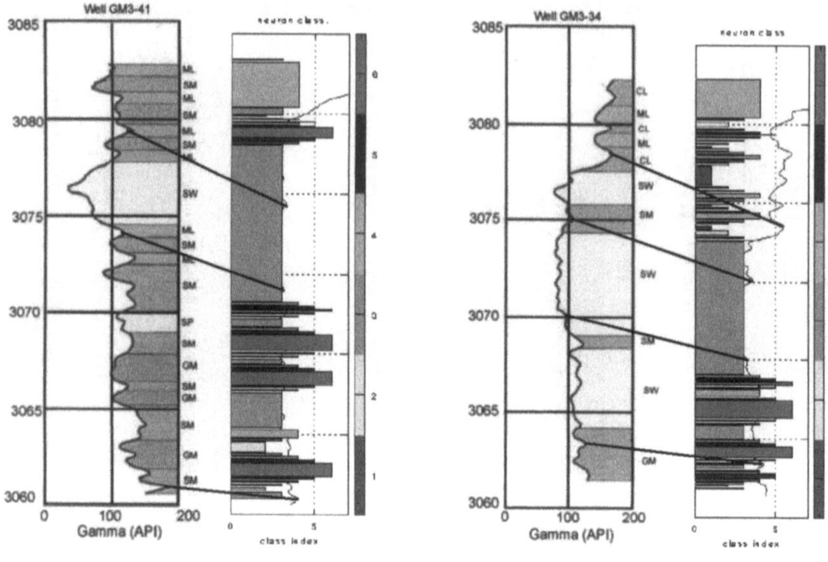

(a) (b)

Figure 5. Floodplain sand deposits. Logs and SOFM classifications for wells GM3-41 (a) and GM3-34 (b). SOFM classifications for 20,000 iterations, 9-sample log window size, 7-point smoothing filter, and 6 classes.

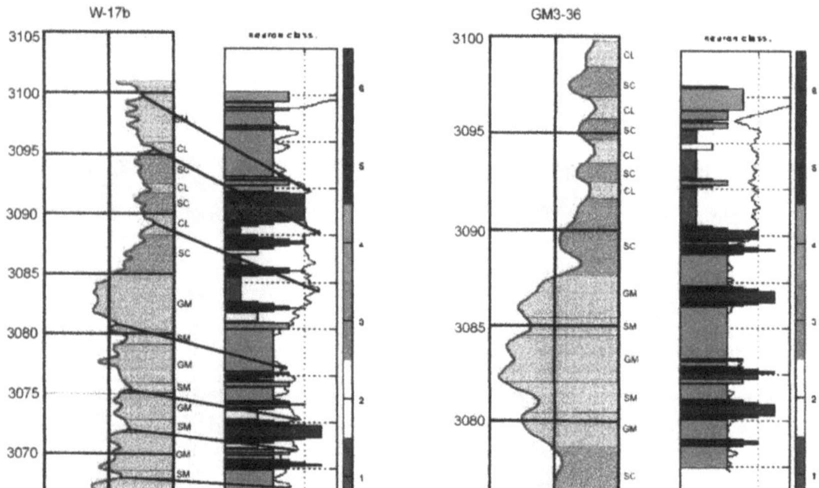

(a) (b)

Figure 6. Clay and silt-rich floodplain sequences. Interpreted logs and SOFM classifications for wells W17-b (a) and GM3-37 (conventional interpretation) GM3-36 (SOFM classification) (b). SOFM classifications for 20,000 iterations, 9-sample window, 7-point smoothing filter, and 6 classes. (Note: GM3-37 and GM3-36 are close together and have similar gamma ray logs.)

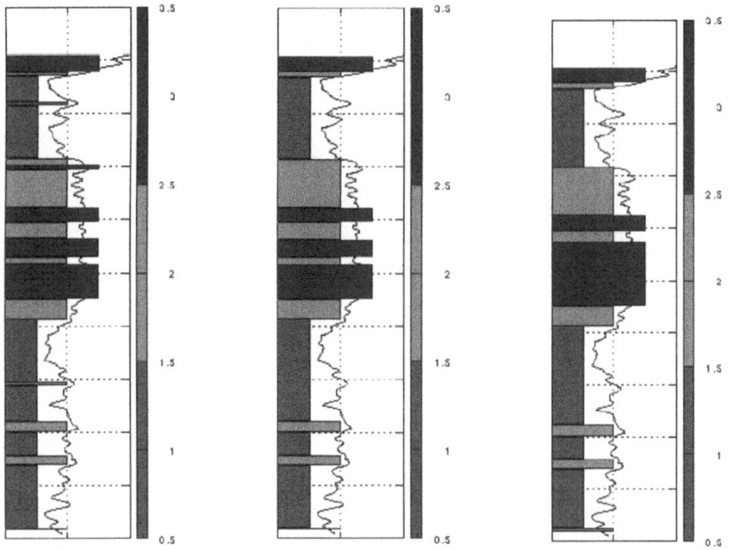

(a) (b) (c)

Figure 7. Comparison of smoothing filter size on 3-neuron classification for well W17-b. The classification is based on the gamma ray log using a sliding window size of 9 samples, 3 output classes, and smoothing filter sizes of (a) 5 points (b) 9 points, and (c) 15 points.

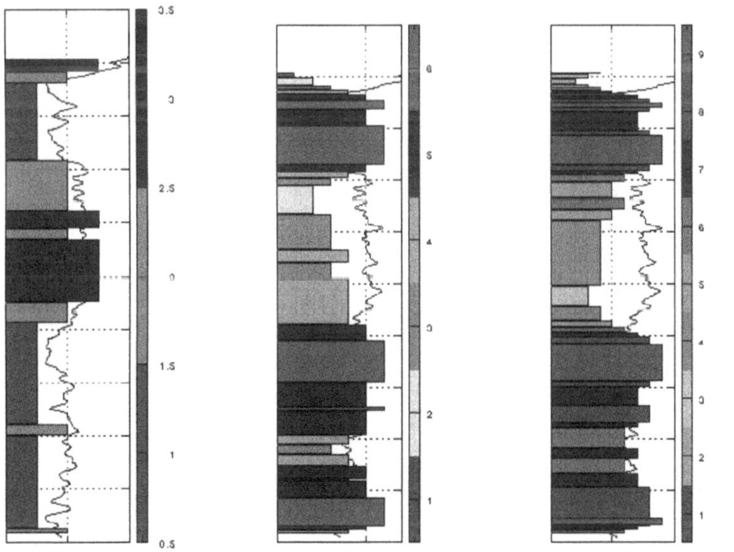

(a) (b) (c)

Figure 8. Comparison of number of classes with 21 point smoothing filter for well W17-b. The classification is based on the gamma ray log using a sliding window size of 9 samples, for (a) 3 output classes, (b) 6 classes, and (c) 9 classes.

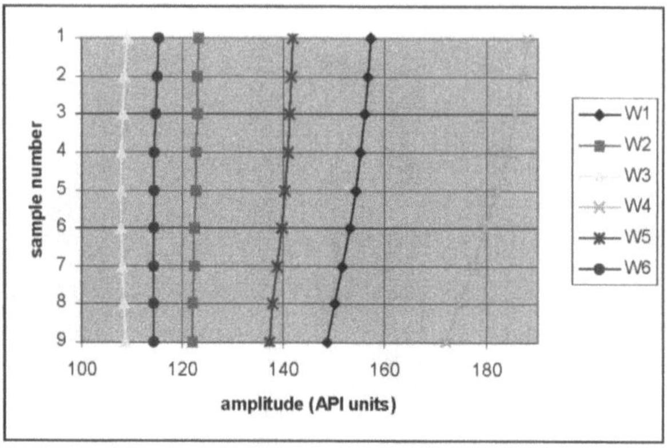

Figure 9. **Plot of weight vectors for classifications resulting from 20,000 iterations, 9-sample log window size, 7 point smoothing filter, and 6 output classes. The weight vectors can be viewed as representative class members for the particular input used, in this case, the gamma log.**

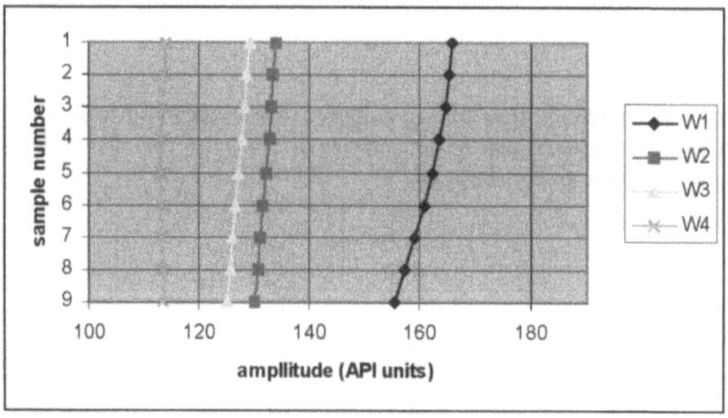

Figure 10. **Prototype weight vectors for classifications resulting from 20,000 iterations, 9-sample log window size, 7-point smoothing filter, and 4 classes.**

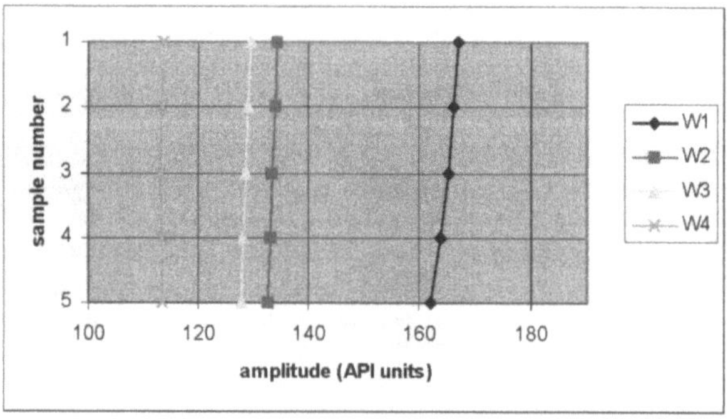

Figure 11. **Prototype weight vectors for classifications from 20,000 iterations, 5-sample log window size, 7-point smoothing filter, and 4 output classes.**

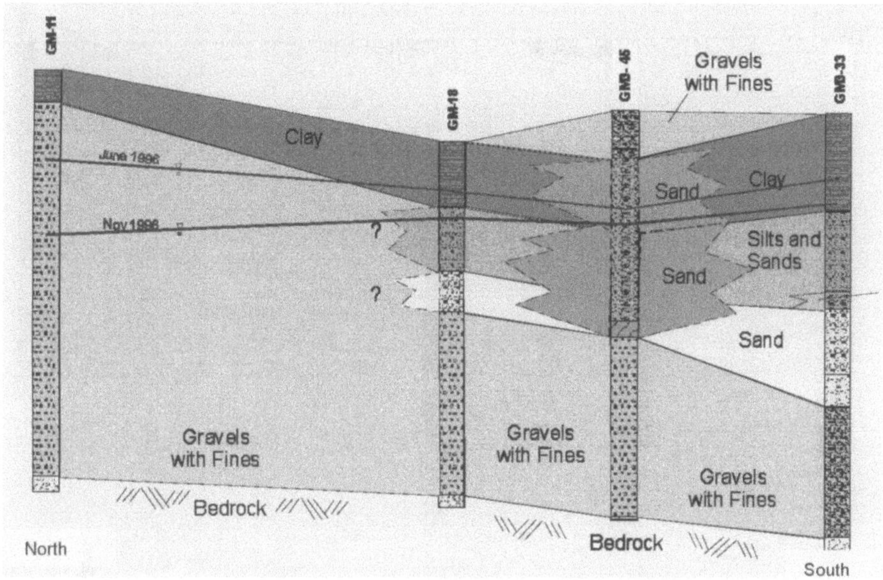

(a)

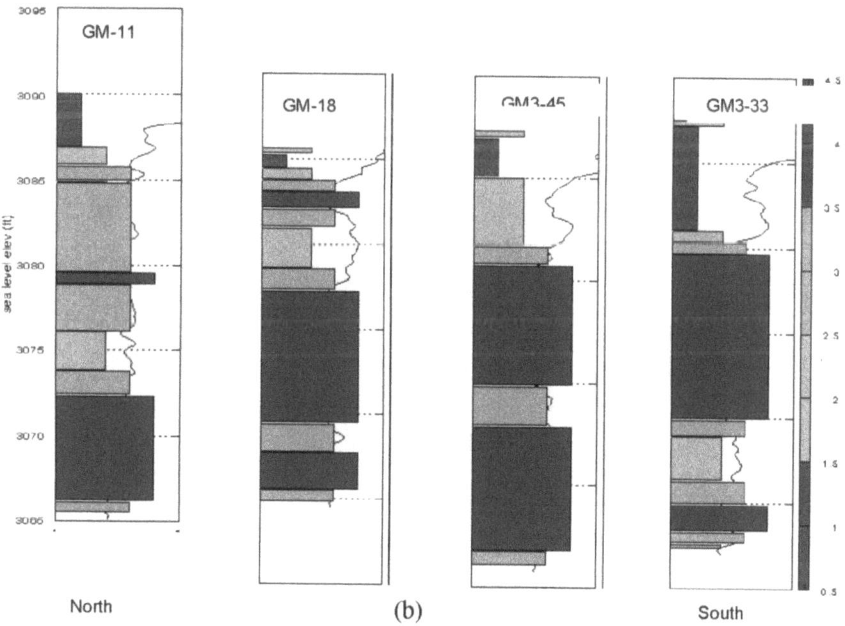

(b)

Figure 12. Conventional (a) and neural network (b) lithologic interpretations for cross-section B-B' (see Figure 1 for location). There is good agreement in stratigraphy for the bottom zones in both interpretations. SOFM output classes are: class 1 - clay, class 2 - silt, class 3 - sand, class 4 - gravel.

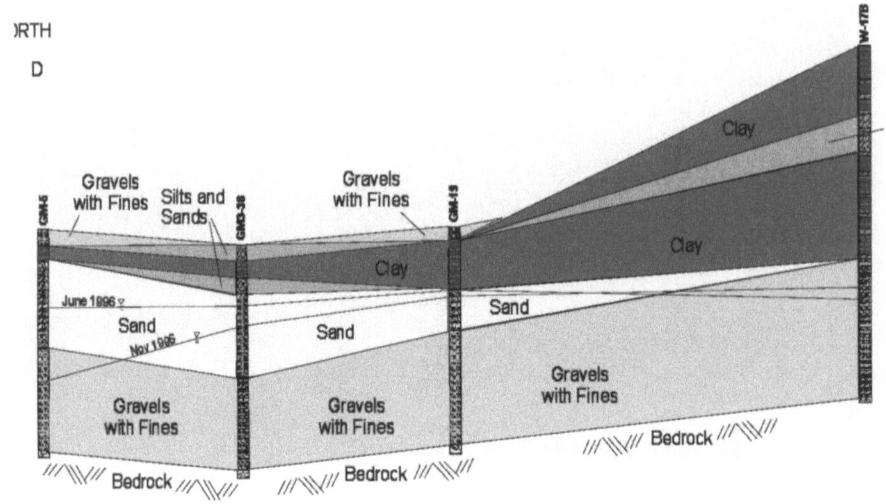

(a)

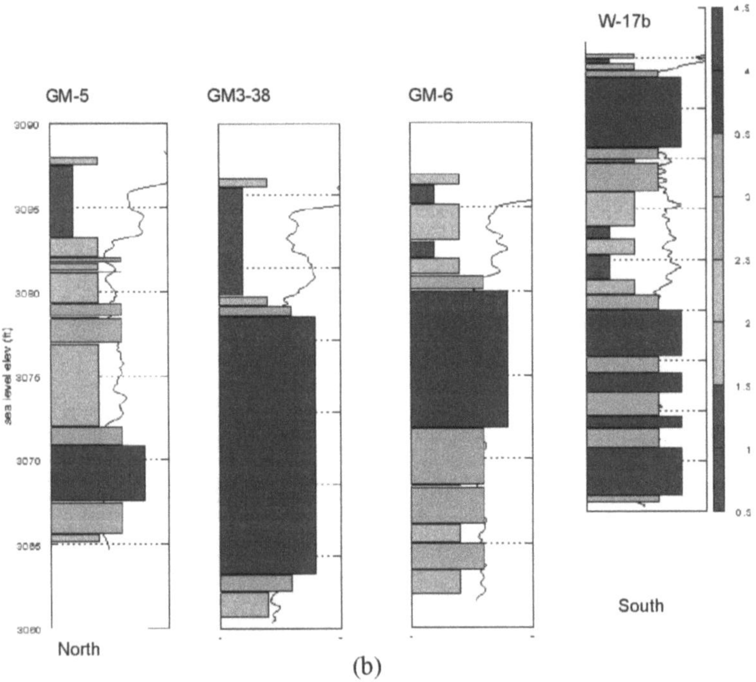

(b)

Figure 13. Conventional (a) and neural network (b) lithologic interpretations for cross-section D-D' (see Figure 1 for location). There is general agreement in stratigraphy (gravels and sands) for the bottom zones in both cross-sections. Also, note the thickening of the clay-rich zones (classes 1 and 2) in both cross-sections although the overlying gravel zone in W-17b does not show up in the conventional interpretation; possibly due to different depth intervals. Output classes are: class 1 - clay, class 2 - silt, class 3 - sand, class 4 - gravel.

NEURAL NETWORK INVERSION OF EM39 INDUCTION LOG DATA

LIN ZHANG[†] and MARY POULTON[‡]

† *Chevron Petroleum Technology Company,*
Houston,
Texas,
USA
E-Mail: linz@chevron.com

‡ *Department of Mining and Geological Engineering,*
University of Arizona,
Tucson,
AZ 85721-0012
USA
E-Mail: mpoulton@u.arizona.edu

Abstract

A modular neural network (MNN) is applied to invert well-logging curves from a Geonics EM39 induction-logging tool. In the interpretation scheme, there are several subsets of networks that depend on the relative resistivities of adjacent layers, e.g. R1>R2>R3, R1<R2>R3, etc. The well-logging curves are subdivided into fixed-length windows by each sub-network. The results are estimations of resistivity and thickness of each layer. The networks are examined for their ability to compute the correct output patterns for the corresponding input patterns of the training set, and the ability to interpret the new patterns that are not present in the training set. The networks are tested using synthetic and field data. The results show that the neural networks can facilitate interpretation of well-log data. When tested on a multi-layer case, the large shoulder effects can make the prediction difficult, especially for the thin resistive and conductive layers. However, the estimates of resistivity and thickness for each layer are sufficiently accurate that an interpreter will not be mislead as to the geological material properties. This approach could also be used for other logging tools or suites of tools.

1. Introduction

The Geonics EM39 induction log is a popular method for mapping the subsurface resistivity structure at different scales for ground water, environmental, and geotechnical applications. The instrument has one coil transmitter and one co-axial coil

W. Sandham and M. Leggett (eds.),
Geophysical Applications of Artificial Neural Networks and Fuzzy Logic, 231-249.

receiver. The separation between the receiver and transmitter is 0.5 m. The transmitter emits electromagnetic energy into the surrounding formation at a constant frequency of 39.2 kHz. The emitted energy induces eddy currents and charges in the formation, which then radiate as secondary sources to the receiver. The intensity of the induced eddy currents and charges depends on the formation resistivity and its variation. Hence, variations in the measured data reflect the formation resistivity structure. The Geonics EM39 instrument outputs an apparent resistivity value based on the measured voltage for every depth-point that is logged. The program EM1DDB2 by Ki Ha Lee is used for modeling the EM39 forward response.

Several studies have been carried out to invert the apparent resistivity curves from a horizontal layered earth model. The generalized inverse theory proposed by Inman et al. (1973) was found to be unsatisfactory if some of the parameters were nearly linearly dependent. Marquardt (1970) and Hoerl et al. (1970a and 1970b) proposed a method called "ridge regression" that yielded a better estimation for the model parameters than the method of least squares. Hence, Inman (1975) applied the ridge regression techniques to invert the apparent resistivity curves. The ridge regression inversion is more stable than the generalized linear inversion. Hoversten et al. (1982) compared the five different least-squares inversion methods for inverting the apparent resistivity curves, and found that the ridge regression algorithm performed best. However, little work has been done on the application of neural networks to invert apparent resistivity curves. Neural networks offer the advantage of fast and accurate inversion but at the cost of some flexibility.

In this chapter, there are four network subsets that depend on the relative resistivities of adjacent layers in a three-layer model, such as R1>R2<R3, R1<R2>R3, R1>R2>R3, and R1<R2<R3. The well-logging curves are then divided into several segments and applied to each sub-network. The outputs from each sub-network are the estimated resistivity and thickness of each layer. The modular neural network (MNN) used in the present application, is first trained using several sets consisting of synthetic data. The trained networks are then tested using "unseen" synthetic and field data. Finally, the speed, accuracy and robustness of the MNN are compared with traditional inversion methods.

2. Data Sets

It is assumed that the subsurface can be described approximately as a layered structure. The goal of this study is to identify each layer's thickness and resistivity from well-log data. In the logging application, many layers may be required to describe the formation resistivity variation with depth. However, training a neural network that covers all the layer combinations would be impossible. Instead, we assume a three-layer model as our basic interpretation unit. The model contains five parameters, the resistivities of the three layers and the thickness of the first and the second layer. The resistivities have four possible type curves, referred to as R1<R2>R3, R1>R2<R3, R1<R2<R3 and R1>R2>R3. Each distinct type is trained using a separate network. A very limited training set is generated to test the general methodology.

The resistivity and thickness values used for training are 1, 10, 20, 30, 40, 50, 60, 80, and 100 Ωm, and 1, 1.5, 2, 2.5, and 3m, respectively. A total of 2,590 synthetic models were generated with different combinations of resistivity and thickness for training (with 1.5% noise added to imitate noise observed in field data). The log10 values of apparent resistivity were supplied as input. The output is the resistivity and thickness for each layer.

There are four types of curve that can be described by the changes in resistivity in a three-layer model. Four training sets were therefore generated according to these four curves, in the interpretation scheme based on the change of formation resistivities in each layer, with each training set corresponding to one type of curve. TABLE 1 lists the parameters of these four training sets.

TABLE 1. The four training sets representing four three-layer resistivity type curves.

Training sets	Set31 R1<R2>R3	Set32 R1>R2<R3	Set33 R1<R2<R3	Set34 R1>R2>R3
# of training patterns	890	900	400	400
Input nodes	41	41	41	41
Output nodes	5	5	5	5

It is important to select the points in the well-logging curve, which represent the continuous nature of the log. In each training pattern, there are 41 inputs and 5 outputs for the three-layer cases. The first 40 input points are log 10-based apparent resistivity values from the logging tool, which are evenly distributed but independent of the sampling interval of the well-log. The 41st input is the depth interval from the first point to the 40th point. The 5 outputs are the log10 resistivity and thickness of each layer. When selecting points, the rate of the sampling is not fixed, and depends on the depth interval. For example, if the depth interval from the first point to the 40th point is 10 m, then the sample interval would be 25 cm. If the depth interval is 4 m, then the sample interval would be 10 cm. The requirement is to select 40 points evenly covering the desired layers. Hence, any section of a logging curve representing three lithologic layers may be extracted and sampled evenly by 40 points, regardless of the original sampling interval of the log.

3. MNN Training

The features and algorithms used for MNNs are described in Haykin (1994). The MNN uses a series of multilayer perceptron (MLP) networks as local experts that compete with each other to learn subsets of the training patterns that have similar features.

The MNN consists of a group of modules (local experts) and a gating network, which combines supervised and unsupervised learning. The gating network learns to break a task into several parts, using unsupervised learning, and each module is assigned to learn one part of the task, using supervised learning. Figure 1 is a block diagram of a modular network (Haykin, 1994). Both the modules and the gating network are fully connected to the input layer. The number of output nodes in the gating network equals the number of modules. The output of each module is connected to the

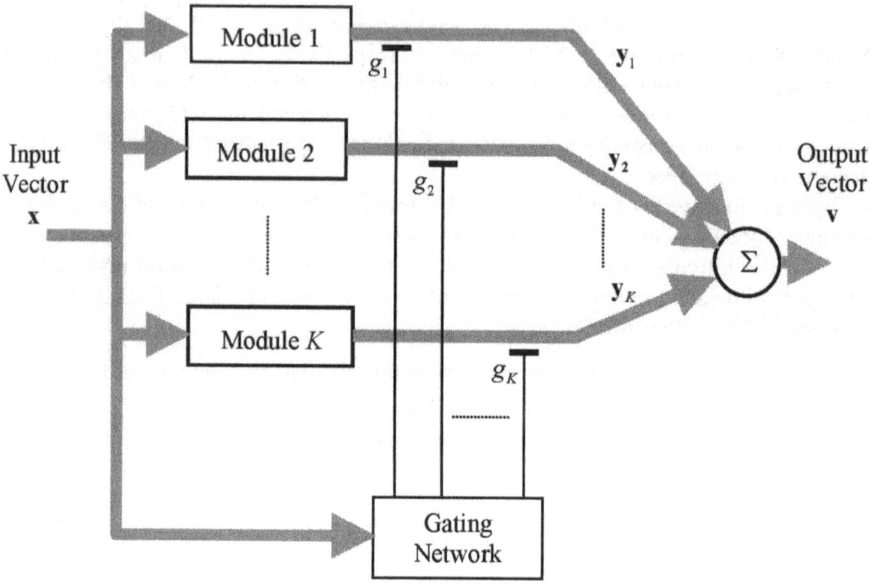

Figure 1. Block diagram of a modular neural network (MNN); the outputs of the expert networks (modules) are mediated by a gating network (after Haykin, 1994).

output layer. The output values of the gating network have been normalized to sum to 1.0 (eq. 4). These normalized output values are used to weight the output vector from the corresponding modules, so that the output from the best module will be passed to the output layer with little change, while the outputs from the other modules will be weighted by a number close to zero and will have little impact on the solution. The final output is the sum of the weighted output vectors (eq. 5).

Poulton and Birken (1998), and Haykin (1994), define the variables used in a modular neural network, as follows:

K Number of modules, also the number of output nodes in gating network
N Number of output nodes in MNN output layer and each module's output layer
M Number of input nodes
Q Number of hidden nodes in each module
P Number of hidden nodes in the gating network
$\mathbf{x} = (x_1, x_2, \ldots, x_M)$ Input training vector
$\mathbf{d} = (d_1, d_2, \ldots, d_N)$ Desired output vector
$\mathbf{u} = (u_1, u_2, \ldots, u_K)$ Output vector of the gating network before normalization to sum to 1
$\mathbf{g} = (g_1, g_2, \ldots, g_K)$ Output vector of the gating network after normalization to sum to 1
$\mathbf{z} = (z_1, z_2, \ldots, z_N)$ Output vector of the whole network
$\mathbf{o}^k = (o_1, o_2, \ldots, o_N)$ Output vector of the k-th module
w_{qm}^k Connection weight between hidden and output layer in k-th module

w^{g}_{pm} Connection weight between hidden and input layer in the gating network

Sum^{k}_{n} Weighted sum for PE n in module k

Each module or local expert, and the gating network, receives the same input pattern from the training set. The gating network and the modules are trained simultaneously. The gating network determines which local expert produces the most accurate response to the training pattern, and the connection weights in that module are updated to increase the probability that the module will respond best to similar input patterns.

The learning algorithm can be summarized as follows:

- Initialize by assigning initial random values to the connection weights in the modules and the gating network.
- Calculate the output for module k,

$$o^{k}_{n} = f\ \mathrm{Sum}^{k}_{n} \tag{1}$$

where
$$\mathrm{Sum}^{k}_{n} = \sum_{q=1}^{Q}\left[f\left(\sum_{m=1}^{M} x_m w^{k}_{qm}\right)\right] w^{k}_{nq} \tag{2}$$

- Calculate the activation for the gating network,

$$u_k = f\left[\sum_{k=1}^{K} f\left(\sum_{p=1}^{P} x_m w^{g}_{pm}\right) w^{g}_{kp}\right] \tag{3}$$

- Calculate the softmax output for the gating network,

$$g_i = \frac{\exp(u_k)}{\sum_{i=1}^{K}\exp(u_i)} \tag{4}$$

- Calculate the network output,

$$z_n = \sum_{i=1}^{K} g_i o_i \tag{5}$$

- Calculate the associative Gaussian mixture model for each output PE,

$$h_k = \frac{g_k \exp\ \frac{-1}{2}\left\|\vec{d} - \vec{o}^{\,k}\right\|^2}{\sum_{j=1}^{K} g_j \exp\ \frac{-1}{2}\left\|\vec{d} - \vec{o}^{\,j}\right\|^2} \tag{6}$$

- Calculate the errors between the desired output and the each module's output,

$$e^{k}_{n} = d^{k}_{n} - o^{k}_{n} \tag{7}$$

- Update the weights for each module. Weights between the output and hidden layer are,

$$w^{k}_{nq}(t+1) = w^{k}_{nq}(t) + \eta h_k \delta^{k}_{nq} act_q \tag{8}$$

where
$$\delta^{k}_{nq} = e^{k}_{n} f'\ \mathrm{Sum}^{k}_{n} \quad \text{and}\ act_q = f\left(\sum_{q=1}^{Q} x_m w^{k}_{qm}\right) \tag{9}$$

- Weights between the input and the hidden layer are,

$$w^k_{qm}(t+1) = w^k_{qm}(t) + \eta h_k \delta^k_{qm} x_m \quad (10)$$

where

$$\delta^k_{qm} = f'(\text{Sum}^k_q) \sum_{n=1}^{N} \delta^k_{nq} w^k_{nq} \quad (11)$$

- Update the weights for the gating network. Weights between the output and hidden layer are,

$$w^g_{kp}(t+1) = w^g_{kp}(t) + \eta \delta^g_{kp} act_p \quad (12)$$

where

$$\delta^g_{kp} = (h_k - g_k) f' \ u_k \quad \text{and } act_p = f\left(\sum x_m w^g_{pm}\right) \quad (13)$$

- Weights between the input and the hidden layer are,

$$w^g_{pm}(t+1) = w^g_{pm}(t) + \eta \delta^g_{pm} x_m \quad (14)$$

where

$$\delta^g_{pm} = f'(\text{Sum}^g_p) \sum_{k=1}^{K} \delta^g_{kp} w^g_{kp} \quad (15)$$

The best architectures of the MNN for each of four training sets are summarized in TABLE 2.

TABLE 2. The best structures for the four training sets.

Training Set	# Local Expert	# Hidden in Gating	# Hidden in Local Expert	Iterations	Learning Rule	Transfer Function	RMS	Mome -ntum
Set31	2	5	12	23,000	Delta-rule	Sigmoid	0.049	0.3
Set32	2	6	12	41,000	Delta-rule	Sigmoid	0.04	0.3
Set33	2	6	10	17,000	Delta-rule	Sigmoid	0.034	0.3
Set34	2	6	8	29,000	Delta-rule	Sigmoid	0.024	0.3

The best structure for training Set31, shown in TABLE 2, uses two local experts. This training set corresponds to a R1<R2>R3 type curve. Each local expert learns a particular shape of a three-layer model (see Figure 2). The 890 training patterns are distributed to the two local experts, with 341 and 549 models, respectively. Local Expert 1 responds to low resistivity models; the resistivities in the forward response increase from 22 to 52 Ωm, then decrease to 35 Ωm. In particular, the first layer's resistivity is less than that in the third layer. Local Expert 2 responds to a relatively higher resistivity model than that in Local Expert 1; the resistivities change from 55 to 70 Ωm, then decrease to 43 Ωm. The first layer's resistivity is larger than that in the third layer. Thus, each local expert learns a special resistivity range and shape of the model.

Training Set32 represents the R1<R2>R3 type curve. The two local experts receive 403 and 497 models, respectively (see Figure 3). Local Expert 1 represents the high resistivity model; the apparent resistivity decreases from 60 to 38 Ωm, then increases to 65 Ωm. In other words, the third layer's resistivity is larger than the first layer's resistivity. Local Expert 2 learns a low resistivity model; the apparent resistivities decrease from 50 to 15 Ωm, then increase to 39 Ωm. The third layer's resistivity is less than that in the first layer.

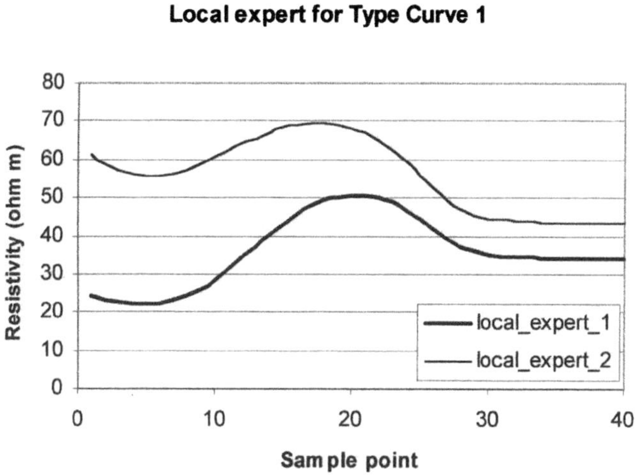

Figure 2. There are two local experts in training Set31 for R1<R2>R3 type curve. Each local expert learns a particular shape of a three-layer model. The plotted curves represent average values for each sample point for models retained by each expert.

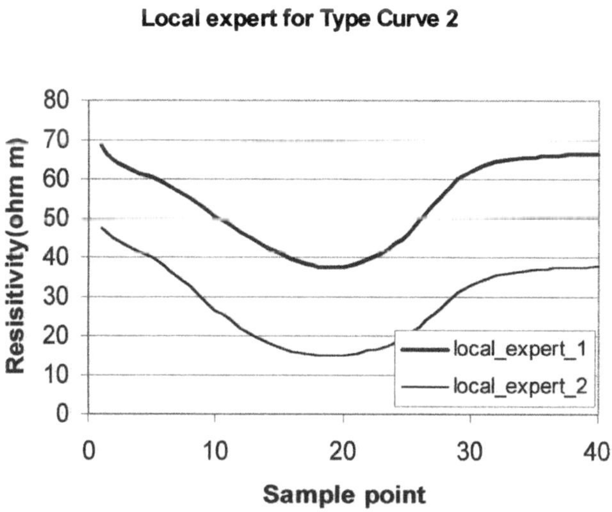

Figure 3. There are two local experts in training Set32 for R1>R2<R3 type curves. Each local expert learns a particular shape of a three-layer model. Curves represent average values for each sample point for models retained by each local expert.

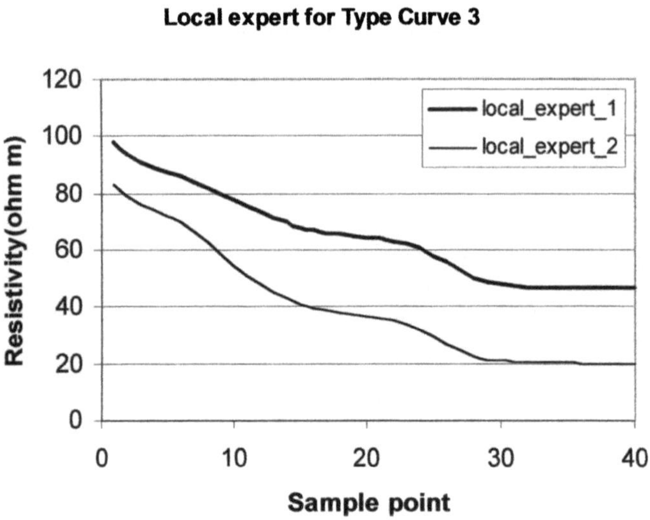

Figure 4. There are two local experts in training Set33 for R1<R2<R3 type curves. Each local expert learns a particular shape of a three-layer model. Curves represent average values for each sample point for models retained by each local expert.

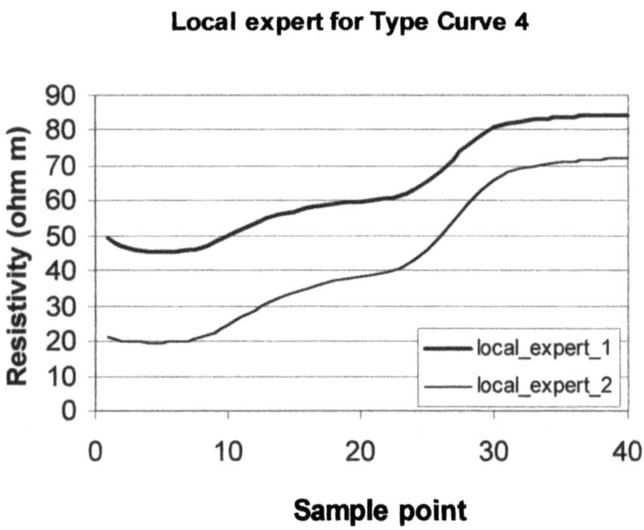

Figure 5. There are two local experts in training Set34 for the R1>R2>R3 type curves. Each local expert learns a particular shape of a three-layer model. Curves represent an average value for each sample point for the training samples retained by each expert.

Training Set33 represents a R1<R2<R3 type curve. The total training patterns are divided between the two local experts, with 262 and 138 models, respectively (see Figure 4). Local Expert 1 corresponds to a high resistivity model, which is from 100 to 50 Ωm. Local Expert 2 corresponds to a low resistivity model, which is from 80 to 20 Ωm.

Training Set34 represents a three-layer model with increasing resistivity, R1>R2>R3. The training patterns are broken into two local experts with 181 and 219 models, respectively (see Figure 5). Local Expert 1 corresponds to a high resistivity model, which is from 45 to 85 Ωm. Local Expert 2 corresponds to a low resistivity model, which is from 20 to 75 Ωm.

4. Testing on Synthetic Data

To test the generalization of the trained neural networks, four testing sets corresponding to the four type curves were used. In the training set, the resistivity and the thickness of each layer ranged from 1-100 Ωm and from 1-3m, respectively. However, all of the testing sets were three-layer resistivity models and the depth range of each model was approximately 7m. Using the traditional inversion method, two hours was required to invert a 7m model. The results show that MNN only takes a second to predict a model. Thus, the MNN has a quick interpretation speed.

The testing methodology involves applying a new example type curve to each network. The resistivity and thickness estimated by the network are input to the EM1DDB2 forward model and a new logging curve is created. This curve is compared to the original log to judge the accuracy of the network output. Figures in this section compare the original resistivity and thickness model with values estimated by the network in addition to the original and estimated logging curves.

Training Set31 (R1<R2>R3 type curves) represents models where the layer resistivity increases and then decreases. The trained neural network was tested using synthetic models. The MNN outputs were then compared with the original synthetic models (see Figure 6). Since the best MNN for training Set31 has two local experts, each local expert responds to one particular type of model (see Figure 2). In case the shape of a new model is not exactly the same as one of the shapes of the local experts, the gating network output weights the output vector from the corresponding local expert. The final output vector is the sum of these weighted output vectors. The weight for each specific local expert depends on how close the new model is to that local expert. The final weighted output vector of each local expert for testing models is analyzed.

The test model (Figure 6) has a similar shape and resistivity range to Local Expert 1, which is a low resistivity model. For the final output, the output vector in Local Expert 1 contributes 90.4% of the solution. Since Local Expert 2 responds to a high resistivity model, only 9.6% of the final output are from the output vector in Local Expert 2. Comparing the MNN output with the synthetic model (Figure 6), it is observed that the MNN provides an excellent prediction in resistivities, especially for the second layer, which is a thin resistive layer. The resistivity and thickness correlate well. The thickness

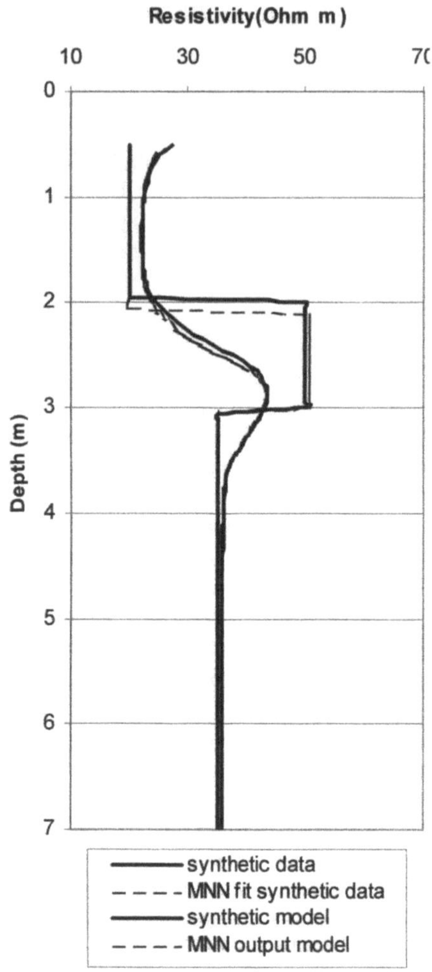

Figure 6. MNN output model fits the Type 1 synthetic model (R1=20 Ωm, t1=2m; R2=50 Ωm, t2=1m; R3= 35 Ωm) quite well. The predicted thickness in the first layer is 9cm deeper than the true thickness. All the resistivity values are estimated very accurately.

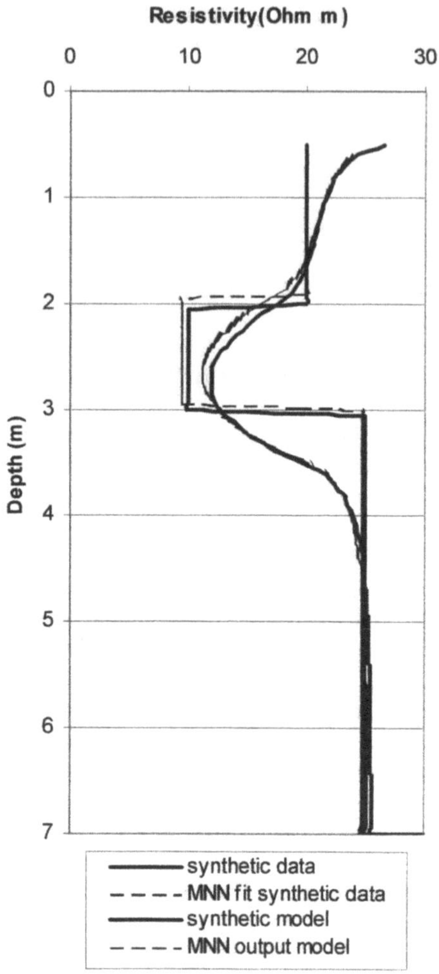

Figure 7. MNN output model fits the Type 2 synthetic model (R1=20 Ωm, t1=2m; R2=10 Ωm, t2=1m; R3= 30 Ωm). The thickness in the first layer is 7cm shallower than the real thickness. The resistivity values are estimated accurately; the resistivity in the second layer is 0.65 Ωm (over 10 Ωm) lower than expected.

in the first layer is predicted 9cm deeper than the true thickness. The accuracy is 89%, which is close to the goal of 90%.

The second network is trained with models that decrease then increase in layer resistivity. The MNN output model is quite comparable with the synthetic models (see Figure 7). The overall accuracy for the test models is above 91%.

The test model (see Figure 7) is a low resistivity model. The resistivity range is similar to the Local Expert 2 prototype curve. The shape of the model is close to the

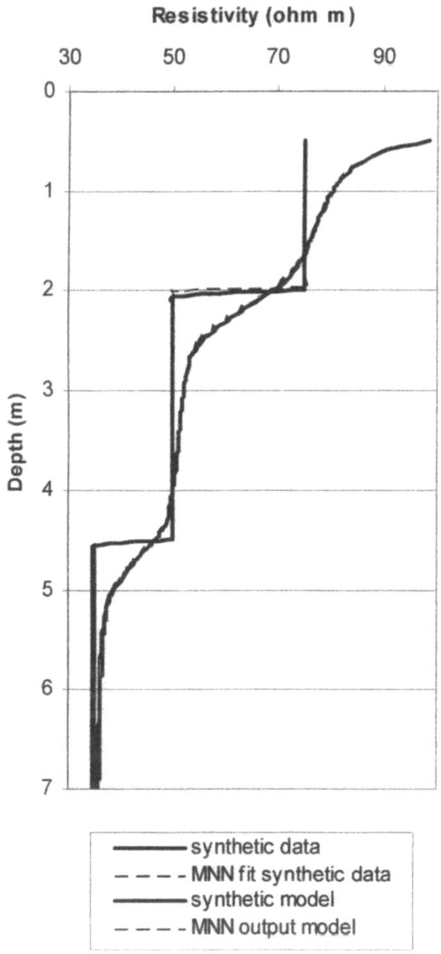

Figure 8. MNN output model fits the synthetic model (R1=70 Ωm, t1=2m; R2=50 Ωm, T2=2.5m; R3= 35 Ωm). The thickness in the second layer was shifted 9cm from the real thickness. The resistivity values are estimated perfectly.

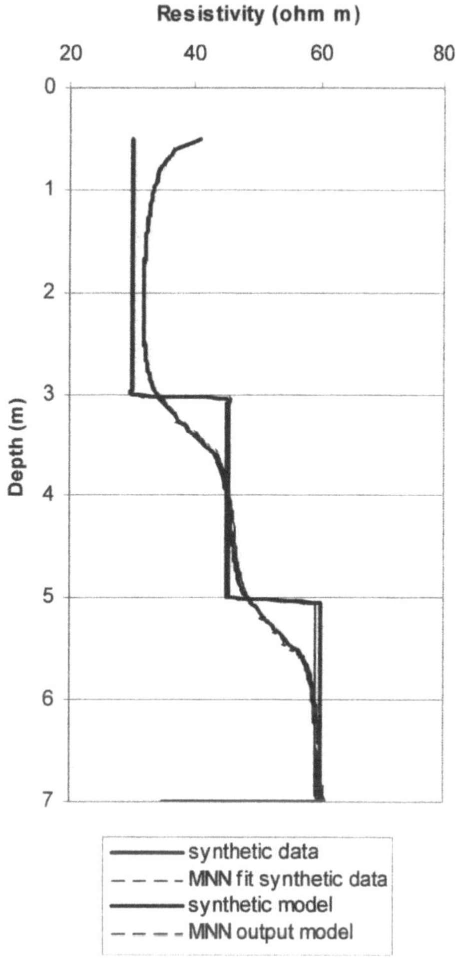

Figure 9. MNN output model fits the synthetic model (R1=30 Ωm, t1=3m; R2=45 Ωm, t2=2m; R3= 60 Ωm) perfectly. There is no shift in thickness. Only the resistivity in the third layer is shifted 0.8 Ωm.

shape of Local Expert 1. For the test output, 76.7% is contributed by Local Expert 2 and 23.3% from Local Expert 1. As seen in Figure 7, the MNN fits the synthetic model quite well. Although the second layer is a thin conductive layer and the thickness is only 1m, which is hard to predict accurately, the MNN fits it well, and there is only a 7cm shift in the location of the predicted first layer. The maximum resistivity shift is 0.65 Ωm (over 10 Ωm) in the conductive layer. The overall accuracy is 91.2%.

Network 3 learned models where the layer resistivities keep decreasing. The MNN output test model is very close to the synthetic models (see Figure 8). All the thickness and resistivity values are predicted exactly. The overall accuracy for the testing models is above 95%.

The resistivity range in the test model lies between that for Local Experts 1 and 2 (see Figure 4), thus, both local experts respond to it; 60% of the final output is from Local Expert 1 and 40% from Local Expert 2. The MNN fits the synthetic model very well for both resistivity and thickness. The accuracy for the whole model is 96%.

The accuracy for test models that continuously decrease in resistivity is much higher than that for alternating resistivity layers.

The fourth network learned to represent models in which the layer resistivity values keep increasing. The MNN output test models fit the synthetic model quite well (see Figure 9); all the thickness and resistivity values are predicted exactly. The overall accuracy for the testing models is above 98%.

The final output for the test model is a combination of the output vectors from Local Experts 1 and 2 (see Figure 5), which contribute 80% and 20% of the solution, respectively. In Figure 9, the MNN fits the synthetic model nearly perfectly. The layer thickness and resistivity values are fit accurately. The predicted resistivity in the third layer is 0.8 Ωm (over 60 Ωm) lower than the true resistivity. The accuracy is above 98%.

The results indicate that neural networks can more easily learn models that continuously increase or decrease in resistivity, as opposed to those that alternate between increasing and decreasing resistivity. The accuracy for testing training Set33 (R1<R2<R3 type curves) and Set34 (R1>R2>R3 type curves), which contain the models that keep increasing or decreasing in resistivity, is above 95%. The accuracy for testing the trained neural networks that have models that alternate between increasing and decreasing resistivity, is above 90%. It is difficult for the neural networks to estimate the models that have a conductive layer embedded in the resistive layers, or a resistive layer embedded in the conductive layers, due to shoulder effects. The overall accuracy, however, even for the difficult models is 90%. Neural networks are shown to have excellent capabilities to invert the three-layer synthetic models with high accuracy and quick interpretation speed.

5. Testing on Field Data

The trained neural networks exhibit good generalization when they are tested using synthetic models. In this section, their performance is evaluated when they are tested using field data. Since all the training patterns in the training sets represent three-layer models and the field data usually contain multiple layers, a new method is carried out to split the field data into segments having three layers.

The interpretation approach is to split the complete well-logging curve into several segments, and analyze each segment using different subsets of networks. The most important step is to determine where to break the curve. Every point in the curve has a slope. These slopes can be connected as a slope-curve. The depth of the local minimum or local maximum slope corresponds to the boundary of the layers (Figure 10). The boundary of the layers could be located depending on these local maximum or local minimum slopes. The local maximum changes of slope correspond to the layer

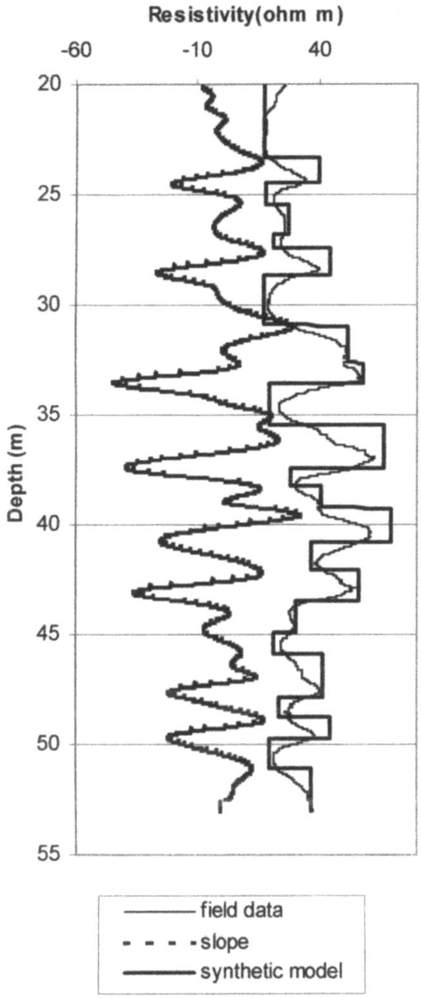

Figure 10. Field data for well Ls2_d1 showing fit of the synthetic model and corresponding slope curve. The local maximum changes of slopes are approximate indicators of the layer boundaries.

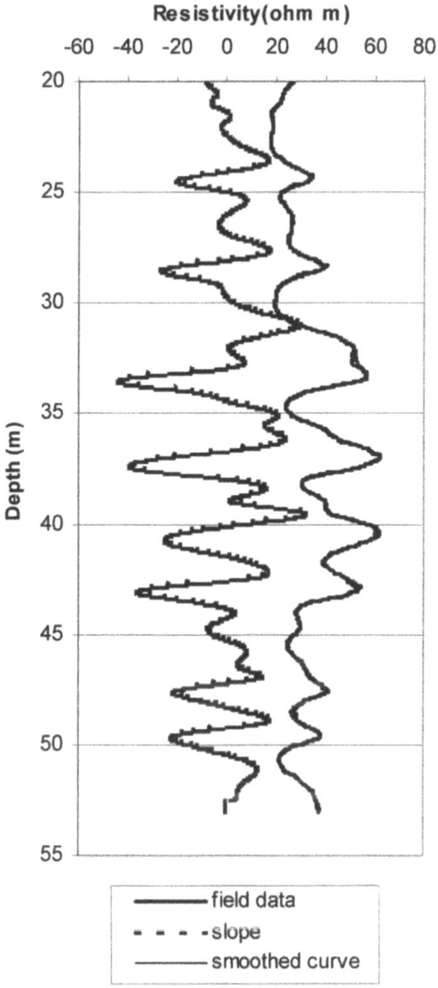

Figure 11. Field data for well Ls2_d1 showing fit of smoothed curve to field data and corresponding slope curve, which is calculated through the smoothed curve.

boundaries. The approximate thickness of each layer can be obtained through these local maximum or local minimum slopes. The thickness is not an exact thickness, just an approximate thickness. Depending on these slopes, the well-logging curve can be segmented and the number of layers in each piece can be counted.

For a smoothed-curve, it is straightforward to obtain the slope points (local minimum or local maximum). However, for field data it is difficult to pick these points due to noise. Thus, the field data are smoothed before processing in order that the

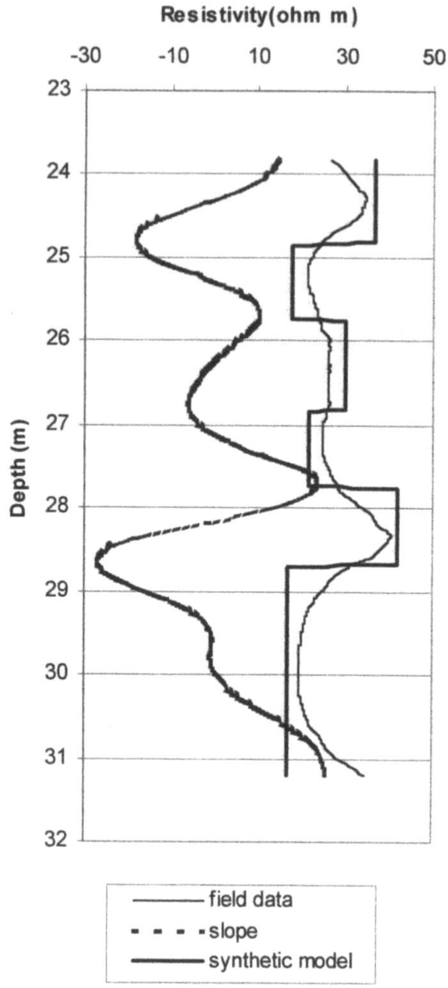

Figure 12. Layers are first picked based on a slope curve calculated from a splined field curve. The local maximum changes of slopes indicate the layer boundaries approximately.

smoothed curve is a good fit to the field data. A cubic spline method is used for smoothing, however this can result in loss of information for very thin layers. In this case, the data are selected from the field well-logging curve with a certain sampling interval. Every 0.5m, a data point is selected for the spline, and then several new points are interpolated in the interval. The number of interpolated data points depends on the sampling interval of the log. The 0.5m interval is found to work best for most of the field data since the spacing between the tool transmitter and receiver is 0.5m; other intervals could also be used. When the field data and the smoothed curve are compared (Figure 11), it is obvious that the smoothed curve is a very close fit to the field data. A slope curve is then calculated, based on the smoothed curve, to obtain an estimate of each layer thickness. The field data curve is broken into segments and each segment represents a three-layer resistivity model. An 8m section of a log is picked to illustrate how to pick points in each segment.

The log in Figure 12 is apparent resistivity (field data) for an 8m section of a well. The bottom curve is the slope curve for the same 8m segment. Six layers are identified in the resistivity curve. An approximate layer thickness is calculated from the slope curve. The layers are at 23.8-24.8m, 24.8-25.6m, 25.6-26.8m, 26.8-27.7m, 27.7-28.65m, and 28.65-31.2m. The logging curve is split into two parts in order to obtain three-layer training patterns. One part is from 23.8-26.8m, the second part is from 26.8-31.2m. For the first part, the depth interval was 3m, and the 40 points are picked in this 3m interval. The first layer is 1 m thick, the second layer is 0.8m thick, and the third layer is 1.2m thick. The 40 points are distributed in proportion to the thickness. Hence, 13 points are picked in the first layer, 11 points in the second layer and 16 points in the last layer. The second part was from 26.8-31.2m, the depth interval was 4.4m. The first layer in this part is 0.9m, the second layer is 0.9m, and the last one is 2.55m. Therefore, there are 8 points in the first layer, 9 points in the second layer, and 23 points in the last layer, based on the layer thickness.

Thus, the field data are smoothed by a cubic spline, then windowed based on the local minimum or maximum slopes. Each window represents three layers and is run through a different network depending on the change of resistivities in the segment. Once the neural networks have tested the segment, the results for each segment are appended together to obtain the predicted model for the complete logging curve.

The original training set had 2,590 training patterns representing three-layer resistivity models from 1-100 Ωm and thickness from 1-3m. The field data, however, have some layer thickness values less than 1m. More thin-layer models are therefore created for thickness values of 0.5, 0.6 and 0.8m. The final training set has 3,491 patterns.

6. Analysis of Results

The MNN was trained using 3,491 input patterns and tested with field data not included in the training sets. The field data are from the San Pedro National Riparian Conservation Area, Cochise County, Arizona. The wells penetrated unindurated Holocene alluvium of the San Pedro River underlain by semi-indurated lower Pleistocene and Pliocene basin-fill alluvium. The data were collected as part of a study of the hydrology of the San Pedro River, one of the few remaining perennial streams in southern Arizona.

Ls2_d1 is a well of 53m depth. The first 20m of the curve are deleted because of excessive noise due to an initial section of steel well casing. The networks were tested on the remaining 33m. There are a total of 23 layers in the log. The resistivity has a range of 17-65 Ωm, and the thickness ranges from 0.5-3.2m. The MNN output model compared well with the synthetic model in Figure 13. The predicted layer thickness and resistivity values are reasonable even though the thin and conductive layers cause large shoulder effects. The trend between the MNN output and the field data is very compatible; the correlation is 0.92. Thus, we are confident that there are no misleading

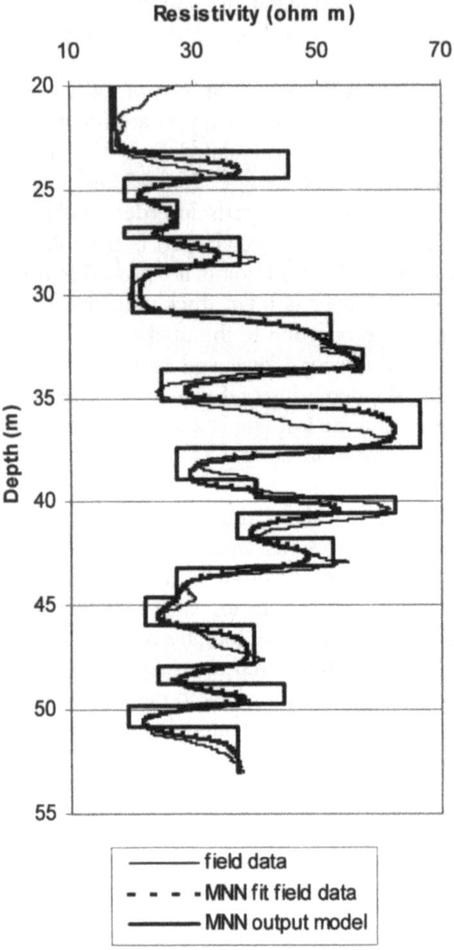

Figure 13. **Comparison of neural network output resistivity and thickness values to the field data (Ls2_d1) and modeled result from neural network output.**

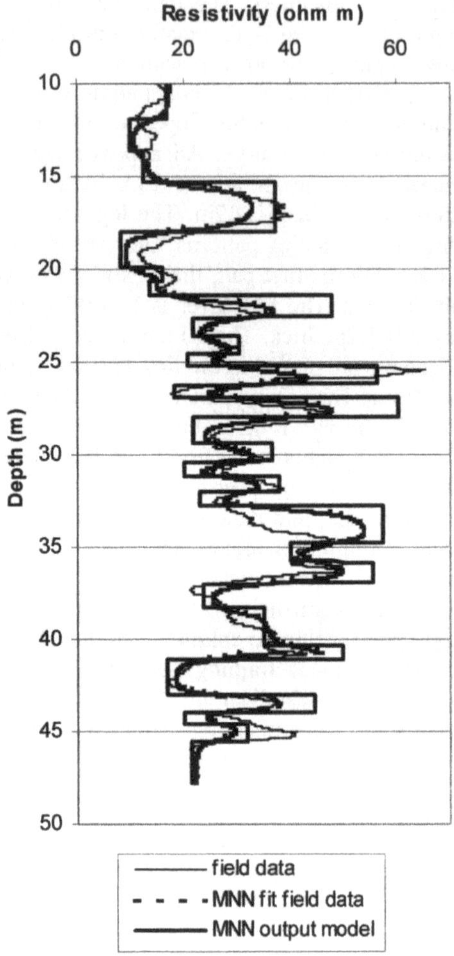

Figure 14. **Comparison of neural network output resistivity and thickness values to the field data (Ls5_d1) and modeled result from neural network output.**

geological material properties in our interpretation. Since Ls2_d1 has some thin layers, the resistivities from MNN could not reach the peaks or troughs because of the shoulder effects. This happens in the 5th, 6th, 7th, 10th, 14th, and 16th layers (see Figure 13).

Ls5_d1 has a total-depth of 48m. The first 10m were removed because of noise. The networks tested the remaining 38m. There were a total of 30 layers interpreted in the log. The resistivity had a range of 10-70 Ωm and the thickness had a range of 0.5-3m. Compared to Ls2_d1, Ls5_d1 (see Figure 14) has more thin layers and sharp resistivity changes. Most of the layers in Ls5_d1 had a thickness of approximately 0.5-1m. Despite the large shoulder effects, the MNN can still fit the field data quite well. Some very thin layers (thickness less than 1 m) are recognized and predicted accurately. The trend between the MNN fit to field data and the field data is very consistent and the geophysicists are not misled as to the interpreted lithology. At layer 12, from 25-26m, the predicted resistivity could not reach the peak since the layer thickness is only 1 m, and both the 11th and 13th layers are thin and conductive. That causes a large shoulder effect. The resistivity predicted by MNN is 56 Ωm, which is 24.5 Ωm lower than expected. Again, at the 13th layer, 26-27m, the MNN output resistivity could not reach the trough. Since the 13th layer is a thin conductive layer, the shoulder effects from the high resistive layers, the 12th and the 14th, are high. The predicted resistivity in this layer is 18.18 Ωm, 6 Ωm higher than the expected resistivities. Similar errors occurred in the 6th, 14th, 16th, 18th, 23rd, 28th, 29th layers as well.

7. Robustness Test

The neural networks described in this chapter are trained to predict three-layer models. Hence, in order to test the robustness of the trained neural networks, two new testing models were generated. The models are 8m long; one model has four layers and another one contains five layers. The trained neural network can only predict three-layer models, hence the purpose of this test is to determine if the neural network predicted three-layer model can characterize the four layer or five layer models, and if the resulting interpretation leads to a misleading lithology.

Figure 15 shows the results for a five-layer model. The neural network output is a three-layer model, however, it can still capture the characters of the five-layer model. From the forward responses, the data generated by the MNN output matches the synthetic data very well and the correlation is 0.95.

Figure 16 shows the results for a four-layer model. The neural network output model captures the features of the four-layer model. From the forward responses, the data generated by MNN output tracks the synthetic data very well and the correlation is 0.92.

While this test is very limited, it appears that the trained neural networks are robust. The MNN output can capture the features of the four or five-layer models and the correlation between the synthetic data and the MNN fit curve is more than 0.9, thus, we are not misled about the lithologic properties.

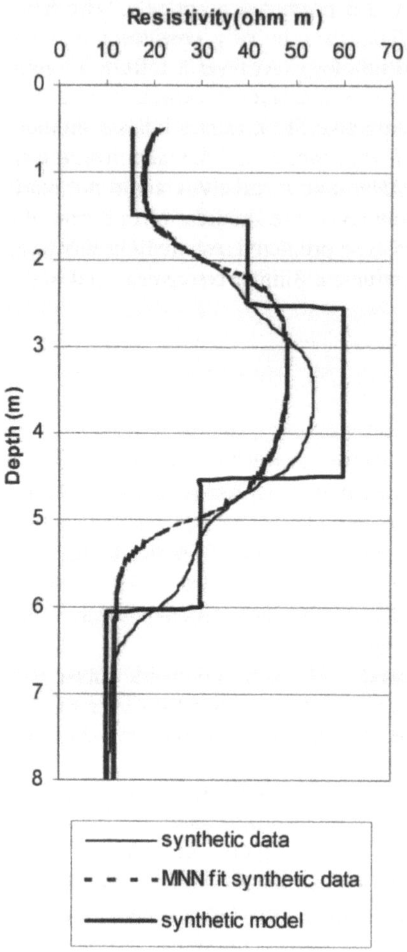

Figure 15. The trained neural network fit to a five-layer model (R1=15 Ωm, t1=1.5m; R2=40 Ωm, t2=1m; R3= 60 Ωm, t3= 2m; R4=30 Ωm, t4=1.5m; and R5=10 Ωm). Although the neural network output is a three-layer model (R1=15.1 Ωm, t1=1.53m; R2=49.6 Ωm, t2=3.49m; and R3= 11.1 Ωm), it can still fit the synthetic data very well and the correlation between the synthetic data and the neural network output data is 0.95.

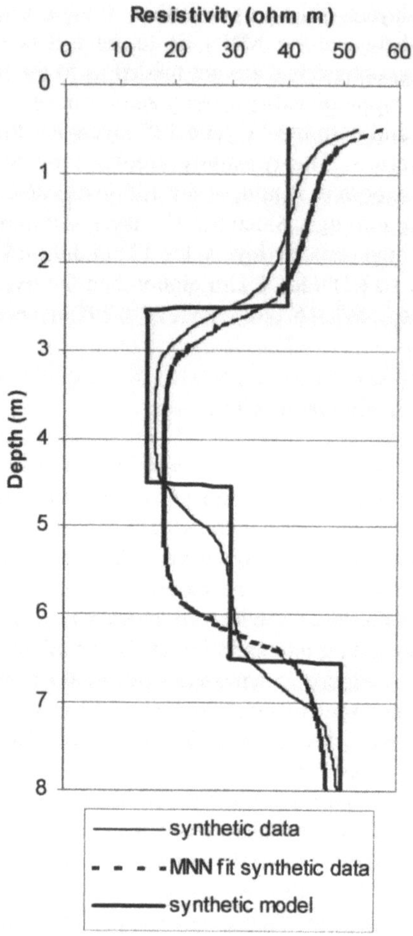

Figure 16. The trained neural network fit to a four-layer model (R1=40 Ωm, t1=2.5m; R2=15 Ωm, t2=2m; R3= 30 Ωm, t3=2m; and R4=50 Ωm). Although the neural network output is a three-layer model (R1=44.5 Ωm, t1=2.3m; R2=18.8 Ωm, t2=3.9m; and R3= 47.5 Ωm), it can still fit the synthetic data and the correlation between the synthetic data and the neural network output data is 0.92.

8. Conclusions

A modular neural network (MNN) has been applied to invert well-logging curves from a Geonics EM39 induction-logging tool. In the interpretation scheme, several subsets of networks depend on the relative resistivities of adjacent layers. The well-logging curves were subdivided into fixed-length windows by each sub-network, to give estimations of resistivity and thickness for each layer. The networks were tested using both synthetic and field data.

The MNN took less than one second to provide a quantitative interpretation for each well-log when run on a 450 MHz Pentium PC. It performs very well on synthetic three-layer models and is robust. For field data with many layers, especially thin conductive or thin resistive layers, the large shoulder effects can make the prediction difficult. However, the estimates of resistivity and thickness for each layer are sufficiently accurate that an interpreter will not be mislead as to the geological material properties. This approach could also be used for other logging tools or suites of tools.

References

Haykin, S., 1994, Neural Networks: A Comprehensive Foundation: Prentice Hall, New York.

Hoerl, A., and Kennard, R., 1970a, Ridge regression: Biased estimation for nonorthogonal problems: Technometrics, **12**, 55-67.

Hoerl, A., and Kennard, R., 1970b, Ridge regression: Applications to nonorthogonal problems: Technometrics, **12**, 69-82.

Hoversten, G., Dey, A., and Morrison, H., 1982, Comparsion of five least-squares inversion techniques in resistivity sounding: Geophys. Prosp., **30**, 688-715.

Inman, J., Ryu, J., and Ward, S., 1973, Resistivity inversion: Geophysics, **83**, 1088-1108.

Inman, J., 1975, Resistivity inversion with ridge regression: Geophysics, **40**, 798-817.

Marquardt, D., 1970, Generalized inverses, ridge regression, biased linear estimation and nonlinear estimation: Technometrics, **12**, 591-612.

Poulton, M. M., and Birken, R. A., 1998, Estimating one-dimensional models from frequency-domain electromagnetic data using modular neural networks: IEEE Trans. Geosci. Remote Sensing, **36**, 547-555.

SECTION C

ELECTROMAGNETIC EXPLORATION

INTERPRETATION OF AIRBORNE ELECTROMAGNETIC DATA WITH NEURAL NETWORKS

EDMUND WINKLER[†], WOLFGANG SEIBERL[‡] and ANDREAS AHL[‡]

† *Geological Survey of Austria,*
Rasumofskygasse 23,
A-1030 Wien,
AUSTRIA
E-Mail: winedm@cc.geolba.ac.at

‡ *Institute of Meteorology and Geophysics,*
Althanstraße 14,
1090 Wien,
AUSTRIA
E-Mail:wolfgang.seiberl@univie.ac.at
E-Mail:andreas.ahl@univie.ac.at

Abstract

Artificial Neural Networks (ANNs) are used for the interpretation of multi-frequency airborne electromagnetic (AEM) data independently of the sensor height, with one-dimensional (1-D) horizontally layered homogeneous earth structures. A divide-and-conquer strategy is applied. One ANN is trained to interpret data, which are best described by homogeneous half-space (HHS) models. A second ANN inverts data from horizontally layered half-space models with two layers (2LHS). Tests have shown that when the 2LHS ANN is applied to data, which are best, described with a HHS-like structure, interpretation errors can become large. Therefore, a third ANN is trained, which classifies the best interpretation of measurements as a HHS model or a 2LHS model. This modular ANN approach shows a good performance on synthetic data. Finally, the interpretation of data from an AEM survey over a tertiary basin structure, shows good accordance with known geological data.

W. Sandham and M. Leggett (eds.),
Geophysical Applications of Artificial Neural Networks and Fuzzy Logic, 253-265.

1. Introduction

The Geological Survey of Austria in cooperation with the Institute of Meteorology and Geophysics of the University of Vienna, has performed multi-frequency airborne electromagnetic measurements with a helicopter towed system (GEOTECH-HUMMINGBIRD). The system operates with four coil configurations: two horizontal co-planar at 434 and 7002 Hz, and two vertical coaxial at 3212 and 34133 Hz (Table 1). The in-phase and quadrature parts of the secondary magnetic field are measured in parts per million (ppm) at these four frequencies, resulting in 8 ppm values per measuring position. Our present interest is to interpret these 8 ppm values by a two-layer half-space model (Figure 1) using ANNs. In the following, the problem is formulated only for the two-layer case (a horizontal layer with resistivity ρ_1 and thickness h_1 over an infinite half-space of resistivity ρ_2), the HHS is a special case of the 2LHS. The inversion of AEM data in the context of the two-layer case is equivalent to finding a function,

$$F:(ppm_1, ppm_2, ppm_3, ppm_4, ppm_5, ppm_6, ppm_7, ppm_8) \rightarrow (\rho_1, \rho_2, h_1) \quad (1)$$

There are analytical solutions to calculate the 8 ppm values from a given set of model parameters ρ_1, ρ_2 and h_1 (Wait, 1982; Anderson, 1979):

$$G:(\rho_1, \rho_2, h_1) \rightarrow (ppm_1, ppm_2, ppm_3, ppm_4, ppm_5, ppm_6, ppm_7, ppm_8) \quad (2)$$

Due to the non-uniqueness of the mapping from the measured data to model parameters, the inversion of AEM data is a complex and often time-consuming task. AEM surveys however require fast processing of very large data volumes. ANNs are best suited for this inversion task, since they can learn a good approximation of the function F from an appropriate set of examples, called training vectors, to enable the mapping of measured data onto model parameters. In addition, the application of an ANN is very fast, due to the small number of simple numeric operations required for the interpretation of a data vector.

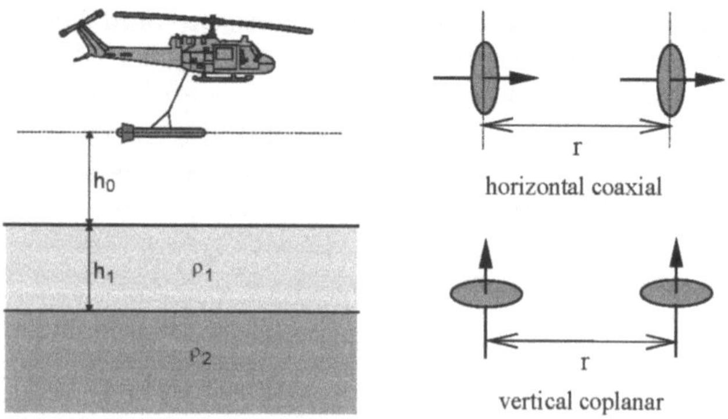

Figure 1. Two layer model parameters and the coil configurations of the AEM system.

TABLE 1. Frequencies and coil separations of the AEM system

	Frequency	Coil Separation (r)	Configuration
f_1	434 Hz	4.53 m	horizontal co-planar
f_2	3212 Hz	4.53 m	vertical coaxial
f_3	7002 Hz	4.49 m	horizontal co-planar
f_4	34133 Hz	4.66 m	vertical coaxial

2. Forward Modeling Two-Layer Models

The helicopter-towed AEM system currently consists of two horizontal co-planar transmitter/receiver coils and two vertical coaxial ones, each coil pair operating at a separate frequency (Table 1). An electromagnetic field is generated by applying a time harmonic voltage of frequency f to the transmitter coil. This primary field directly induces a voltage U_p in the receiver coil. Flying above the earth's surface, the primary electromagnetic field induces currents in the earth's crustal layers. These currents are the source of a secondary electromagnetic field, which generates the voltage U_s in the receiver coil. Hence, the total voltage induced in the receiver coil is $U_p + U_s$. The result of a measurement process is the complex value U_s/U_p in ppm units. The real and imaginary parts of U_s/U_p are called in-phase and quadrature (out-phase). In the following we will restrict ourselves to the HHS and 2LHS models as a description for the geological situation. From electromagnetic theory (Wait, 1982) we can derive the measured ppm value as a function of the frequency f, the coil separation r, the sensor height h_0 and the earth layer resistivities ρ_1, ρ_2 and first layer thickness h_1. Here μ_0 and ε_0 denote the magnetic and the dielectric permittivity in free space, respectively:

$$\text{Vertical coaxial configuration: } \left(\frac{U_s}{U_p}\right)(r,\omega,\rho_1,\rho_2,h_o,h_1) = 1 + B^3 T_o \tag{3}$$

Horizontal co-planar configuration:

$$\left(\frac{U_s}{U_p}\right)(r,\omega,\rho_1,\rho_2,h_o,h_1) = -2\left(1 + \frac{B^2}{2}(T_2 - BT_o)\right) \tag{4}$$

$$\delta = \sqrt{\rho_1/\pi \, \mathrm{f} \, \mu_0} \tag{5}$$

T_0 and T_2 are integrals defined as follows,

$$T_0 = -\delta^3 \int_0^{+\infty} \lambda^2 \, R_{TE}(\lambda)\, e^{-2\lambda h_0} \, J_0(\lambda r)\, d\lambda$$

$$T_2 = -\delta^2 \int_0^{+\infty} \lambda \, R_{TE}(\lambda)\, e^{-2\lambda h_0} \, J_1(\lambda r)\, d\lambda$$

where J_0 and J_1 are Bessel functions of order 1 and 2. T_0 and T_2 are so called Hankel transforms, which can be evaluated numerically by means of linear digital filters as described in Anderson (1979). For a 2LHS, the coefficient R_{TE} in integrals T_0 and T_2 is,

$$R_{TE} = \frac{\alpha_0 - \alpha_1 \dfrac{\alpha_2 + \alpha_1 \tanh(\alpha_1 h_1)}{\alpha_1 + \alpha_2 \tanh(\alpha_1 h_1)}}{\alpha_0 + \alpha_1 \dfrac{\alpha_2 + \alpha_1 \tanh(\alpha_1 h_1)}{\alpha_1 + \alpha_2 \tanh(\alpha_1 h_1)}}$$

where $\alpha_i = \sqrt{\lambda^2 + j\omega\mu_0 (1/\rho_i + j\omega\varepsilon_0)}$, $i = 0,1,2$, $j = \sqrt{-1}$ and $\omega = 2\pi f$.

For a special half-space case where $\rho_1 = \rho_2$, we have,

$$R_{TE} = \frac{\alpha_0 - \alpha_1}{\alpha_0 + \alpha_1}$$

So, the result of one measurement process is a vector of four complex values,

$$\left(\frac{U_s}{U_p}\right)_{\text{coplanar}}^{434\,\text{Hz}}, \left(\frac{U_s}{U_p}\right)_{\text{coaxial}}^{3212\,\text{Hz}}, \left(\frac{U_s}{U_p}\right)_{\text{coplanar}}^{7002\,\text{Hz}}, \left(\frac{U_s}{U_p}\right)_{\text{coaxial}}^{34133\,\text{Hz}} \qquad (6)$$

which are actually 8 ppm values $(ppm_1, ppm_2, ppm_3, ppm_4, ppm_5, ppm_6, ppm_7, ppm_8)$.

3. Feedforward Artificial Neural Networks

Among the great variety of ANN paradigms, we have chosen multilayer perceptrons (MLPs). These are normally used for classification and function approximation problems, where sufficient examples are available to empirically describe a task for which hard and fast rules cannot easily be applied. As an extension to the standard architecture of MLPs, units in one layer may be connected directly to neurons in subsequent layers as shown in Figure 2.

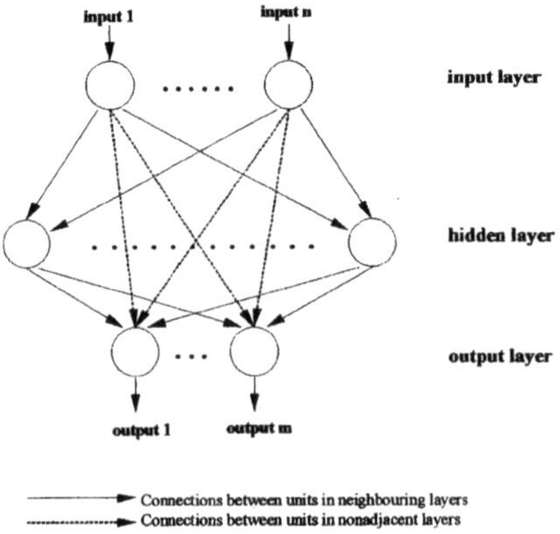

Figure 2. The architecture of a multilayer perceptron

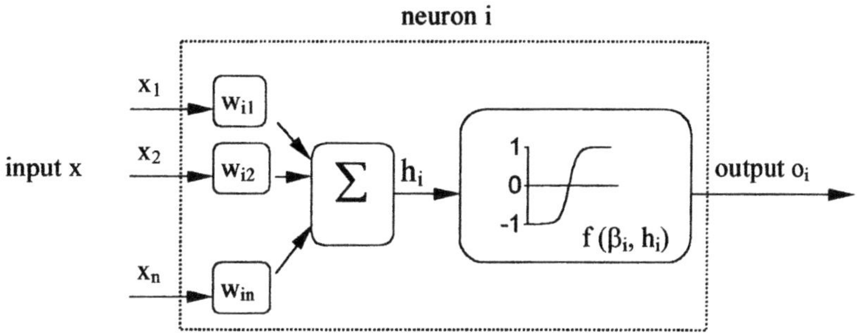

Figure 3. The functionality of a neuron is described by the weights w_{ij}, gain β_i, activation level $h_i=\sum_j w_{ij} \cdot x_j$, activation function f and neuron output $o_i = f(\beta_i, h_i)$.

The activation function of the neurons was modified slightly by keeping the gain parameter variable (Figure 3).

To save CPU time, the hyperbolic tangent $f(\beta,h)=(1-e^{-2\beta h})/(1+e^{-2\beta h})$, the Elliot sigmoid $f(\beta,h)=\beta h/(1+|\beta h|)$ (Elliot, 1993) and a piecewise approximation of the hyperbolic tangent with straight lines and polynomials (Anguita et al., 1993), were used as activation functions in our MLP implementations, instead of the logistic function $f(\beta,h)=1/(1+e^{-\beta h})$.

Learning was directed by the "descending epsilon technique" (Yu and Simmons, 1990). At iteration $t=0$, all weights and gains are initialized to be uniformly distributed within 0.01 and 0.1, and an initial error threshold of 0.4 for the output units is set. The order of the training patterns is changed after each presentation of the complete training set. Each pattern in the training set is presented at the input layer and the corresponding error is calculated at each output node. If the error at any output unit is larger than the current error threshold, backpropagation, as described in eq.(7) is performed for this pattern. w_{ij} is the connection from unit i to unit j, β_i is the gain of unit i, $\eta_w(m_w)$ and $\eta_\beta(m_\beta)$ are the learning rates (momentum terms) for weight and gain update. An error function E was used as proposed in Fahlman (1988),

$$\Delta w_{ij}(t) = -\eta_w \frac{\partial E}{\partial w_{ij}(t)} + m_w \cdot \Delta w_{ij}(t-1)$$
$$\Delta \beta_i(t) = -\eta_\beta \frac{\partial E}{\partial \beta_i(t)} + m_\beta \cdot \Delta \beta_i(t-1) \qquad (7)$$

$$w_{ij}(t+1) = w_{ij}(t) + \Delta w_{ij}(t)$$
$$\beta_i(t+1) = \beta_i(t) + \Delta \beta_i(t)$$

This procedure continues until all patterns produce an error less than the current threshold. After saving weights and gains the threshold is reduced and the sequence starts again. Parameters to this algorithm are the initial and final output error threshold and the error threshold stepsize.

4. ANN Inversion and Classification of a 2LHS

4.1. AEM INVERSION OF A 2LHS

In the following, only the techniques to train an ANN for a 2LHS are described, as they contain everything to handle the simpler HHS. An ANN is trained to interpret AEM data with a two-layer model independently from the sensor height h_0, where h_0 varies from 40m to 70m. If we denote the input vector with,

$$\mathbf{X}=\left(\frac{U_s}{U_p}\right)^{434\,\text{Hz}}_{\text{coplanar}},\left(\frac{U_s}{U_p}\right)^{3212\,\text{Hz}}_{\text{coaxial}},\left(\frac{U_s}{U_p}\right)^{7002\,\text{Hz}}_{\text{coplanar}},\left(\frac{U_s}{U_p}\right)^{34133\text{Hz}}_{\text{coaxial}}$$

and the output vector $\mathbf{Y}=(\rho_1, \rho_2, h_1)$, the function to be approximated is F:X→Y. To learn this mapping, an ANN needs a training set with examples giving a good representation of the mapping's behavior. With the numerical implementation of eqs. (3) and (4) a set of (X_i,Y_i) pairs was generated by varying the model parameters (ρ_1, ρ_2, h_0, h_1). The values for the resistivities were in the range 1 to 4000Ωm, sensor height h_0 was between 40m and 70m, and the thickness of the first layer ranged from 1m to 80m. The raw training set $T_0=\{(X_i,Y_i),i=1\ldots n_0\}$ consists of n_0=11800 elements. The approximation problem F:X→Y is split into three simpler sub-problems: $F_{\rho 1}:X\rightarrow\rho_1$, $F_{\rho 2}:X\rightarrow\rho_2$ and $F_{h1}:X\rightarrow h_1$, by creating three training sets $T_0^{\rho 1}=\{(X_i,\rho_{1i}), i=1\ldots n_0\}$, $T_0^{\rho 2}=\{(X_i,\rho_{2i}), i=1\ldots n_0\}$ and $T_0^{h1}=\{(X_i,h_{1i}), i=1\ldots n_0\}$.

Because the error in the measured data is at least ± 2ppm and since there are situations where the mapping G in eq.(2) computes the same eight data points for different two-layer models, we often find sets of different model vectors Y belonging to the same data vectors X ± 1ppm. Therefore in a preprocessing stage, clustering of the raw training sets $T_0^{\rho 1}$, $T_0^{\rho 2}$ and T_0^{h1} was performed with respect to the X_i component of each training sample and the resulting centroid sets were used for MLP training. The distance measure between the 8-dimensional X_i vectors is the maximum norm.

Starting with a random permutation of the training set, the first training vector is chosen as the centroid of the first cluster. All other training vectors, which are within a given radius RC, are added to the cluster. After moving a new pattern to a cluster the centroid is updated. If no more patterns can be found close enough to the first cluster, this process is repeated with the next free pattern as the centroid of the next cluster, until all patterns have been assigned to a cluster. In a second phase, patterns are exchanged between clusters if the within-cluster variations decrease. To check the influence of the initial ordering of the patterns on the final result, clustering was performed for 20 different initial random permutations of the patterns in each training set. Results for different R_C values are shown in Figure 4.

Taking the clustering with the smallest within-cluster variance as the final result, 520 clusters were obtained for R_C=2ppm, 834 for R_C=1.5ppm and 1506 for R_C=1ppm. The logarithm of ppm values and model parameters was linearly transformed to lie within −1 and +1 in the case of the ppm components. Due to the nature of the sigmoid functions in the output layer neurons, the range of scaled model parameters is within 0.1 and 0.9. In a first step backpropagation training with the descending epsilon technique, the centroid set resulting from clustering with R_C=2ppm was employed initially. In a second step, training of the network with the best performance on the centroids from clustering with R_C=1.5ppm, was continued with the centroids from R_C=1.5ppm clustering. From these trained networks, the network was selected which performed best on the centroids from R_C=1.0ppm clustering, and training continued with this set.

Networks resulting from this last training step were tested using a set, which was independent of the raw training set. This process was repeated for $T_0^{\rho 1}$, $T_0^{\rho 2}$ and T_0^{h1} starting with a feedforward network with two hidden layers and connections from each layer to all the following layers. Each hidden layer consisted of 40 sigmoid units (Figure 5). To demonstrate the performance of the ANN inversion, the three final ANNs were used to interpret a synthetic AEM anomaly of a basin structure. In Figure 5 it can be seen that a good fit of the ANN results with the theoretical basin structure, and occurs only in regions with sufficient information regarding both layers in the AEM data.

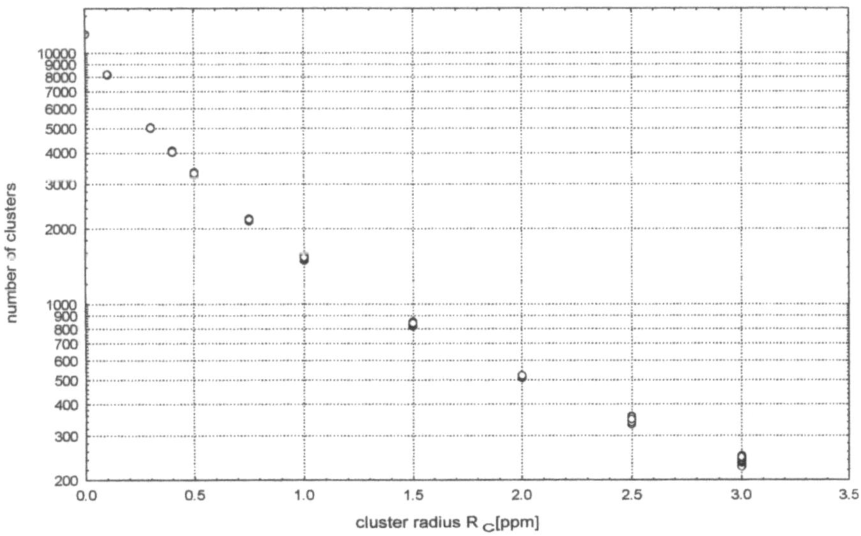

Figure 4. Number of clusters in training set $T_0^{\rho 1}$ as a function of the given radius R_C. For every value of R_C, 20 clusterings, each with an initial random permutation of patterns in the training set, were performed.

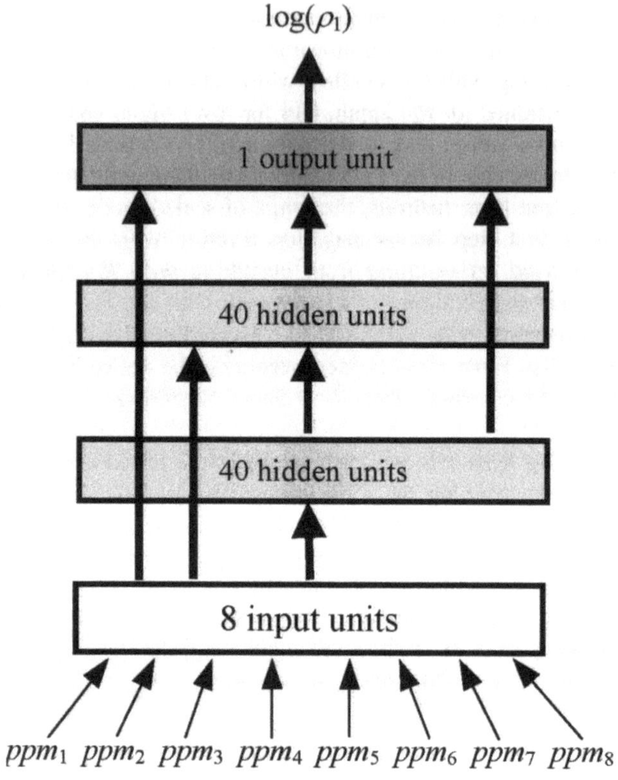

Figure 5. Architecture of a feed forward neural network with two hidden layers. Arrows show the flow of information within the network. The input layer with 8 units receives the 8 ppm data values. At the output layer the network delivers an estimate for the resistivity of the first layer. Similar networks were trained to estimate the model parameters ρ_2 and h_1 at the output.

In the transition zones from HHS to 2LHS (small h_1 values), and when the layer thickness grows larger than 80m, the results of a two-layer inversion become unreliable. Albeit patterns in the raw training sets from these two regions show a great variation in the model parameter component. They are accumulated in large clusters since they have almost identical ppm components, due to the limited parameter resolution capability of the AEM system. Thus, the replacement of all elements within a cluster by the centroid introduces a large error in the model parameter estimation. This is because the network has learned to approximate the centroid value of the model parameter. Since the interpretation with a HHS reduces the error in the ρ_1 parameter in these two cases, a separate classifier is necessary to decide when a 2LHS is reasonable and when it is more appropriate to apply a HHS model.

4.2. CLASSIFICATION OF AEM DATA

Eq.(5) for the skin depth shows that the depth information content in AEM data is reciprocal to the frequency. To employ this fact for each of the three frequency pairs f_1 and f_2, f_2 and f_3, f_3 and f_4, an ANN was trained with synthetic HHS data to determine the sensor height above ground h_0 and the HHS resistivity ρ_1. The outputs from each ANN, when applied to data resulting from 2LHS models, represent an average of the resistivity distribution in subsoil. Due to the fact that the penetration depths of the frequency pairs decrease with increasing frequency, each ANN obtains knowledge regarding the subsoil for different depths.

A fourth ANN, whose inputs are the outputs of the three ANNs for the HHS, decides if the measured data are best represented by a HHS or a 2LHS model (Figure 7). If this ANN decides in favor of a HHS, the final interpretation is a HHS with resistivity ρ_1 from the HHS1 ANN, since the lowest frequency pair has the greatest penetration depth. Figure 6(c) demonstrates that with the additional information from the 1-D Classification ANN, we can decide which model is the best interpretation for a ppm data vector.

4.3. ANN INVERSION OF AEM SURVEY DATA

In the following, an application of the ANN inversion scheme (Figure 6) to AEM data is illustrated. The survey area is situated in Austria, northwest of Vienna in the range of the Bohemian Massif. There are several tertiary basins on the high metamorphic crystalline. One of them, with a hook like contour, is the so-called Horn basin. Drillings down to 140m have shown that this basin is partially filled with clay of resistivity about 20 Ωm. The results of inverting AEM data from this survey area using ANNs, is shown in Figure 8. The output from the inversion is a uniform subsoil resistivity ρ_1 (HHS) or a resistivity ρ_1 for the first h_1 meters followed by ρ_2 for greater depths (2LHS). The distribution of the subsoil resistivities is displayed at four depth levels: 0m, -15m, -50m and -80m. This is in good accordance with the asymmetrical form of the basin, which is slightly deeper towards the east.

5. Conclusions

Theoretical and practical tests have proven that ANNs can perform inversion of AEM data very efficiently, in the context of 2LHS or HHS models. The problem is split into a number of sub-tasks. First, depending on the information content in the AEM data, an ANN decides if a AEM measurement is best explained using a HHS or a 2LHS model. For each model class, ANNs are trained. In the 2LHS case, the thickness of the first layer is limited to 80m. In both cases the ANNs interpret data independently of the sensor height within 40m and 70m.

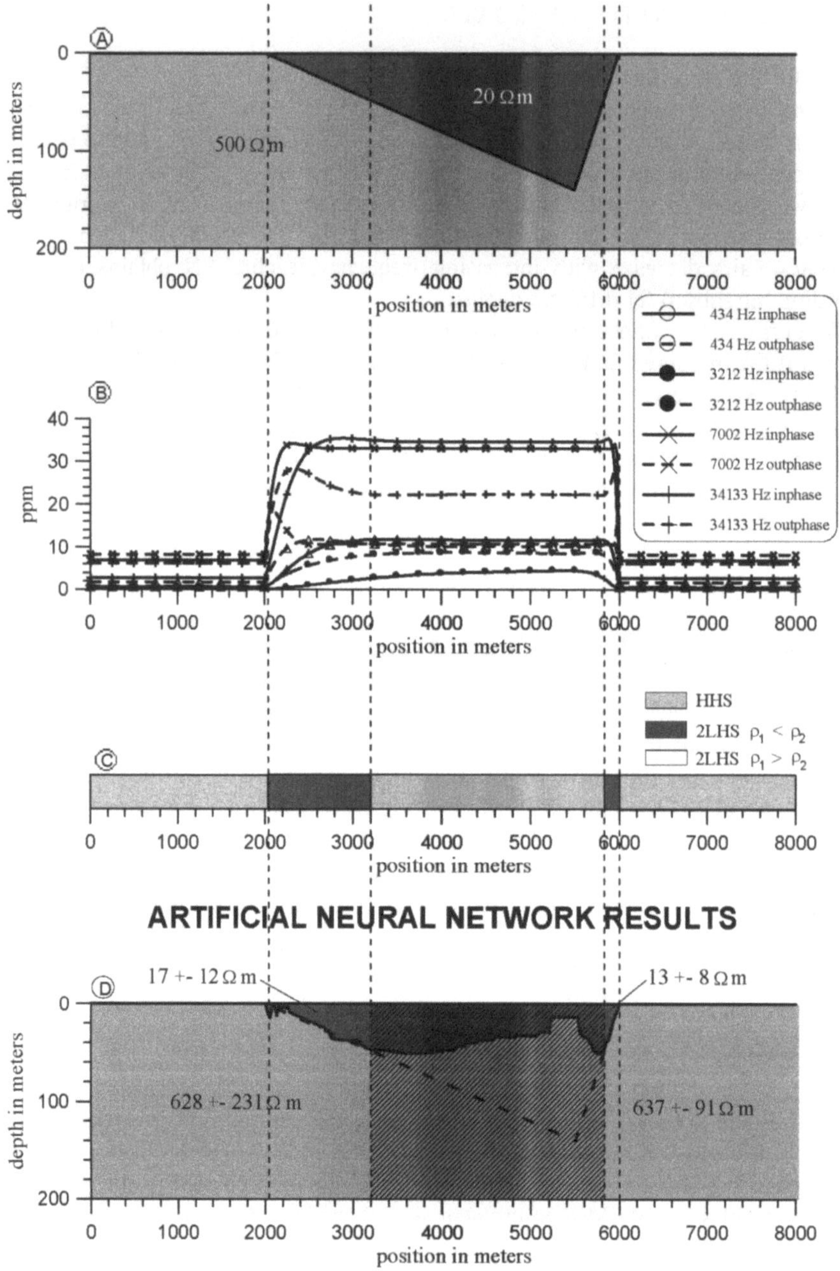

Figure 6. **ANN interpretation of AEM data (A : synthetic basin structure; B: AEM-traces; C: classification to distinguish between a two-layer homogeneous half-space (2LHS) and a homogeneous half-space(HHS) model; D: 2LHS ANN-inversion).**

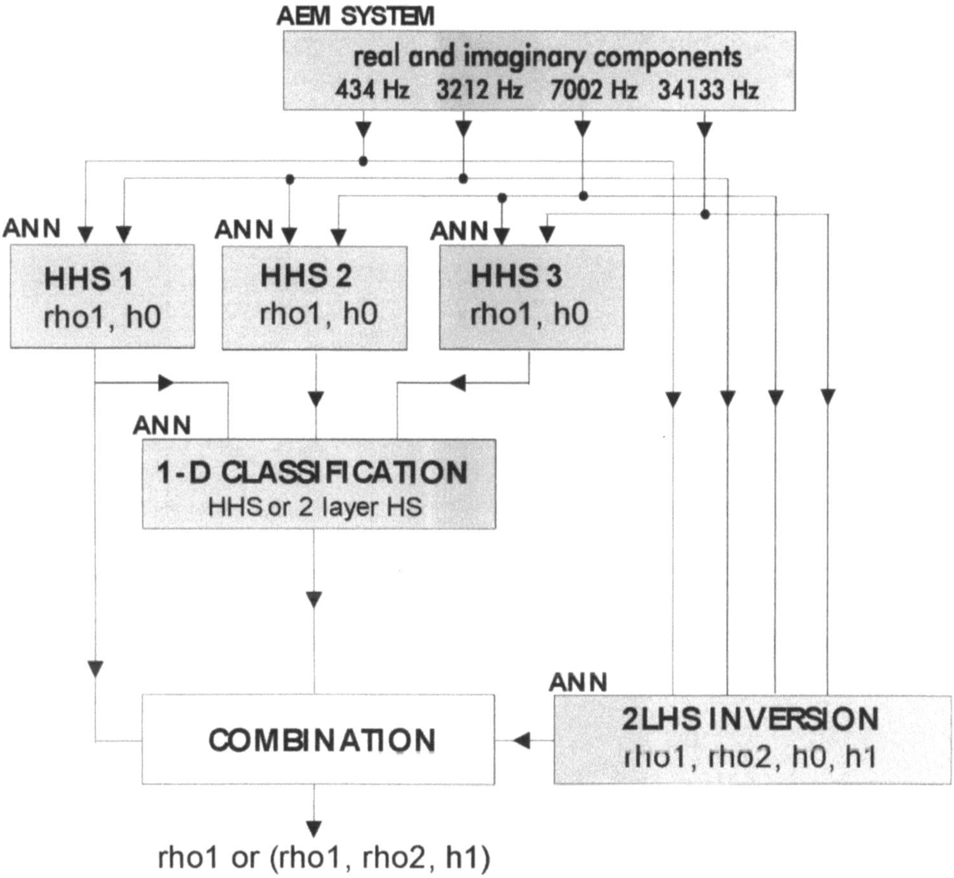

Figure 7. Modular structure of the ANN interpretation scheme for AEM data (ANN = artificial neural network, HS = homogeneous half-space, 2LHS = two-layer homogeneous half-space).

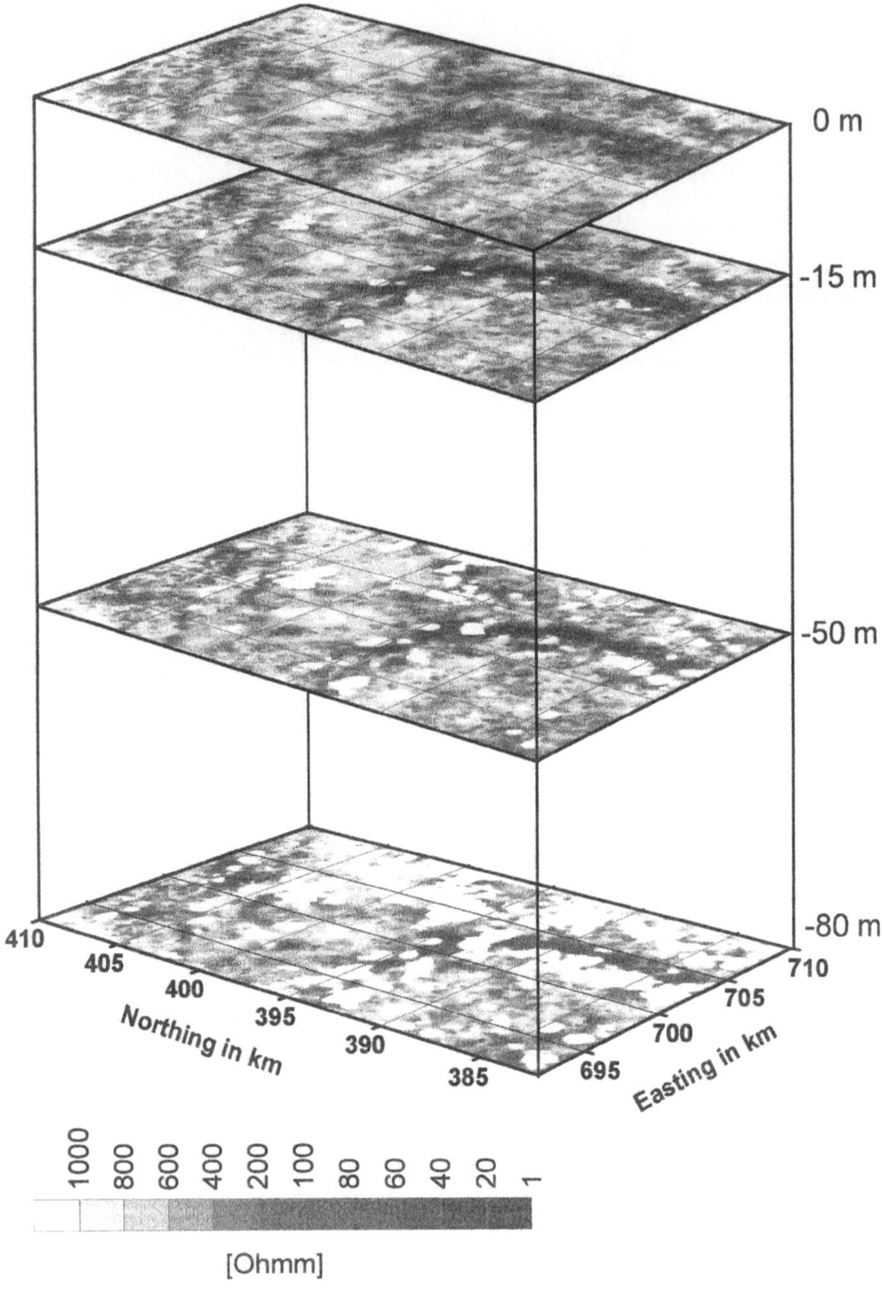

Figure 8. ANN inversion of AEM data from an area NW of Vienna (Horn basin). The resistivity distributions at four depth levels are shown.

Acknowledgement

We gratefully acknowledge the support of this work by the Austrian Scientific Foundation (FWF–Grant No. P11833-GEO).

References

Anderson, W., 1979, Numerical integration of related Hankel transforms of order 0 and 1 by adaptive digital filtering : Geophysics, **44**, 1287-1305.

Anguita, D., Parodi, G., and Zunino, R., 1993, Speed Improvement of the back-propagation on current generation workstations: Proceedings of the World Congress on Neural Networking, Portland, Oregon, 1993, vol. **1**, Lawrence Erlbaum/INNS Press, 165-168.

Elliott, D., 1993, A better activation function for artificial neural networks: ISR Technical Report TR93-8, Institute for System Research, University of Maryland.

Fahlman, S. E., 1988, An empirical study of learning speed in back-propagation networks: CMU-CS-88-162, September 1988.

Wait, J. R., 1982, Geo-electromagnetism: Academic Press Inc., 121-124.

Yu, Y., and Simmons, R., 1990, Descending epsilon in backpropagation: a technique for better generalization: Proc. Internat. Joint Conf. on Neural Networks, 167-172.

SECTION D

OTHER GEOPHYSICAL APPLICATIONS

INTEGRATED PROCESSING AND IMAGING OF EXPLORATION DATA: AN APPLICATION OF FUZZY LOGIC

WOOIL M. MOON† and PING AN‡

† *Dept. of Earth Sciences,*
University of Manitoba,
Winnipeg MB R3T 2N2,
CANADA
E-Mail: Wmoon@cc.umanitoba.ca

‡ *Schlumberger-Geoquest,*
5599 San Felipe, Suite 1700,
Houston TX 77056-2722,
USA
E-Mail: pan@houston.geoquest.slb.com

Abstract

Recently, Earth observation tools and geological exploration techniques have been developing rapidly, alongside significant advancements in computer technology and data processing capabilities. These developments encourage integrated multi-sensor, and multi-target observation of the Earth System environment, resulting in massive volumes of Earth science and geological exploration data. A number of new approaches for handling such large volumes of data have been proposed, including those based on classical Bayesian statistics and probability theory. However, fuzzy logic approximation of the spatial information (Earth science and geological exploration data) has been quick and effective for efficient and accurate analysis of complex geological exploration data and associated Earth system information. For more complex spatial reasoning, a combined single-stage fuzzy neural network approach has also been successfully applied, with future prospects of cascaded multi-stage fuzzy neural network systems. This chapter reviews some of the recent developments in the theoretical aspects of fuzzy logic and neural network models in Earth System science and geological application problems.

1. Introduction

In geological resource exploration, it has been customary for a geologist to estimate and combine field survey information qualitatively, based purely on his/her knowledge and

W. Sandham and M. Leggett (eds.),
Geophysical Applications of Artificial Neural Networks and Fuzzy Logic, 269-285.

experience. In information processing theory, however, there have been serious attempts to quantify the observed information and associated uncertainties using entropy, as demonstrated by Shannon (1948). Many scientific disciplines, including geophysical inversion theory, have not adopted Shannon's concepts until recently. Instead, they have attempted to define the available information from various laboratory experiments and field observations, and then drawn semi-quantitative conclusions.

In geology and resource exploration, many field surveys are limited to near-surface regions and produce data sets often with incomplete spatial coverage even at the surface, due to logistical constraints. As a consequence, geologists routinely use intuitive expertise in geological reasoning and decision-making processes. Many traditional data processing techniques adopted in geophysics have focused on inversion and statistical approaches (Mathai and Rathie, 1975), however these are very limited in handling situations lacking full spatial coverage and which introduce considerable uncertainties.

Since the early days of geological science, geologists, and Earth scientists in general, have recognized and accepted situations involving incomplete data sets and imperfect knowledge. Indeed, the problem of imperfect knowledge has been rigorously tackled by philosophers, logicians and mathematicians for more than 2000 years. Although increasing numbers of geological scientists now use mathematical analysis and numerical techniques that are based on statistics and probability, most applications rely on classical set theory, established by Georg Cantor (1845-1918). However, following recent developments in computer science, and the necessity of analysing increasing volumes and varieties of exploration data, it has become necessary for Earth scientists to accept new approaches in reasoning using imperfect knowledge from incomplete data sets, and to develop more accurate and reliable decision making processes.

There have been many approaches to the problem of imperfect geological knowledge: traditional statistical theory, evidential belief function theory, Boolean reasoning, rough set theory (Pawlak, 1996), and fuzzy set theory proposed by Zadeh (1968, 2001). Fuzzy logic is also utilized in classification of spatial themes in space-borne sensor images and surface geophysical data interpretation (An et al., 1991; Ishibuchi et al., 1996).

In this chapter, the basic theory of the fuzzy logic approach in geological resource exploration applications is reviewed and illustrated with examples. Neural network reasoning of the information represented using fuzzy logic is also discussed along with fuzzy logic spatial data fusion.

2. Classical Set Theory

Traditionally, geological scientists have relied on the classical set as developed by Georg Cantor. A set is a collection of objects or thoughts, within a certain realm, taken as a whole. Each object in the collection is called an element (or member) of the set. The notation,

$$a \in A$$

denotes that a is an element of the set A. In this case, it is understood that a is a member of A or a belongs to A. This method of describing an object is comparable to the way a geologist maps an outcrop in the field and determines to which rock type the outcrop belongs,

$$\text{outcrop } a \in \text{rock type } A$$

In general, this is a definite statement that the outcrop a belongs to a rock type A as determined by the geologist.

Now consider the same geologist taking a magnetometer to the same outcrop and acquiring a reading in the hope of determining the relative content of magnetic minerals in the outcrop rock. The magnetometer reading X, taken by the geologist, is a number calibrated with respect to the magnetometer design parameters. If this number is properly calibrated, it can indeed provide the geologist with a definite percentage of the magnetic mineral. However, if the geologist wishes to connect the magnetic survey data (i.e. magnetometer reading) with a specific mineral deposit type, the magnetic survey data X cannot provide 100% evidence that a specific mineral deposit exists at the location. The magnetometer reading X can, at best, provide an indirect indication of the target mineral deposit. In this context, the exploration data (i.e. magnetometer reading) represents partial information or "fuzzy" information about the exploration target and the mineral deposit hypothesis.

The classical set satisfies algebraic rules, such as the commutative, associative, distributive, and absorption laws. Operations that assign each element of a set A to each element of a set B are said to be mapping or transforming A to B. Union and intersection of sets A and B are also defined in the classical set theory (Iyanaga and Kawada, 1980). The first publications in fuzzy set theory were by Zadeh (1965) and Goguen (1967, 1969) and included the generalization of classical definitions of a set and propositions to accommodate fuzziness of the newly defined sets.

3. The Basic Concept of a Fuzzy Set

Consider the characteristic features of a real exploration process; real exploration data are very often uncertain or vague in many ways. Geological mapping of rocks in the field can be viewed as a definitive and deterministic step, even when the transitional change from one rock type to another can be clearly seen, because each rock type is defined by a certain range of component minerals. However, in many real cases, the problem is intensified due to many factors that cannot be controlled by a geologist. In most exploration areas, bedrock is at least partially covered by thick overburden material and/or heavy vegetation. In such cases, even a sharp geological boundary has to be extrapolated under the overburden cover, and the exact spatial extent of the geological unit becomes vague.

Most geophysical survey data pose a similar problem to exploration hypotheses. The design of most geophysical instrumentation is precise, and calibration is also exact with respect to chosen references. However, the information content of each data set is vague and non-deterministic with respect to exploration hypotheses (e.g. lithologies, mineral deposit models, complex geological factors associated with hydrocarbon traps). Zadeh (1987b) wrote appropriately: "As the complexity of a system increases, our ability to make a precise and yet significant statement about its behavior diminishes until a threshold is reached beyond which precision and significance become almost mutually exclusive characteristics".

Since information is incomplete, the "final outcome" of an exploration task cannot be known exactly. This type of inherent uncertainty, in the theoretical classical set sense, has often been handled using probability and statistical methods. In this situation, the events (elements of sets) or the statements are well defined, and this uncertainty or vagueness is called "stochastic uncertainty." In contrast, vagueness concerning the description of the semantic meaning of the events, phenomena or statements themselves, is called "fuzziness." Fuzziness can be found in many areas of geological exploration, such as porosity of formation rocks, electrical capacitance of disseminated mineral ores, reflectance of outcrop rocks, and other deposit model parameters in exploration problems.

Let X be a collection of objects denoted generically by x, then a fuzzy set $\tilde{A}$ in X is a collection of ordered pairs,

$$\tilde{A} = \{(x, \mu_{\tilde{A}}(x)) | x \in X\}$$

where $\mu_{\tilde{A}}(x)$ is called the membership function or grade of membership (or degree of truth) of x in $\tilde{A}$ which maps X to the membership space M. (If M contains only two points 0 and 1, $\tilde{A}$ is a non-fuzzy classical set). The range of the membership function is a subset of the non-negative real numbers whose supremum is finite (Zadeh, 1965, 1987a, 1987c; Zimmermann, 1984).

4. Classical Measures of Fuzzy Events

Zadeh (1968) published technical literature regarding the probability of a fuzzy event and the axiomatic definition of fuzzy probability measures. He demonstrated that fuzzy probability measures can be characterized uniquely, either by a probability measure or a Markoff kernel. He also defined the probability $\tilde{P}$ of a fuzzy event $\tilde{A}$ or a fuzzy set $\tilde{A}$ with membership function μ_A, as follows,

$$\tilde{P}(\tilde{A}) = \int \mu_A(x) dp(x)$$

where p is some classical probability measure (Zadeh, 1987; Klement, 1980). The probability $\tilde{P}$ has the following properties:

(i) $\tilde{P}(\tilde{0}) = 0$ and $\tilde{P}(\tilde{1}) = 1$

(ii) $\tilde{P}(\tilde{A} \cup \tilde{B}) + \tilde{P}(\tilde{A} \cap \tilde{B}) = \tilde{P}(\tilde{A}) + \tilde{P}(\tilde{B})$

(iii) If $(\tilde{A}_n)_{n \in IN}$ is an increasing sequence of fuzzy events then,

$$\tilde{P}(\bigcup_{n \in IN} \tilde{A}_n) = \sup_{n \in IN} \tilde{P}(\tilde{A})$$

The union and intersection of fuzzy sets are used in the usual sense here i.e. union and intersection are expressed by the maximum and minimum of the membership functions, respectively.

A fuzzy probability measure is a function,

$$m:\tilde{A} \rightarrow [0,1]$$

which fulfills the above properties (i)-(iii), amongst others that are satisfied by many classical probability spaces (Klement, 1980; Wang and Chang, 1980).

5. Fuzzy Operator

The membership function is the fundamental component of a fuzzy set, and operations with fuzzy sets are defined using their membership functions (Zadeh, 1965). The membership function $\mu_{MIN}(x)$ of intersection of a fuzzy set $\tilde{A}$ is defined as follows,

$$\mu_{MIN}(x) = \min \ \mu_{\tilde{A}}(x), \mu_{\tilde{B}}(x) \ , \ x \in X$$

This intersection operator is also called the "fuzzy AND" operator or "MIN operator" in some publications (An et al., 1991; Bonham-Carter, 1994).

The membership function $\mu_{MAX}(x)$ of the union $\tilde{D} = \tilde{A} \cup \tilde{B}$ is point-wise defined by,

$$\mu_{MAX}(x) = \max\{\mu_{\tilde{A}}(x), \mu_{\tilde{B}}(x)\}, \qquad x \in X$$

As above, this operation is called the "fuzzy OR" or "MAX operator". This operator behaves like the Boolean OR and the output membership values are controlled by the maximum values of the input information.

The membership function of the complement of a fuzzy set $\tilde{A}$, $\mu_c(x)$ is defined by,

$$\mu_c(x) = 1 - \mu_{\tilde{A}}(x), \qquad x \in X$$

In addition to the above basic operators, there are a large number of operators which are specially designed for specific applications in complicated engineering problems. Some of these operators are also used effectively in the spatial data fusion applications of geological exploration data (An, 1992; Bonham-Carter, 1994). A combined membership function $\mu_{comb}(x)$ of a fuzzy set $\tilde{A}$ is defined as,

$$\mu_{comb}(x) = \prod_{i=1}^{n} \mu_i(x)$$

where $\mu_i(x)$ is the fuzzy membership function for the *i-th* subset of $\tilde{A}$ (in a geological exploration application, the subscript i may represent individual map layers to be combined). In this case, the combined fuzzy membership values tend to be very small because of the effect of multiplying several numbers smaller than 1.0. This fuzzy operator is also called a fuzzy algebraic product operator.

The next useful fuzzy operator is the algebraic sum operator, defined as follows,

$$\mu_{A-SUM}(x) = 1 - \prod_{i=1}^{n} (1 - \mu_i(x))$$

The algebraic sum operator is complementary to the algebraic product operator and the result is always greater than or equal to the largest contributing fuzzy membership value. This operator can be used when there is evidence that supports the chosen exploration hypothesis and the combined evidence is more supportive than individual pieces of evidence.

Another useful fuzzy operator is the γ (gamma) operator, which was proposed by Zimmermann and Zysno (1980). The membership function $\mu_\gamma(x)$ of the combined fuzzy algebraic product and fuzzy algebraic sum operation is defined as follows,

$$\mu_\gamma(x) = (\text{Fuzzy algebraic sum})^\gamma * (\text{Fuzzy algebraic product})^{(1-\gamma)}$$

where γ is a parameter chosen in the range (0, 1). When γ is 1, the combination is the same as the fuzzy algebraic sum, and when γ is 0, the combination is the same as the fuzzy algebraic product (An, 1992). Therefore, careful choice of γ can produce output membership values that can ensure a flexible compromise.

In geological applications of digital data fusion, the available data or spatial data have, in most cases, varying degrees of information content with respect to the chosen exploration hypothesis. In such cases, it becomes necessary to use several different fuzzy operators separately or a combination of selected operators depending on the characteristics of each data layer (Moon and Jiang, 1995; Moon et al., 2001). The situation is similar for oil and gas exploration applications utilizing a fuzzy logic approach. If an attempt is made to integrate imaging of well–log data, surface seismic and VSP data, and associated reservoir characteristics, the information content of each data set will be considerably different with respect to the hydrocarbon deposit model of the study site. In such a case, the relationship between each data set with respect to the specific hydrocarbon deposit model will favor a certain fuzzy operator or a combination of several fuzzy operators.

6. Fuzzy Information Representation

Most geological exploration tasks include multiple sets of spatial data layers. These include geological maps, geochemical data, geophysical data and other auxiliary information. A systematic integration (or fusion) of these different types of spatial data towards a chosen exploration hypothesis involves several steps: pre-processing of individual data layers, information representation, integration (digital fusion), visualization, and decision making (Moon, 1995). Assuming that all exploration data are properly digitized, the pre-processing and geocoding steps, with the exception of certain specialized processing tasks, can be carried out utilizing any of a number of available commercial and public domain geographic information systems (GIS).

6.1. EXPLORATION DATA: SPATIAL DATA

The definition of spatial data may vary slightly depending on the application. In resource exploration, the data-gathering step usually comes first along with the field work. Information layers obtained from field observation or field measurements using specific instrumentation are called "spatial data," which are often symbolic models or a collection of symbols. As an example, a geological map is a symbolic model that includes all the geological symbols mapped in the field. Most survey data for non-renewable resource exploration are thus spatial data, whereas many of the data sets used in renewable resource studies also include (multiple) temporal information. In hydrocarbon and mineral exploration, the most important and basic information layer is the geological map. The next most important spatial data are the topographic maps, including both paper and digital maps, which are necessary as base maps.

Other exploration data include geological structure maps, geochemical survey data, well-log data and geophysical survey data, including various types of seismic survey sections (see Figure 1).

These real-world observations and measurements are characteristic of the instrument adopted and the personnel involved in the survey, and are usually represented by field variables and spatial objects. Definitions and a detailed description of spatial objects are given in many textbooks (Bonham-Carter, 1994; Burrough, 1986) and will not be repeated here.

Spatial objects are usually described in terms of dimensions (0-D, 1-D, 2-D, and 3-D), and in terms of mode of occurrence at the time of observation. Observed gravity values at each gravity station and geochemical analysis results of samples at each sampling site, represent point data and they are typical of 0-D data. Lithological boundaries of geological formations are 1-D spatial objects and anomalous zones on a geochemical map represent 2-D spatial objects. Similarly, a geometric description of a subsurface ore body constitutes a 3-D spatial object. Salt domes and underground cavities are also 3-D spatial objects. In applications where the geometric dimension *n* is more than 3, such as multi-spectral remote sensing data, a feature defined by each spectral window datum can be described as a hyperspectral or n-D spatial object (Benediktsson et al., 1997).

Recently, the concept of the elementary entropy function of a fuzzy set has been introduced (Rudas et al, 1998), together with new sets of generalized fuzzy operators. There have also been new attempts to develop non-parametric models for fusing heterogeneous fuzzy models (Pedrycz et al., 1998). Digital fusion of the heterogeneous fuzzy information is a very important new development because there are more multi-

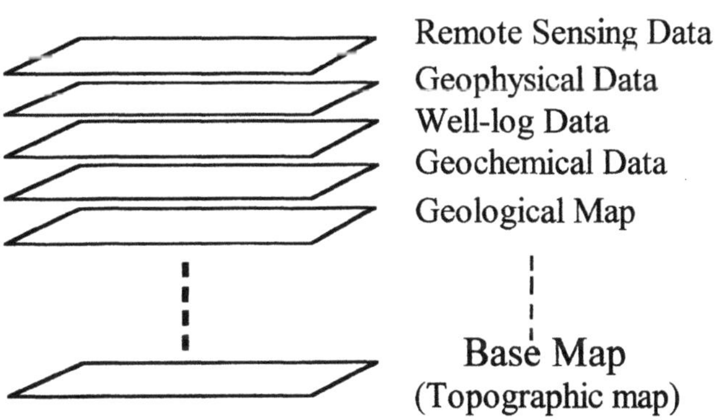

Figure 1. Typical spatial data layers in resource exploration.

sensor information becoming available in Earth science research and applications. These new concepts have not been tested yet, and so we will focus here on previously established well test operators (An et al., 1991).

6.2. SPATIAL REASONING IN NON-RENEWABLE RESOURCE EXPLORATION

An exploration project involves several steps: planning, field survey, data processing, digital data integration or data fusion, visualization of the fused information with respect to the target proposition, interpretation and decision making. Among these steps, information representation, digital data fusion, visualization, interpretation and decision making, are often collectively referred to as the spatial reasoning process, a term probably derived from the discipline of artificial intelligence (AI) and expert system study in computer engineering (see Figure 2).

Mathematically, the spatial data fusion steps may be explained as a mapping or a transformation between the raw field data and the final, digitally fused information with respect to a chosen exploration hypothesis. Suppose that n maps represent various exploration data, such as geology and specific geochemical survey data in a prospecting area A. Evidence contained in the *i-th* layer of the n map databases may be denoted by E_k $(k = 1, 2, 3, \ldots, n)$. The complete set of the n-layer databases in A may be written as follows,

$$E = \{E_1, E_2, \ldots, E_n\}$$

For each layer of exploration data, information representation can be scaled with respect to the target hypothesis in such a way that,

$$g_k : E_k \Rightarrow [0,1]$$

where $g_k(e)$ represents the actual information content with respect to the exploration target model. Here, $g_k(e)$ also defines a mapping $G(E)$ from the observed data space to the exploration information space.

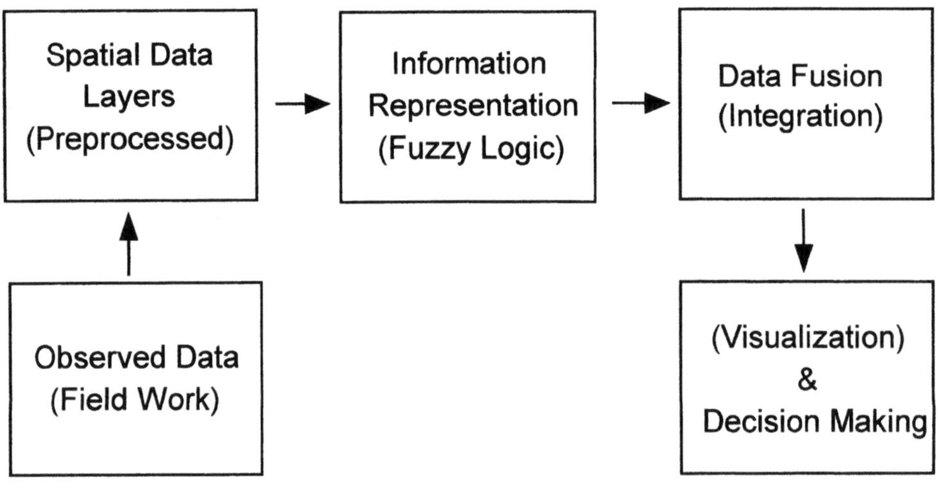

Figure 2. Block diagram illustrating basic fuzzy spatial data processing.

Suppose that $d_k(e)$ is defined as the "probability" that e of E_k is related to having an exploration target at a point p in A where the observation e is made. Here, the word "probability" does not necessarily have the same meaning as in "probability and statistics". The mapping d_k for each layer of exploration evidence E_k, represents the degree of "compatibility" of the observation e at p in A (Moon, 1993). A proposition exploration target (ET) may be defined such that a point p in A belongs to a deposit of the target type. A membership function,

$$U_k(ET|e) = d_k(e)$$

then represents the degree of certainty that p is a member of the set of points that belongs to a deposit, given e in A.

Now consider a set of n observed values $\{e_1, e_2, \ldots, e_n\}$ of pieces of evidence $\{E_1, E_2, \ldots, E_n\}$ at a point p in A. Suppose $\{d_1, d_2, \ldots, d_n\}$ are defined on $\{E_1, E_2, \ldots, E_n\}$. We then have representations $\{d_1(e_1), d_2(e_2), \ldots, d_n(e_n)\}$ at the point p where the observations are made for ET. We now wish to integrate these n representations into one single function. The integration rules depend on the interpretation of the mapping used (Moon, 1993). In general, a traditional probabilistic approach has been used most frequently, however an evidential belief function approach has also been used (Moon, 1990; An, 1992; Bonham-Carter, 1994). Although the basic concept is very similar, the mathematical details and interpretation of the results are quite different.

Suppose we now have the fuzzy membership function, $U_k(ET|e) = d_k(e)$ for all $k = 1, 2, \ldots, n$ for each layer of evidence $\{e_1, e_2, \ldots, e_n\}$ at a point p in A. Now, our task is to define a membership function $U(ET|e_1, e_2, \ldots, e_n)$ from n membership functions $U_k(ET|e_k)$. This can be accomplished using many different types of operator available in fuzzy set theory as discussed above. The choice of operator (or operators) depends on the exploration hypotheses (model) and the data sets available. Among the fuzzy operators listed above, the fused membership function using the algebraic sum operator takes the following form,

$$U(ET|e_1, e_2, \ldots, e_n) = \sum_{k=1}^{n} U_k(ET \mid e_k) + \sum_{k=1}^{n} \sum_{j=k+1}^{n} U_k(ET \mid e_k) U_j(ET \mid e_j) + \ldots$$

$$\ldots + (-1)^n U_1(ET \mid e_1) \ldots U_n(ET \mid e_n)$$

Similarly, the fused membership function using the Π-operator is,

$$U(ET|e_1, e_2, \ldots, e_n) = \left[\prod_{k=1}^{n} U_k(ET|e_k)\right]^{(1-\gamma)} - \left[1 - \prod_{k=1}^{n} (1 - U_k(ET|e_k))\right]^{\gamma}$$

where $0 \le \gamma \le 1$.

Spatial information representation and fusion procedures using classical probability and evidential probability are similar in their basic approach. The data layers collected from an exploration project are represented with conditional probabilities defined as

$\text{Prob}_k(e \mid ET)$, where the observation e of E_k is made at a point p, conditional on the location p containing the target deposit. The individual target probability contributions can then be combined using Bayes theorem (Moon, 1993).

Similarly, if an evidential belief function is used for representing multiple layers of exploration data, Dempster's rule can be employed for fusion of the multiple layers of the exploration data (Moon, 1990, 1993). If there are unsurveyed or missing sections of data, an evidential belief function approach is particularly effective, since further quantization of spatial information (or unknowns) may be performed in terms of plausibility and ignorance functions. These two additional mappings of plausibility and ignorance along with the weighted evidential belief function representation, make the final decision making process considerably more accurate and precise. This unique feature can be crucially important with exploration projects exhibiting sparse survey coverage.

7. Application Formulation

Digital data fusion of multiple layers of exploration data is, in fact, a mapping process, which transforms multiple layers of unweighted survey data into a single fused layer of information tuned towards the exploration target. Let [E, $U(ET|E)$] be the multiple sensor input and output vector pairs where $E = \{E_1, E_2, \ldots, E_n\}$ are the input exploration data, and $U(ET|e_1, e_2, \ldots, e_n)$ is the final fused information. The mapping $\tilde{G}$ can then be defined such that,

$$U(ET|e_1, e_2, \ldots, e_n) = \tilde{G}(w, E) + \varepsilon$$

where the final fused information layer $U(ET|e_1, e_2, \ldots, e_n)$ is the final m-dimensional output, depending on the problem formulation, w represents the weighting for each exploration data layer with respect to the chosen exploration model (or target hypothesis), and $\varepsilon = \{\varepsilon_1, \varepsilon_2, \ldots, \varepsilon_m\}$ is an error vector associated with $U(ET|e_1, e_2, \ldots, e_n)$. Even in a single exploration area with a given set of survey data, w can be quite different depending on the target deposition model; therefore, the mapping $\tilde{G}$ includes two-stage mapping: information representation and digital fusion (Moon, 1993).

8. Fuzzy-Neural Network

Fuzzy set theory is relatively well understood for approximate reasoning and fusion of geological exploration data (An, 1992; An et al., 1991; Moon, 1993; Bonham-Carter, 1994). Although fuzzy set theory has also been applied successfully in various fields of science and engineering, there still exists two difficulties: lack of guidelines to decide or adjust the membership functions of fuzzy variables, and lack of algorithms for automatic rule generation. Therefore, many investigators have recently tried to combine the concepts of neural networks with fuzzy set theory to overcome these problems (Takagi and Hayashi, 1991; Yager, 1992).

Generally, neural networks are good at approximating nonlinear systems, whereas fuzzy set theory is useful for simplifying nonlinear information structure. Learning ability, parallel processing and distributed knowledge representation are the major features of neural networks. On the other hand, neural networks suffer from unstructured knowledge representation, in contrast to the structured knowledge representation in fuzzy systems. Therefore, the combination of the two techniques should provide a new paradigm to model realistic, often nonlinear exploration models with high degrees of uncertainty. Fuzzy neural networks are being used increasingly in many other disciplines, particularly in engineering control system applications (Yager, 1992; Hauptmann and Heesche, 1995; Yao and Kuo, 1996).

A schematic diagram of a three-layer fuzzy neural network structure, representing a multiple data set exploration project, is shown in Figure 3. In many engineering problems, the hidden layer may be adequately trained with the known input and an expected or desired output. However, many geological exploration tasks depend on either a selected deposit model or models, in which case the hidden layer will closely

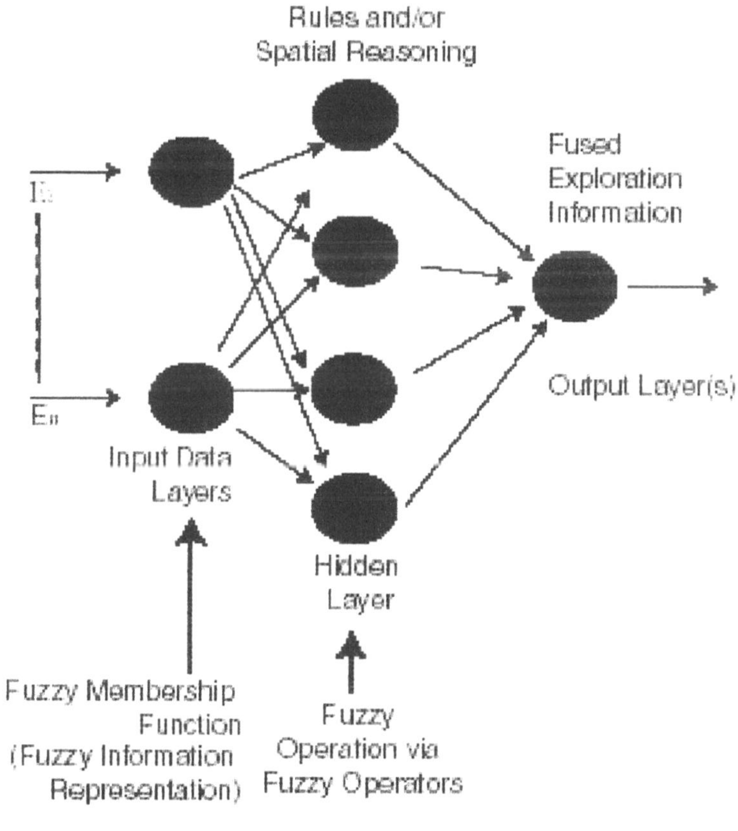

Figure 3. Structure of a simple three layer fuzzy neural network for exploration data fusion.

represent the deposit model and the exploration reasoning process will depend largely on the model parameters. If a data driven exploration approach is adopted, the hidden layer can pose problems, because the final output is completely unknown in many resource exploration problems and the hidden layer cannot be trained adequately.

With the evolution of second-generation fuzzy systems, a considerable amount of research has focused on the integration of fuzzy logic and artificial neural networks, giving birth to neuro-fuzzy systems or an integrated neuro-fuzzy system (INFS). The term neuro-fuzzy generally refers to the fusion of fuzzy systems and neural networks with the aim of combining the advantages of both paradigms and at the same time compensating for the inadequacies with respect to the target hypothesis. In many exploration problems, the deposit models are neither robust nor universal and application of a neural network approach to data fusion does not appear to be suitable. Some of the data layers, such as a set of geophysical data, can be modeled using a specific mathematical model, in which case an optimized inversion of limited amount of observation can be used for training of the hidden layer.

In the data driven exploration model, there is usually no background information with respect to the exploration hypothesis, and the information content of the observed input data is speculative. Consequently, the choice of fuzzy operators that will replace some of the hidden elements of the neural network to be adopted can be somewhat arbitrary.

With increasing interest in the fusion of neural networks and fuzzy logic, there are more attempts to integrate the fuzzy reasoning with neural network modeling. The simple fuzzy neural network reasoning discussed above is the most basic among various types of reasoning mechanisms. Syllogistic fuzzy reasoning, where the consequence of a rule in one reasoning stage is passed to the next stage as a new fact is essential, in certain cases, to effectively build a large scale system with high level of intelligence. A cascaded fuzzy neural network (CFNN) is proposed, which is based on the syllogistic fuzzy logic and neural network (Duan and Chung, 2001). The cascaded fussy neural network (CFNN) can generate an appropriate syllogistic fuzzy rule set through structural learning and back propagating parameter learning. This type of cascaded fuzzy neural network can be useful in the spatial data fusion with several heterogeneous fuzzy data layers.

9. Error and Uncertainty Analysis

There are, in general, two types of uncertainty and error in integration and fusion of exploration data sets. The first type is introduced to the data at the beginning, when experiments are carried out or when the field surveys are performed. Analysis of this type of error and uncertainty is relatively well established and there are several methods for estimating and evaluating them. The second type are the ones introduced during the information representation and digital fusion of the pre-processed information layers, because of vagueness of the original information contained in the data sets with respect to the exploration target. The impreciseness of mathematical operators employed during the digital fusion stage introduces additional uncertainty (An et al., 1994). This has been an inherent problem with data fusion using both fuzzy logic and evidential belief function approaches. In a neural network approach, the design of the network and training of the neurons using both the input and output information can minimize this type of error or uncertainty.

The fusion of exploration data includes independent interpretation of each data layer $E_k (k=1,2,\ldots,n)$, and mapping of the data into a fuzzy space, assigning membership functions to each data object. These errors and uncertainties associated in the fuzzy membership functions, subsequently propagate through the entire reasoning process towards the final membership function (An, 1992). In the above, the algebraic product of n membership functions is defined as,

$$\mu_{comb}(x) = \prod_{i=1}^{n} \mu_i$$

Let μ_{-i} be the algebraic product with the *i-th* term deleted,

$$\mu_{-i} = \mu_1, \mu_2, \ldots, \mu_{i-1}, \mu_{i+1}, \ldots, \mu_n$$

Then,

$$\frac{\partial \mu_p}{\partial \mu_i} = \mu_{-i}$$

Let ε_i be the error term associated with the membership function from the *i-th* E_i. The error ε_p associated with the algebraic product can be estimated using a Taylor expansion, with second- and higher-order terms ignored,

$$\varepsilon_{comb} = \sum_{i=1}^{n} \mu_{-i}\varepsilon_i$$

$$= \mu_p \sum_{i=1}^{n} \frac{\varepsilon_i}{\mu_i}$$

ε_p may be further normalized using the resulting membership function μ. The normalized error is, in fact, the relative error. Similarly, if the algebraic sum discussed above is denoted by,

$$\mu_{A-SUM} = 1 - \prod_{i=1}^{n}(1-\mu_i)$$

$$= 1 - \nu$$

where $\nu = \prod_{i=1}^{n}(1-\mu_i)$, the relative error associated with the algebraic sum operation becomes,

$$\varepsilon_{A-SUM} = \frac{\nu}{1-\nu} \sum_{i=1}^{n} \frac{\varepsilon_i}{1-\mu_i}$$

In the case of the γ-operator, the relative error in the final fused information becomes (An, 1992; Moon, 1993),

$$\varepsilon_{\gamma} = (1-\gamma)\varepsilon_{comb} + \varepsilon_{A-SUM}$$

Numerical computation of the errors that propagate through the fusion (or aggregation) processes is a relatively easy task, whereas quantitative estimation of the errors and uncertainties introduced through information representation and data fusion operators is considerably more difficult.

Another further complication is the fact that, as discussed in Section 7, the multiple layers of data sets collected in the field for a specific exploration project are often inter-linked through a very complex depositional model. Some of these data sets are also dependent on each other, making the whole data fusion process even more complicated (Jiang et al., 1997).

As discussed at the beginning of this chapter, membership functions are relative. A fuzzy membership function close to unity does not necessarily imply certainty with respect to the target hypothesis. If, however, the final fused membership value is higher at one location than at another, it does mean that the possibility for the target proposition is higher at the former location than at the latter. The errors or uncertainties propagated through the fusion of various exploration data layers do not necessarily have the same precise meaning. In this sense, the interpretation of the errors and uncertainties in the final results is more vague than the error analyses of many physical experiments.

10. Conclusions

Exploration strategies for non-renewable resources have changed rapidly over the years, and have been complemented with accelerating innovations in computer hardware and information processing technology. Although the basic philosophy of geological exploration remains the same, the methodology has been changing to the extent that multiple sets of exploration data are now utilized simultaneously, and the final target information is visualized in an optimum manner, with minimized uncertainties.

There are still a number of areas for further investigation, and many of the mathematical tools we employ require further research. The main problem which may be identified at this point is the exact quantization of the field survey data. There are several ambiguities. First, most exploration data collected in the field are 1-D, 2-D, and rarely 3-D, whereas almost all exploration targets are 3-D. This requires interpolation of the available information by a geologist, utilizing his or her own expert intuition, but utilizing a more quantitative and sophisticated mathematical approach. Second, most exploration equipment used routinely is designed on simple principles with a limited number of parameters representing the overall physical characteristics. Therefore, the survey data collected using such equipment can provide us with only partial information.

To overcome the inherent ambiguities in modern exploration techniques, exploration geologists have used several mathematical tools, such as statistical and probability theory, fuzzy logic approaches for quantizing vagueness, and evidential belief function methods, in addition to conventional data processing techniques.

The fuzzy set approach of converting exploration field data into exploration information is not exact in the mathematical sense, but it provides us with an opportunity to carry out a spatial reasoning process with less subjective human intervention. It also improves the consistency of exploration data processing and reduces errors considerably. However, one of the problems of using a fuzzy logic approach to processing exploration data lies in the availability of optimum operators with respect to the target proposition. Many of the simple fuzzy operators being used by geologists today are designed by engineers with different application objectives, and there is no one fuzzy operator that can be adequately utilized for all deposit models (Jiang et al., 1997). In fact it may be impossible to expect one fuzzy operator to integrate and fuse multiple sets of exploration data. Even for one type of base metal exploration, several fuzzy operators are used in parallel and some in combination (Moon and Jiang, 1995; Jiang et al., 1997). Application of fuzzy operators, both in parallel and/or in combination, can adequately fuse partial information contained in each data layer with respect to the target hypothesis. The problem of conditional dependency of certain data layers can also be resolved in this approach. A marriage of fuzzy information representation and fuzzy reasoning with neural networks can provide a more effective and accurate solution. However, development of a neuro-fuzzy system for resource exploration problems requires further research and investigation.

There are several major converging trends today:

(1) Major advances in geophysical exploration techniques.
(2) Major advances in Earth observation (or remote sensing) sensor technology.
(3) Major advances in geophysical data processing and inversion technology.
(4) Development of soft computing techniques to infer a conclusion from multi-source observation data sets.

In view of these developments, Professor Lofti Zadeh (2001) argues that the principal components of information representation and digital fusion such as fuzzy logic, neurocomputing, probabilistic computing and chaotic computing should all be parts of the principal (digital) brainwares. Although a single spatial data fusion stage is often sufficient, we should research new tools for Earth observation data processing and more accurate imaging of target information in the Earth sciences.

Error and uncertainty analysis is an important step in any integrated exploration data processing task. There are two types of error and uncertainty in fuzzy data fusion and information imaging of exploration data. The errors and uncertainties originating from direct experiments and field surveys can be adequately estimated following the traditional approach. However, the errors and uncertainties introduced from fuzzy membership assignment and during application of the fuzzy operator(s) are considerably more complicated, and require further investigation and research.

Acknowledgments

This research is supported by an NSERC of Canada operating grant #7400 (to W.M. Moon). Some of the research presented in this paper was carried out by Wei-Wei Jiang, Bo Li, and Sung-Won Choi.

References

An, P., 1992, Spatial reasoning and integration techniques for geophysical and geological exploration data: Unpublished Ph.D. thesis, Univ. of Manitoba, Canada.

An, P., Moon, W. M., and Rencz, A., 1991, Integration of geological, geophysical and remote sensing data using fuzzy logic: J. Can. Soc. Expl. Geophys., **27**, 1-11.

An, P., Moon, W. M., and Bonham-Carter, G. F., 1994, Uncertainty management of exploration data using the belief function: Nonrenewable Resources, **3**, 60-71.

Benediktsson, J. A., Ingimundarson, J. I., and Sveinsson, J. R., 1997, Classification and feature extraction of hyperdimensional data using LOOC covariance estimation: IGARSS'97 Proceedings, B10.03.

Bonham-Carter, G. F., 1994, Geographic information systems for geoscientists: Pergamon Press.

Burrough, P. A., 1986, Principles of geographic information systems for land resources assessment: Clarendon Press.

Duan, J. C., and Chung, F. L., 2001, Cascaded fuzzy neural network model based on syllogistic fuzzy reasoning, IEEE Trans. Fuzzy Systems, **9**, 293-306.

Goguen, J. A., 1967, L-fuzzy sets: J. Math. A. A., **18**, 145-174.

Goguen, J. A., 1969, The logic of inexact concepts: Synthese, **19**, 325-373.

Hauptmann, W., and Heesche, K., 1995, A neural net topology for bidirectional fuzzy-neuro transformation: Proc. IEEE Internat. Conf. on Fuzzy Systems, vol. III, 1511-1518.

Ishibuchi, H., Morisawa, T., and Nakashima, T., 1996, Voting schemes for fuzzy rule-based classification systems: Proc. IEEE Internat. Conf. on Fuzzy Systems, vol. I, 614-620.

Iyanaga, S., and Kawada, Y., Eds., 1980, Encyclopedic dictionary of mathematics: MIT Press, 1167 – 1170.

Jiang, W. W., Moon, W. M., and Feng, L.N., 1997, Data fusion research for mineral exploration data from Hwanggang-ri region, Korea: Technical Report # 17, Dept. of Earth Sciences, University of Manitoba, Winnipeg, Canada, 1 - 62.

Klement, E. P., 1980, Fuzzy measures by classical measures, *in* Wang, P. P., and Chang, S. K., Eds., Fuzzy sets: 25 – 33.

Mathai, A. M., and Rathie, P. N., 1975, Information theory and statistics: John Wiley and Sons, 1 – 137.

Moon, W. M., 1990, Integration of multiple sets of geophysical information using evidential belief function: IEEE Trans. Geosci. Remote Sensing, **28**, 272-278.

Moon, W. M., 1993, Mathematical basis for geophysical information representation and integration: Can. J. Remote Sensing, **19**, 63-67.

Moon, W. M., 1995, Information representation and integration of multiple sets of spatial geoscience data: IGARSS'95 Proceedings.

Moon, W. M., and Jiang, W. W., 1995, Integrated exploration of base metal deposits in East Java, Indonesia: Technical Report #14, The University of Manitoba, Canada.

Moon, W. M., Jiang, W. W., Yun, S. G., and So, C. S., 2001, Data fusion modeling of the Hwang-Gang-Ri mineralization zone: in preparation.

Pawlak, Z., 1996, Why rough sets ?: Proc. IEEE Int. Conf. on Fuzzy Systems, vol. 2, 738-743.

Pedrycz, W., Bezdek, J. C., Hathway, R. J., and Rogers, G. W., 1998, Two non-parametric models for fusing heterogeneous fuzzy data, IEEE Trans. Fuzzy Systems, **6**, 411-424.

Rudas, I.J., and Kaynak, M. Okay, 1998, Entropy-based operations on fuzzy sets, IEEE Trans. Fuzzy systems, **6**, 33-40.

Shannon, C. E., 1948, A mathematical theory of communication: Bell System Tech. J., **27**, 379-423, 623-656.

Takagi, H., and Hayashi, I., 1991, NN-driven fuzzy reasoning: J. Approx. Reasoning, **5**, 191-212.

Wang, P. P., and Chang, S. K., Eds., 1980, Fuzzy sets: Plenum Press.

Yager, R. R., 1992, Implementing fuzzy logic controllers using a neural network framework: Fuzzy Sets and Systems, **48**, 53-64.

Yao, C. C., and Kuo, Y. H., 1996, A fuzzy neural network model with three-layered structure: Proc. IEEE Internat. Conf. on Fuzzy Systems, Vol. III, 1503-1510.

Zadeh, L. A., 1965, Fuzzy sets: Information and Control, **8**, 338-353.

Zadeh, L. A., 1968, Probability measures of fuzzy events: J. Math. Anal. Appl., **23**, 421-427.

Zadeh, L. A., 1987a, Fuzzy sets, *in* Yager, R. R., Ovichinikov, S., Tong, R. M., and Nguyon, H. T., Eds., Fuzzy sets and applications: Selected papers by L.A. Zadeh: John Wiley, 29-44.

Zadeh, L. A., 1987b, Fuzzy sets as a basis for a theory of possibility, *in* Yager, R. R., Ovichinikov, S., Tong, R. M., and Nguyon, H. T., Eds., Fuzzy sets and applications: Selected papers by L.A. Zadeh: John Wiley, 193-218.

Zadeh, L. A., 1987c, A theory of approximate reasoning, *in* Yager, R. R., Ovichinikov, S., Tong, R. M., and Nguyon, H. T., Eds., Fuzzy sets and applications: Selected papers by L.A. Zadeh: John Wiley, 368-412.
Zadeh, L.A., 2001, Recognition technology and fuzzy logic, IEEE Trans. Fuzzy Systems, **9**, 3-4.
Zimmermann, H. J., 1984, Fuzzy set theory – and its applications: Kluwer-Nijhoff Publishing.
Zimmermann, H. J., and Zysno, P., 1980, Latent connectives in human decision making: Fuzzy Sets and Systems, **4**, 37-51.

APPLICATION OF MULTILAYER PERCEPTRONS TO EARTHQUAKE SEISMIC ANALYSIS

HENGCHANG DAI

British Geological Survey,
Murchison House,
West Mains Road,
Edinburgh EH9 3LA,
SCOTLAND
E-Mail: hcd@mail.nmh.ac.uk

Abstract

Two multilayer perceptron (MLP) methodologies are developed to tackle the problems of earthquake seismic arrival picking and identification. The arrival picking is achieved using the vector modulus of three-component (3-C) seismic recording or the absolute value of amplitude of single-component recording, as the input to an MLP. A discriminate function, determined from the output of the trained MLP, is then employed to define the arrival onset. Arrival identification is achieved using the degree of polarization of the arrival segment of the 3-C recordings as the input to an MLP; the output of the trained MLP indicates the arrival type. The results are encouraging. This work demonstrates significant potential in the application of MLPs to automatic earthquake analysis and other geophysical problems. In principle, it should be possible to develop a complete automatic seismic analysis system based on the two methodologies described herein.

1. Introduction

The aim of earthquake seismic analysis is to extract information on earthquake events from seismograms, the recorded seismic signals, as soon as possible after they occur. Such information includes the event position, origin time and magnitude, which can then be accessed for further scientific analysis or public inquiry. Seismograms are usually contaminated by various kinds of noise, such as artificial vibrations, traffic noise, and background noise. Traditionally, they have been handled using empirical methods based on the expertise of the human operator to discriminate genuine seismic signals from the various

W. Sandham and M. Leggett (eds.),
Geophysical Applications of Artificial Neural Networks and Fuzzy Logic, 287-304.

sources of noise. Once an event has occurred, analysts must visually check the seismograms which contain the event, and pick the onset time of possible seismic arrivals, for example primary (*P*) and secondary (*S*) wave arrivals, and then to identify their types. Both tasks involve the implementation of extensive pattern recognition algorithms, conventional methods of which are usually incapable of dealing with such a complex problem, and which are time-consuming and subjective. There is an urgent need to find an alternative pattern recognition approach to tackle the problem. The multilayer perceptron (MLP) provides a natural alternative for automation of seismic analysis, as it has been successful at handling complicated pattern recognition problems due to its learning ability. A wide range of applications emphasizes the particular strengths of MLPs for solving a particular problem, compared to conventional methods incorporating a fixed algorithm

Two MLP approaches which automatically pick and identify seismic arrivals, are presented in this chapter. The details of the two approaches can be found in Dai (1995) and Dai and MacBeth (1995, 1997a, 1997b). The data set used in this work is from the TDP3 local seismic network, which is located near the North Anatolian Fault in Turkey, and which is part of the third Turkish Dilatancy Project (Lovell, 1989). Between March and November 1984, many hundreds of local earthquakes were recorded. Among them, 371 and 391 recordings from stations DP and AY, respectively, were selected for further analysis. All recorded events were local with depths between 2km and 14km, epicentral distances less than 30km, and magnitudes (M_L) between -0.3 and 2.0. The most prominent *P*- and *S*-waves were manually identified from the seismograms as a reference.

2. Automatic Picking of Seismic Arrivals Using MLPs

The aim of arrival picking is to estimate the onset time of a predominant seismic arrival, reliably and accurately. There is no shortage of techniques which profess to tackle this task (see Dai, 1995; Dai and MacBeth, 1995, for an extensive reference list). However, they do tend to be data specific and are not generally applicable. Analysts still need to pick arrivals visually using interactive methods. The task performed by the trained analyst in manually picking arrival onset times involves an intensive amount of pattern recognition. Experience provides a judicious balancing of wave characteristics such as amplitude, frequency and polarization from previous records, to determine the most likely onset time. If questioned about a particular decision, however, the analyst may offer a few rules for guidance but can often give no obvious systematic reasoning, because the decision is partly subjective and based upon his past experience. Consequently, different trained analysts may give different answers, and the same analyst may even choose a different interpretation after some time has elapsed. Since MLPs utilize a learning scheme to develop an appropriate solution, they provide a natural alternative to this type of earthquake analysis.

Here, an approach is developed using an MLP to emulate this manual analysis procedure, but overcome its weaknesses. The MLP is designed to be similar in operation to an analyst, and is trained by presenting many different seismic arrivals. After training is accomplished, the MLP remembers these arrivals and should then be able to recognize arrivals in a variety of new seismograms.

2.1. ARRIVAL DETECTION AND PICKING

2.1.1. *Methodology*

Generally speaking, a method for interpreting seismograms can be regarded as using a filter to operate on a characteristic function, which is either extracted from the seismogram or consists of the seismogram itself. The filter will produce output features, which indicate the signal properties such as the arrival onset of a wave, its polarization, its propagation direction, and its type. The characteristic function is a time series, which is obtained by performing a transformation on a seismogram. It might enhance desired signals from the seismogram or separate different signals according to their characteristics, so that the subsequent "filtering" procedure can be performed easily and efficiently.

For arrival picking, an MLP is used as a filter to act on the characteristic function of the seismogram. This approach includes two phases: the training (learning) phase and the testing (generating) phase. The MLP is trained using the back-propagation training algorithm (Rumelhart et al., 1986). During the training phase, a training data set is selected by an analyst using his geophysical knowledge and experience. After training, the MLP is applied to the seismogram by using a window which slides along the entire modulus trace (Figure 1). The output of the MLP then yields a time series which can then be interrogated for a decision regarding the seismic arrivals. The arrival onset times are defined by a discriminant function, determined from the output of the trained MLP. Some post-processing procedures are used to improve the performance in conjunction with this MLP.

2.1.2. *The Input Characteristic Function F(t).*

The input function $F(t)$ to the MLP is the vector modulus of three-component (3-C) recordings, or absolute values of one-component (1-C) recordings. The input is presented to the MLP in segments which are selected from a sliding window which passes across the entire input seismogram (Figure 1). This function is strongly dependent on the magnitude and epicentral distance of the earthquake. For the data used here, the modulus of a large event is more than 3600 counts in scale, and the modulus of a small event is just above the background noise (about 20-30 counts). If such large variations of modulus are fed directly into the MLP for training, a large training data set is required to cover the range, with a

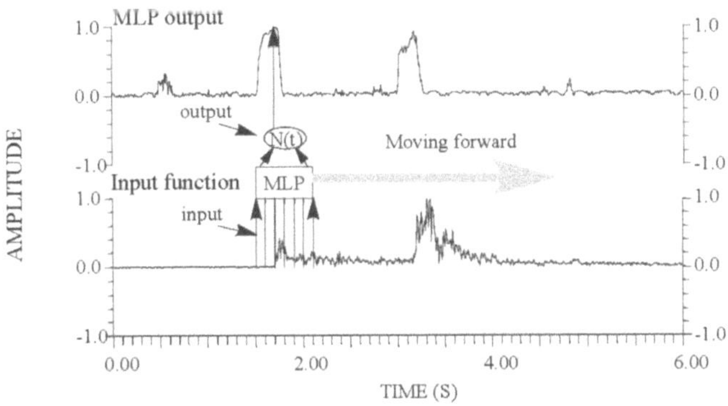

Figure 1. Schematic diagram illustrating the use of an MLP to pick seismic arrivals.

consequent increase in training time and a larger MLP structure. Otherwise, large changes in modulus may bias the estimates. This problem may be overcome by using the relative characteristic function in which each segment is individually normalized so that it is not dependent upon the magnitude and distance of an earthquake. This reduces the number of training examples required.

2.1.3. *The MLP Structure*

The MLP used in this work has an input layer, a single hidden layer and an output layer (Figure 2). The input nodes correspond to the input segment of modulus which is chosen to include several complete cycles of a wave in a sliding window with fixed length. This number should be chosen after a process of trial-and-error with different training runs.

There are two nodes in the output layer flagging the result; (0,1) for an arrival and (1,0) for pure background noise. It is possible to use only one output node to flag the two states i.e. '1' for arrivals and '0' for pure background noise. However, a single output node cannot properly flag more than two output states, and so for this kind of pattern recognition, the output node should equal the desired number of output states (Haykin, 1994). Hence, two output nodes are used in the MLP for the work described here.

The number of nodes in the hidden layer was finally chosen after a process of trial-and-error using different training runs. Although this solution is considered optimum for the current application, further architecture optimization could undoubtedly be achieved by a more exhaustive search procedure using a more powerful computing system.

2.1.4. *Trained MLP Output Function*

The present analysis involves determining $F(t)$, acquiring a windowed segment of it, normalizing the segment, and finally applying it to the trained MLP. The output of the MLP is stored for subsequent analysis. The window is then shifted by one sample and the process repeated, until the end of the seismogram is reached. If a segment is the same as the training *P*-arrival segment or background noise segment, the output should be (0,1) or (1,0),

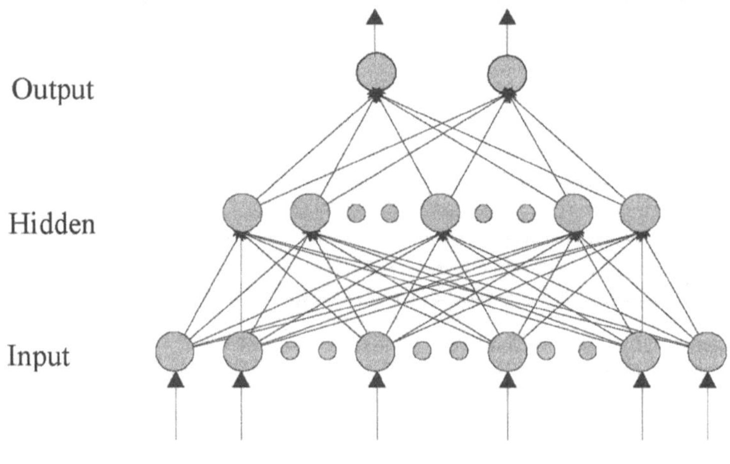

Figure 2. MLP structure for picking seismic arrivals.

respectively. In general, the output $[o_1(t), o_2(t)]$ lies between the ideal for a signal or for background noise. To provide a single indication of the onset, we define a function $N(t)$ which highlights the difference between the actual output and ideal noise,

$$N(t) = \frac{1}{2}\{[1 - o_1(t)]^2 + o_2^2(t)\} \tag{1}$$

$N(t) = 1.0$ or 0.0 if an input segment is the same as the training *P*-arrival segments or the training background noise segment, respectively. In reality, $N(t)$ is a measure of the similarity between the input segment and the training segments. A time series of $N(t)$ is obtained as the input window slides across the entire seismogram. The significant feature of the $N(t)$ function is that some peaks emerge from the smooth background. These peaks correspond to changes in $F(t)$, with magnitudes reflecting the level of change. A high-value peak indicates an abrupt change and a low-value peak indicates a smooth change, regarded as a noise burst. From the transformation point of view, the trained MLP transforms the rate of change of $F(t)$ into the value of $N(t)$, enhancing the changes and making them more easily detectable.

2.1.5. *Rules of Arrival Detection and Picking of Arrival Onset Times*

Usually, most arrivals correspond to high-value peaks of $N(t)$, and it is relatively simple to detect and pick the arrivals, and to measure their onset times. In this case, a threshold rule is sufficient to detect the arrival; if the value of $N(t)$ exceeds a threshold, it flags an arrival.

Note that as the segments of $F(t)$ are fed into the trained MLP, $N(t)$ reaches a maximum when the arrival time is at a point of the input window which corresponds to the onset point of the training *P*-arrival segment. Either side of this maximum, $F(t)$ is shifted and the MLP output $N(t)$ decreases. This implies that the onset time may be estimated by searching for a local maximum after $N(t)$ exceeds the threshold.

2.2. AUTOMATIC PICKING OF SEISMIC ARRIVALS FROM 3-C RECORDINGS

2.2.1. *Input Characteristics of the Data*

The characteristic function of seismograms for the detection and picking of arrivals from 3-C recordings, is the instantaneous vector modulus, defined as follows:

$$M(t) = \sqrt{x^2(t) + y^2(t) + z^2(t)} \tag{2}$$

where $x(t)$, $y(t)$, and $z(t)$ are the 3-C recordings. This is a useful measure, since the signals in 1-C recordings are strongly dependent on the source position and ray direction, which may otherwise give rise to a misleading interpretation. The instantaneous vector modulus $M(t)$, calculated at each individual 3-C sample along the traces, separates the geometric dependency from the seismic vector motion whilst retaining its energy characteristics for picking. $M(t)$ is then fed directly into an MLP. It is believed that this attribute facilitates an easier identification, regardless of the polarization of the wave (Lomax and Michelini, 1988).

2.2.2. *Selection of Training Data set and MLP Structure*

In general, for every problem there exists an optimum MLP, however it is usually difficult to establish. Typically, there is no fixed rule to select the structure, and it must be determined by a lengthy process of trial-and-error. For the problem presented here, the optimum MLP consisted of 30 input nodes, 10 hidden nodes and 2 output nodes. The input to the MLP was a normalized segment of vector modulus of the 3-C recordings with length 290ms (30 samples).

The most important task in training an MLP is the selection of a suitable training data set specific to its purpose. The MLP is employed to pick seismic arrivals from the background noise, so the training data should include both the seismic arrival signal and background noise signal. Although the relative modulus $M(t)$ is independent of the source position, the modulus patterns of seismic arrivals are still quite varied. Different patterns are required to train the MLP. For training, the $M(t)$ segments include either the pure background noise or the *P*-arrival with some early background noise. The *P*-arrival training segments are chosen to include wavelets with different characteristics. *S*-arrivals are not included in the training data set because $M(t)$ for these arrivals appears to exhibit similar characteristics to *P*-arrivals. Background noise segments are extracted prior to the *P*-arrivals in the same seismograms. These segments are arranged so that the predicted onset time of every *P*-arrival lies at the 11th sample. The MLP then produces a (0,1) pattern at its output nodes to flag a *P*-arrival.

The number of training patterns required is determined by the nature of the problem and the anticipated performance of the trained MLP. Although some rules give guidance on this (Fausett, 1994), experience still plays an important role in selecting a suitable number. Too small a number may result in a poorly trained MLP, and too large a number may result in long training times. In practice, it is better to train an MLP with a small training data set first, and then improve its performance by subsequently adding more training data sets. In order to balance the performance of a trained MLP for both *P*-arrivals and background noise, the same number of *P*-arrivals and background noise segments are required. The training sample should also contain the general characteristics of *P*-arrivals. Figure 3 shows the nine pairs of *P*-arrival and background noise segments used for training the MLP in this present study.

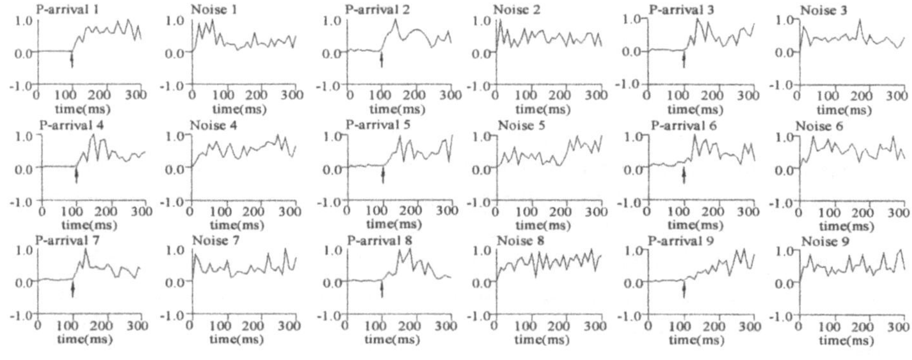

Figure 3. Nine pairs of *P*-arrivals and background noise segments used for training. Arrows on *P*-arrivals, all at the 11th sample, indicate arrival onset times. These segments are individually normalized before being fed into the MLP.

2.2.3. *Selection of Training Parameters and Initial Weights*
The second task involved in the training procedure is two-fold. First, selection of training parameters for the back-propagation training algorithm, including learning rate, momentum rate, system error threshold and pattern error threshold, and second, initialization of the weights and thresholds of the MLP. These are somewhat undetermined and their values depend on the nature of the problem. This has been investigated and reported in various publications (Dai, 1995; Dai and MacBeth, 1995,1997a,1997b). Values used in the current work are reproduced below:

Learning Rate = 0.7
Momentum Rate = 0.9
System Error Threshold = 0.00001
Pattern Error Threshold = 0.0001
Range of Initial Weights: -0.5 to +0.5.

An MLP was trained using the above selected MLP structure, training data set and training parameters. After training, the MLP was used to pick arrivals in the complete data set.

2.2.4. *Test Results*
The trained MLP was tested on 371 recordings from station DP and 391 recordings from station AY. Unfortunately, not all of them include valid seismic signals, so these data were reduced further as manual picks for comparison with the MLP results are only possible for 356 *P*-arrivals and 342 *S*-arrivals from station DP, and 300 *P*-arrivals and 285 *S*-arrivals from station AY. Although this MLP is trained only with *P*-arrivals and background noise from station DP, it can also be applied to station AY, and to pick *S*-arrivals as well, although the picking ability varies as the data quality is changed. This shows that seismic arrivals have some general characteristics i.e. this MLP, trained using *P*-arrivals has learned these general characteristics and can use them to pick other kinds of arrival, even from other stations. Figure 4 shows a comparison of the trained MLP results with the manual picks,

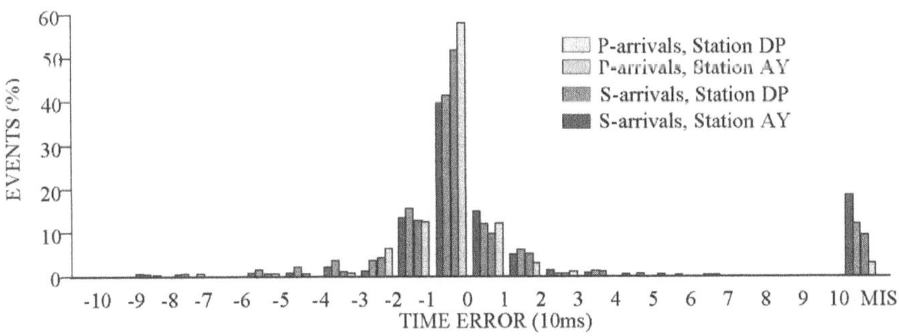

Figure 4. Statistical comparison of *P*- and *S*-wave arrivals picked using the MLP, and picked manually. A negative value of time error indicates a later pick using the MLP than that given by the manual method, whereas positive values indicate an earlier pick. MIS refers to unpicked arrivals, defined as picks with an error greater than a 10 sample (100ms) increment. The success rate of the trained MLP, relative to the manual reference picks, is quoted as a percentage.

using 0.6 as the $N(t)$ threshold.

There are in total 6% of *P*-arrivals and 14% of *S*-arrivals overlooked by the trained MLP. These usually have indistinct first-break motions so that their $N(t)$ value is lower than 0.6. However, most of them have clear peaks in $N(t)$ corresponding to onsets from a smooth background. Reducing the picking threshold of $N(t)$ might enable the picking of more arrivals. For example, as the threshold is reduced to 0.5, the MLP can pick 10 more *P*-arrivals and 32 more *S*-arrivals. Note that once picked, the arrival onset time is insensitive to the $N(t)$ threshold.

2.3. AUTOMATIC PICKING OF SEISMIC ARRIVALS FROM 1-C RECORDINGS

The above MLP approach showed a good performance in picking seismic arrivals from 3-C recordings. Unfortunately, not all seismic stations have 3-C seismometers, or can provide consistently high quality 3-C recordings, due to the failure of one or more components. Hence, there is a practical requirement for picking seismic arrivals from 1-C recordings. We adapted the above MLP approach to pick seismic arrivals from either vertical or horizontal recordings. This was achieved using the absolute value of 1-C recordings as the MLP input.

2.3.1. *Input Characteristics of 1-C Seismic Recordings*

Despite the fact that complete polarization and propagation information cannot be obtained from 1-C recordings which are strongly dependent on the source position and ray direction, the arrival of a seismic signal may still be observed on each separate 1-C recording by changes in the amplitude, frequency, and polarization characteristics. These can be obtained by determining different characteristic functions. Several basic characteristic functions of 1-C recordings can be used for this purpose. Here we chose the relative absolute value function (Anderson, 1978) as the input to the MLP, since it has high-fidelity and processing speed and is objective. The 1-C recording was not used itself, because the first motion of an arrival has two directions (up and down) and is source-dependent. Usually, the vertical component (V-C) recordings are used for picking arrivals in many other methods, so these were used here as a first attempt.

2.3.2. *MLP Structure and Training Data Set*

After a process of trial-and-error, a three-layer MLP was selected with 40 nodes in the input layer, 10 nodes in the hidden layer, and two nodes in the output layer. The input was a normalized segment of absolute values of the V-C recordings of local earthquake data, with length 390ms (40 samples).

The training data set was re-selected from the V-C recordings of local seismic data because the onset times of *P*-arrivals in V-C recordings are sometimes different from those of the vector modulus. Finally, 10 pairs of *P*-arrival and background noise segments (Figure 5) from station DP were used to train the MLP.

2.3.3. *Test Results on a Complete Data Set*

After optimizing the MLP's structure and training data set, the trained MLP was applied to the same data set. Due to the ray paths involved, some *P*-arrivals and some *S*-arrivals were

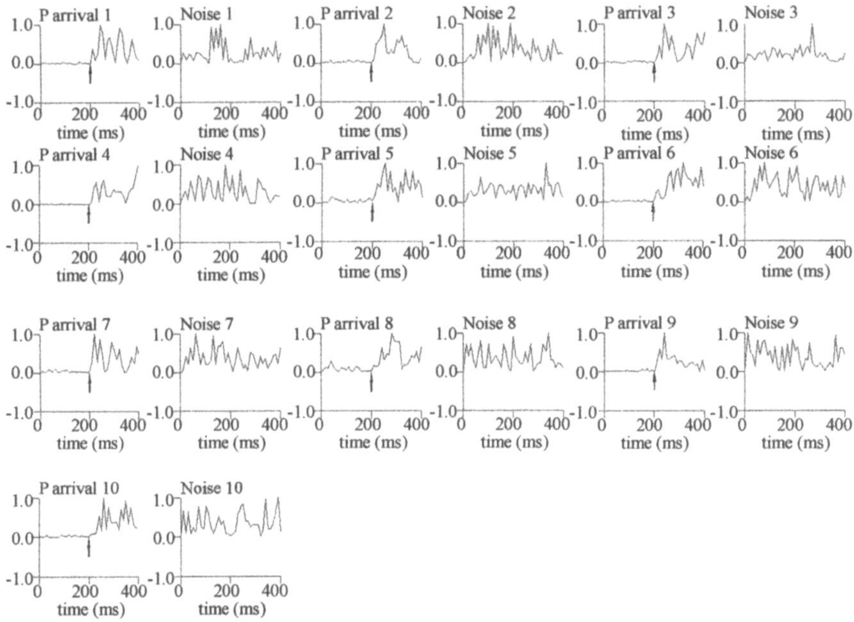

Figure 5. Ten pairs of *P*-arrivals and background noise segments selected from vertical components for training. Arrows, all at the 21st sample, indicate arrival onset times.

low amplitude or absent completely from 1-C recordings, especially for the smaller events, and some arrivals had different onset times for different 1-C recordings due to phase changes. Manual picking of the *P*- and *S*-arrival onset times was carried out for three 1-C recordings, and these were used as a reference to compare with the MLP's picking results. The trained MLP was used to deal with three 1-C recordings, respectively.

Figure 6 shows the statistics of the time accuracy of the MLP-picked onset times compared with manual picks. Since picked onset times are insensitive to the picking threshold, only the results with a threshold value of 0.6 are shown. These figures show that stations DP and AY have a different performance. Detection and picking results from station AY are poorer than those from station DP. This is largely a consequence of the data quality and different characteristics. The recordings for station DP usually have stronger energy and higher SNR than those for station AY. The arrivals for station DP are much clearer than for station AY, so the arrival times picked by the MLP for station DP are more accurate than those for station AY. The source orientation has a larger effect on the picking of *S*-arrivals compared to *P*-arrivals. It is difficult to say which component produces the best performance. The above results suggest that a trained MLP can be applied to all three components individually. It should then be possible to combine and compare results from each component, for which some arrivals may be absent, to obtain both *P*- and *S*-arrival onset times. Even if one of the three components fails, the other two generally give useful results. The arrival times picked by the MLP for three 1-C recordings are usually different, which is thought to be due to the effect of ray path or variations in the inhomogeneous upper crust.

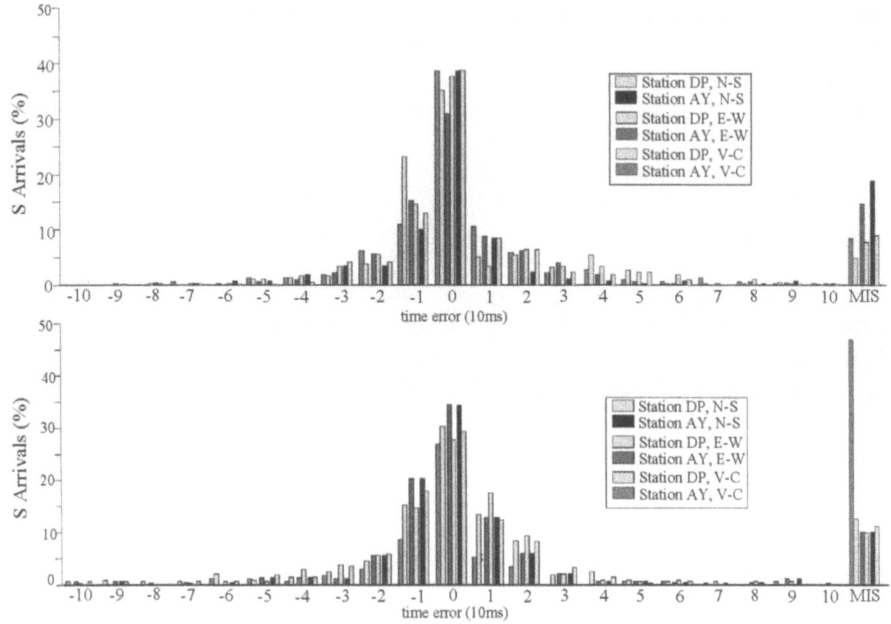

Figure 6. Statistical comparison of *P*- and *S*-wave arrivals for V-C, N-S and E-W components (upper and lower diagram, respectively) picked using the MLP, and picked manually. Notation as in Figure 4.

2.4. DISCUSSION

2.4.1. *Comparison of Results from 3-C and 1-C Recordings*

Comparing the picking results from the 3-C and three 1-C recordings, it can be seen that the 3-C picking has obviously produced a better performance, being more accurate than the 1-C picking. Visually checking the seismograms shows that the arrival onset times picked from the 3-C modulus are closer to the manual picks than those from the 1-C recordings.

2.4.2. *Sensitivity to Segment Length*

The performance and time duration of the MLP approach depends on segment length, which relates to the number of nodes in the input layer. Reducing the number of input nodes to 20 or increasing them to 40 dramatically affects the performance, and there appears to be an optimum number of input nodes for the particular configuration reported here. It is suggested that the segment length should include several complete cycles of a wavelet. This reflects the general observation that the MLP architecture must be tailored specifically to individual applications.

2.4.3. *Effect of Post-Processing*

In order to reduce false picks, post-processing was employed to discard small noise bursts and spikes (Dai, 1995; Dai and MacBeth, 1995, 1997a). This post-processing discarded the majority of small noise bursts and spikes. Although post-processing does not directly improve the ability to pick *P*- and *S*-arrivals, $N(t)$ thresholding can be decreased so that more seismic arrivals are picked, with most false alarms being discarded using post-processing. Other types of noise can be discarded using similar procedures if their particular characteristics can be adequately described.

3. Identification of Seismic Arrival Types Using MLPs

Following arrival picking is arrival identification, which aims to classify individual arrivals into categories related to their amplitude, frequency, polarization and propagation direction. This is more difficult than arrival picking due to the complexity of the signals. Although some methods have been developed (see Dai, 1995; Dai and MacBeth, 1997b, for extensive list), there is no general method to identify *P*- and *S*-arrivals simultaneously, and automation is still an unresolved issue. Due to the intensive amount of pattern recognition involved, a natural candidate for automated identification is the artificial neural network method discussed in this section.

3.1. DEGREE OF POLARIZATION

One of the obvious features of distinguishing *P*- and *S*-arrivals is their polarization direction, which is either parallel or perpendicular to the propagation direction, respectively, for *P*- or *S*-arrivals in an isotropic medium. It would appear simple to identify *P*- and *S*-arrivals using 3-C recordings. However, it is impractical to use this directly since it is related to the event location, which is generally unavailable prior to analysis. Another feature of distinguishing *P*- and *S*-arrivals is their polarization state. In general, it is observed that the direct *P*-arrival is predominantly linearly polarized, whilst arrivals following such as *S*-arrivals, have considerably more complicated polarization patterns involving phase shifts (Basham and Ellis, 1969). The polarization state can be measured using the degree of polarization (DOP), which is independent of the source-receiver direction. DOP is calculated from the covariance of the 3-C recordings (Samson, 1977; Cichowicz, 1993). The covariance matrix is defined as,

$$\mathbf{C} = \begin{Bmatrix} Cov(x,x) & Cov(x,y) & Cov(x,z) \\ Cov(y,x) & Cov(y,y) & Cov(y,z) \\ Cov(z,x) & Cov(z,y) & Cov(z,z) \end{Bmatrix} \tag{3}$$

where the covariance is measured for N samples as follows,

$$Cov(x,y) = \frac{1}{N}\sum_{i=1}^{N}(x_i - \bar{x})(y_i - \bar{y}) \tag{4}$$

where $\bar{x}$ and $\bar{y}$ are the average values of x and y. N is determined from the predominant signal frequency (Cichowicz, 1993), which is ten samples in this case.

Samson (1977) defines DOP as follows,

$$\mathrm{DOP} = \frac{(\lambda_1-\lambda_2)^2+(\lambda_2-\lambda_3)^2+(\lambda_3-\lambda_1)^2}{2(\lambda_1+\lambda_2+\lambda_3)^2} = \frac{3trS^2-(trS)^2}{2(trS)^2} \tag{5}$$

where λ_1, λ_2, and λ_3 are eigenvalues of the covariance matrix of a moving window; *trS* is defined as $\lambda_1 + \lambda_2 + \lambda_3$ (i.e. the trace of **C**), and trS^2 is defined as $\lambda_1{}^2 + \lambda_2{}^2 + \lambda_3{}^2$ (i.e. the trace of $\mathbf{C}^2$). DOP is independent of the co-ordinate system.

For a linearly polarized wave, DOP is unity, and for a completely unpolarized or circularly polarized wave it is zero. Cichowicz (1993) pointed out that both *P*- and *S*-wave arrivals exhibit a high degree of linear polarization, but the *P*-wave coda manifests itself as generally elliptical polarization with a significantly lower value of DOP. For real data, although the *S*-arrival is usually associated with a far larger value of DOP than that of the *P*-wave coda, it does not reach a value of one. The variation of this quantity along the seismogram forms the pattern which may indicate the wave type. Figure 7 shows an example in which DOP is of high amplitude for a *P*-arrival, a median-level value for an *S*-wave, and a low-level pattern for a noise burst.

It is important to note that this particular definition of DOP does not consider the signal amplitude. To accommodate amplitude information, the following modified function is defined,

$$MF(t) = DOP(t) \times \overline{M}(t) \tag{6}$$

where $\overline{M}(t)$ is a smoothed relative function of the modulus $M(t)$ of the 3-C recording

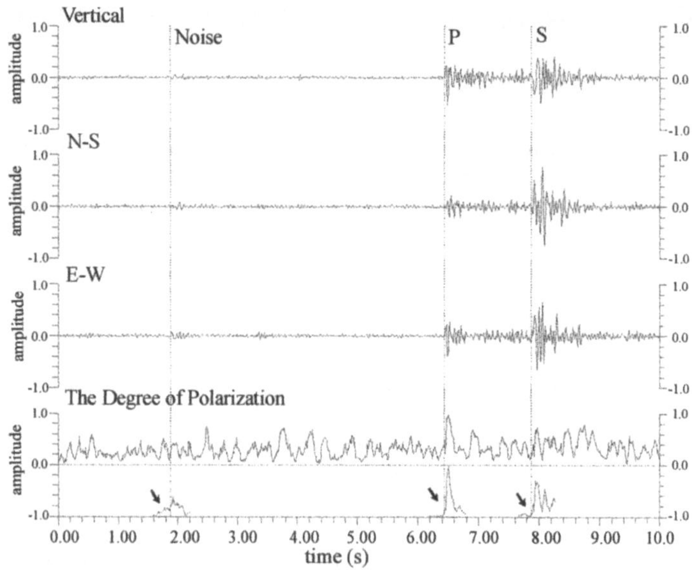

Figure 7. The degree of polarization (lowest trace) determined from a 3-C seismogram (three upper traces). Three vertical lines indicate the arrival onset times of the noise burst, *P*-arrival and *S*-arrival. The degree of polarization (DOP) has a lower level for the noise burst, a higher level for the *P*-arrival and a mid-range value for the *S*-arrival. Their modified DOPs are shown below them (arrowed).

within a window. This value is also independent of the source-receive direction. The normalization factor is calculated using the window between the onset point and the following ten points. Note that *MF*(*t*) and *DOP(t)* have slightly different patterns (Figure 7). The *MF*(*t*) patterns are complex and identifying them requires an intensive amount of pattern recognition for which the MLP is suitable. *MF(t)* is presented to the MLP in a segment selected from a window in which the arrival onset-time is at the centre.

3.2. METHODOLOGY

Figure 8 illustrates a flow chart for identifying arrival types using an MLP. This stage follows arrival picking. Unlike the MLP picking approach which uses an MLP as a filter to deal with the entire seismic trace, only arrival segments are input into the identification MLP. An arrival segment is selected by its onset time, positioned at the centre of the segment, and this *MF*(t) segment is then fed into an MLP for training and testing.

Due to picking errors, the onset time may not be exactly at the centre of the segment and this can affect the MLP output. An adjustment of the onset time is necessary to avoid this effect. For each *MF*(t) segment, the first local maximum after the onset time is set at the centre of the segment. The *MF*(t) segment has a length of 60 samples (590*ms*) which is chosen to include the complete *MF*(*t*) pattern of an arrival.

The MLP used in this approach has three layers. The input layer consists of 60 nodes, corresponding to an *MF*(t) segment with a fixed length of 60 samples. There are three nodes in the output layer to flag the result: (1,0,0) for a noise burst, (0,1,0) for a *P*-arrival, and (0,0,1) for an *S*-arrival (used during training). Ten nodes were chosen for the hidden layer, after a process of trial-and-error with different training runs.

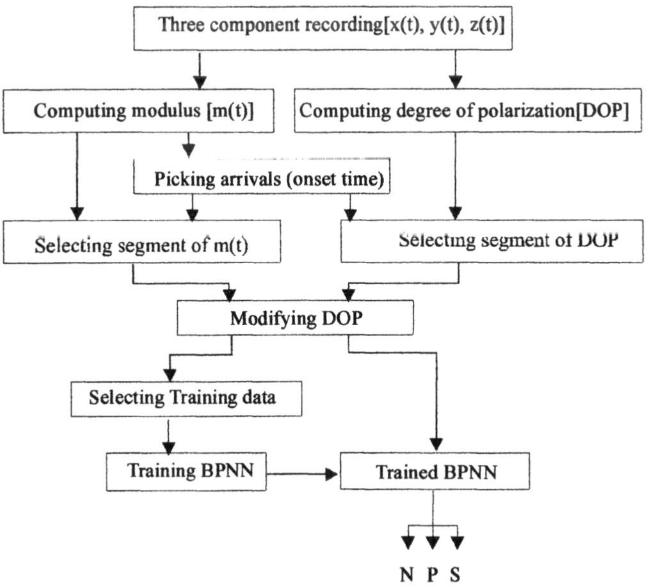

Figure 8. Flowchart illustrating the methodology for identifying arrival type using an MLP.

Using this procedure, only a small number of recordings from DP were used to train the MLP. *P*- and *S*-waves with similar *MF*(*t*) patterns were avoided in the training data set because the MLP cannot distinguish between them and the training procedure might not converge. At the commencement of training, only three *MF*(*t*) segments were selected, representing each of the three categories (noise burst, *P*-arrival and *S*-arrival), with the desired output (1,0,0; 0,1,0; 0,0,1), respectively. After training, this MLP was used to process other data. Using manual analysis results, another three segments incorrectly identified by the MLP were combined with the former training segments, to re-train it. This procedure was repeated until the MLP performance could not be improved by further additions to the training data set.

As each *MF*(*t*) segment is fed into the trained MLP, it outputs three values: o_1, o_2, and o_3. For training segments, the output should be equal to the desired ones i.e. (1,0,0) for a noise burst, (0,1,0) for a *P*-arrival, and (0,0,1) for an *S*-arrival. For non-training segments, the output (o_1, o_2, and o_3) is a measure of similarity between a new segment and the group characteristics defined by the training segments. If non-training and training segments are similar, the MLP output will be close to the corresponding desired output. For identification of segment types, the maximum of the three outputs (o_1, o_2, o_3) was employed. Normally, the trained MLP should have a high output and two low outputs. However, sometimes two or even three high outputs or three low outputs are obtained. This implies that these segments are markedly different from the training segments, and in such cases, the maximum of the three outputs is used to identify the corresponding wave type.

3.3. TEST RESULTS

3.3.1. *Station DP*

For the data from station DP, 345 *P*-arrivals, 302 *S*-arrivals and 174 noise bursts were picked using the MLP arrival picker applied to 371 recordings. An MLP for arrival identification was then trained using nine groups of *MF*(t) segments from noise bursts, *P*- and *S*-arrivals (Figure 9). Table 1a displays the overall performance of this trained MLP, which has a better performance in identifying *P*-arrivals than *S*-arrivals and noise bursts.

3.3.2. *Station AY.*

For the data from station AY, 274 *P*-arrivals, 240 *S*-arrivals and 28 noise bursts were automatically picked from 391 recordings. The MLP described in the above section was applied to them. Table 1a displays the overall performance. Unfortunately, it only correctly identifies 42% of the *P*-arrivals, 39% of the *S*-arrivals and 32% of the noise bursts. This is due to the *MF*(*t*) patterns of seismic arrivals being different between data from stations DP and AY, even for the same arrivals from the same event. Such differences cause the MLP to fail in correctly identifying these. For example, some recordings have *P*-arrivals with low to mid-level values of *MF*(*t*), and *S*-arrivals with high-level values of *MF*(t), which are quite opposite to the trained waves. Hence, to deal successfully with the data from station AY, it is necessary to train the MLP using data from the station itself.

The best performance is obtained when the MLP is trained for station AY using the five groups shown in Figure 10. A much improved performance is obtained with this MLP, as

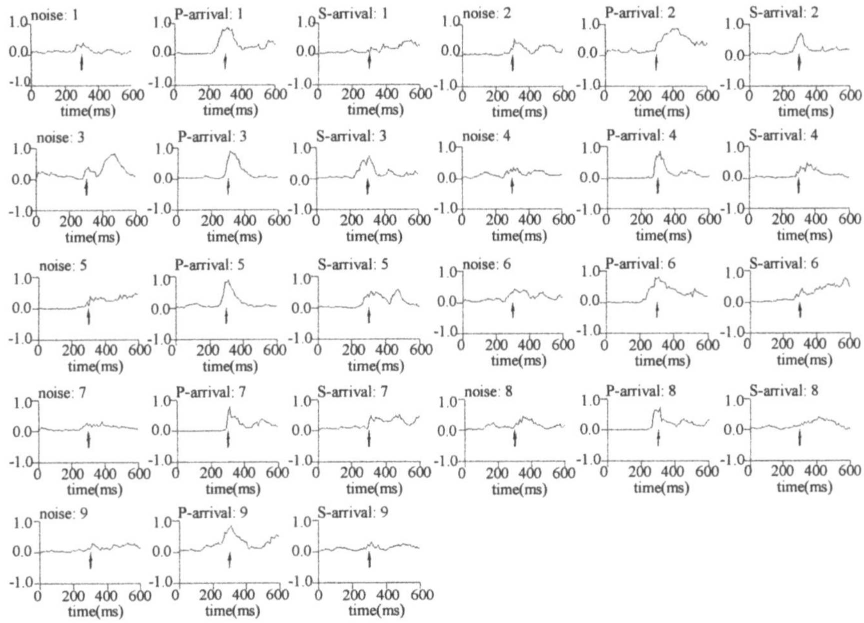

Figure 9. Nine groups of *MF*(t) segments of noise bursts, *P*-arrivals and *S*-arrivals, used for training an MLP for arrival identificiation. Arrows on segments indicate the pre-picked onset times for those arrivals which all lie at the 31st sample.

evidenced by Table 1b. This MLP was also used to test the data from station DP. However, although it can identify 85% of the *P*-arrivals, the 36% success rate for the *S*-arrivals is quite low. Note that the figures for identification of noise bursts may be statistically insignificant, since only 28 bursts were tested.

TABLE 1. Performance of the trained MLP for arrival identification.

(a) MLP (60 input nodes) trained using nine groups of training segments from station DP.

	DP			AY		
	P(345)	S (302)	Noise (174)	P (274)	S (240)	Noise(28)
Identified P	82% (284)	22% (67)	9% (16)	42% (115)	49% (117)	43% (12)
Identified S	11% (36)	63% (189)	42% (75)	44% (120)	39% (93)	25% (7)
Identified N	7% (25)	15% (46)	48% (83)	14% (39)	12% (30)	32% (9)

(b) MLP (60 input nodes) trained using five groups of training segments from station AY.

	DP			AY		
	P(345)	S (302)	Noise (174)	P (274)	S (240)	Noise(28)
Identified P	85% (293)	45% (137)	30% (52)	77% (210)	22% (53)	21% (6)
Identified S	8% (28)	36% (109)	18% (31)	13% (35)	60% (145)	47% (13)
Identified N	7% (24)	19% (56)	52% (91)	11% (31)	18% (42)	32% (9)

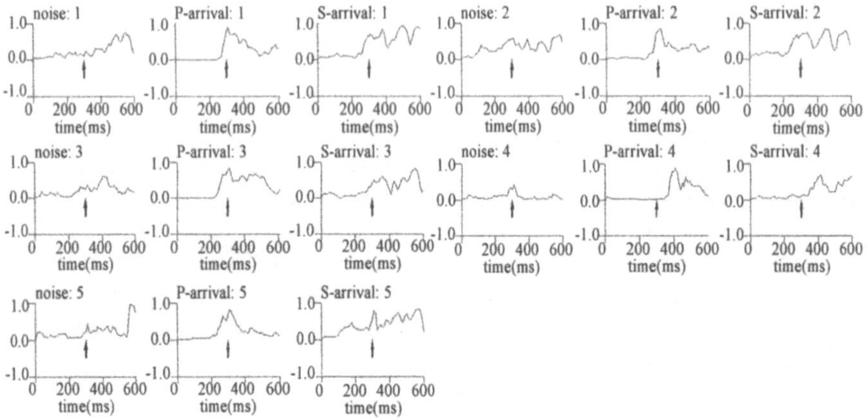

Figure 10. Five groups of *MF*(*t*) segments of noise bursts, *P*-arrivals and *S*-arrivals from station AY, used for training. Arrows on segments indicate the pre-picked onset times for those arrivals which all lie at the 31st sample.

3.4. DISCUSSION

3.4.1. *Effect of Changing Training Data Set*

Since the performance of the MLP is critically dependent on the training data set, an investigation of the sensitivity to the training data set from station DP, is obviously beneficial. As the number of training segments increases, identification of *P*-arrivals improves, but identification of *S*-arrivals and noise bursts becomes worse. This appears consistent with the observation that the *MF*(*t*) patterns for *P*-arrivals are more typically alike, but *MF*(*t*) patterns for *S*-arrivals are often quite different. In addition, it is noted that some *P*-arrivals, *S*-arrivals and noise bursts also have similar *MF*(*t*) patterns. If such a *P*-arrival pattern is used to train the MLP, the network will classify all of them as *P*-arrivals, irrespective of their actual status.

3.4.2. *Effect of Changing Input Nodes.*

The sensitivity to input segment length (number of input nodes) was also investigated. Between 50 and 70 input nodes were tested, but with the same number of hidden and output nodes. The training procedure was identical: one group of training segments, increasing finally to nine groups. The training segments were different due to their different performance at every training stage, but all were from station DP. The MLP with 60 input nodes had the optimum performance. This suggests that segments should include appropriate information from an arrival, otherwise too much or too little information will degrade the MLP performance.

4. Summary and Conclusions

Two MLP methodologies have been developed to pick and identify seismic arrivals automatically for local earthquake data. Compared to other conventional methods (Dai and MacBeth, 1995, 1997a, 1997b), both approaches show encouraging results.

A significant feature of using MLPs is their adaptiveness. The same programs have been used to pick arrivals from 3-C and 1-C recordings and to identify arrival types, with only minor changes of the MLP structure and the input/output. Once the generic MLP paradigms have been developed, it is a simple matter to apply them to the particular problem in hand. The only necessary step is to select suitable training examples for any new application. A trained MLP's performance is objective and consistent and can be easily improved by adjusting the training data set.

The limitation of using MLPs is in finding an optimum architecture for a particular application, because no appropriate theory is currently available. An MLP's performance depends implicitly upon the training data, and its interpretation ability cannot lie too far outside its experience. This approach works well if testing segments are similar to the training segments, otherwise it fails. In fact, this limitation is a consequence of the supervised learning scheme which is employed.

The MLP has great potential for analysing earthquake seismology data automatically. In principle, there is essentially no difficulty in developing a complete automatic seismic analysis system based on the two approaches described here.

Acknowledgements

This work is published with the approval of the Director of the British Geological Survey. Thanks are due to Blackwell Science Ltd, The American Geophysical Union, and Elsevier Sciences B.V., for granting permission to re-use materials from Dai and MacBeth (1995, 1997a, 1997b).

References

Anderson, K. R., 1978, Automatic processing of local earthquake data: PhD thesis, Massachusetts Institute of Technology, Cambridge, Massachusetts, USA,

Basham, P. W., and Ellis, R. M., 1969, The composition of P-code using magnetic tape seismograms: Bull., Seis. Soc. Am., **59**, 473-486.

Cichowicz, A., 1993, An automatic S-phase picker: Bull., Seis. Soc. Am., **83**, 180-189.

Dai, H. C., 1995, Analysis of local earthquake data using artificial neural networks: Ph.D. thesis, Univ. of Edinburgh, Scotland.

Dai, H. C., and MacBeth, C., 1995, Automatic picking of seismic arrivals in local earthquake data using an artificial neural network: Geophys. J. Internat., **120**, 758-774.

Dai, H. C., and MacBeth, C., 1997a, The application of back-propagation neural network to automatic picking seismic arrivals from single-component recordings: J. Geophys. Res., **102**, 15105-15113.

Dai, H. C., and MacBeth, C., 1997b, Application of back-propagation neural network to identification of seismic arrival types: Phys. Earth Plan. Int., **101**,177-188.

Fausett, L. V., 1994, Fundamentals of neural networks: architectures, algorithms and applications: Prentice-Hall, Inc.

Haykin, S., 1994, Neural networks, a comprehensive foundation: Macmillan College Publ. Co.

Lomax, A. J., and Michelini, A., 1988, The use of spherical coordinates in the interpretation of seismograms: Geophys. J. Internat., **93**, 405-412.

Lovell, J. H., 1989, Source parameters of a microearthquake swarm in Turkey: M.Sc. thesis, Univ. of Edinburgh, Scotland.

Rumelhart, D., Hinton, G. E., and Williams, R. J., 1986, Learning representations by back-propagating errors: Nature, **323**, 533-536.

Samson, J. C., 1977, Matrix and Stokes vector representations of detectors for polarized waveforms: theory, with some applications to teleseismic waves: Geophys. J. Roy. Astr. Soc., **51**, 433-452.

BIBLIOGRAPHY

A. ARTIFICIAL NEURAL NETWORKS AND FUZZY LOGIC

1. BOOKS

Abbott, L., and Sejnowski, T. J., 1999, Neural codes and distributed representations: foundations of neural computation: M.I.T. Press.

Anderson, J. A., 1995, An introduction to neural networks: M.I.T. Press.

Anderson, J., and Rosenfeld, E., 1989, Neurocomputing: Foundations of Research: M.I.T. Press.

Bischof, H., 1995, Pyramidal Neural Networks: Lawrence Erlbaum.

Bishop, C. M., 1996, Neural networks for pattern recognition: Oxford Univ. Press, Inc.

Bojadziev, G., and Bojadziev, M., 1996, Fuzzy sets, fuzzy logic, applications: World Scientific.

Bouchon-Meunier, B., Yager, R. R., and Zadeh, L. A., 1995, Fuzzy logic and soft computing: World Scientific.

Cichocki, A., and Unbehauen, R., 1993, Neural Networks for Optimization and Signal Processing: John Wiley & Sons, Inc.

Cloete, I., and Zurada, J. M., 2000, Knowledge-based neurocomputing: M.I.T. Press.

Dayhoff, J. E., 1990, Neural Network Architectures: An Introduction: Van Nostrand Reinhold.

Fausett, L. V., 1994, Fundamentals of neural networks; architectures, algorithms and applications: Prentice-Hall, Inc.

Freeman, J. A., 1994, Simulating Neural Networks: Addison-Wesley Publ. Co.

Freeman, J. A., and Skapura, D. M., 1991, Neural Networks, Algorithms, Applications and Programming Techniques: Addison-Wesley Publ. Co.

Girolami, M., 1999, Self-organising neural networks: independent component analysis and blind source separation: Springer Pub. Co., Inc.

Golden, R. M., 1997, Mathematical methods for neural network analysis and design: M.I.T. Press.

Grossberg, S., Kawato, M., and Taylor, J., 1994, Neural networks: Elsevier Science Publ. Co., Inc.

Hagan, M. T., Demuth, H. B., and Beale, M., 1995, Neural Network Design: Prindle, Webber & Schmidt.

Hassoun, M. H., 1995, Fundamentals of artificial neural networks: M.I.T. Press.

Haykin, S., 1999, Neural networks; a comprehensive foundation: Prentice-Hall, Inc.

Hecht-Nielson, R., 1990, Neurocomputing: Addison-Wesley Publ. Co.

Heinke, D., Humphreys, G. W., and Olson, A., 1999, Connectionist models in cognitive neuroscience: Springer Pub. Co., Inc.

Hertz, J., Krogh, A., and Palmer, R. G., 1991, Introduction to the Theory of Neural Computation: Addison-Wesley Publ. Co.

Hinton, G., and Sejnowski, T. J., 1999, Unsupervised learning:foundations of neural computation: M.I.T. Press.

Hirota, K., and Sugeno, M., 1995, Industrial applications of fuzzy technology in the world: World Scientific.

Hu, Y. H., and Hwang, J. N., 2001, Handbook of neural network signal processing: CRC Press.

Husmeier, D., 1999, Neural networks for conditional probability estimation: forecasting beyond point predictions: Springer Pub. Co., Inc.

International Fuzzy Systems Association (IFSA), 1978, Fuzzy Sets & Systems: Elsevier Science Publ. Co., Inc.

Jain, L. C., 1999, Fusion of neural networks, fuzzy systems and genetic algorithms: CRC Press.

Jain, L. C., and Fanelli, A., 2000, Recent advances in artificial neural networks: design and applications: CRC Press.

Jang, J. S. R., Sun, C. T., and Mizutani, E., 1997, Neuro-fuzzy and soft computing: Prentice-Hall, Inc.

Kasabov, N. K., 1996, Foundations of neural networks, fuzzy logic, and knowledge engineering: M.I.T. Press.

Kay, J. W., and Titterington, D. M., 2000, Statistics and neural networks: advances at the interface: Oxford Univ. Press, Inc.

Khanna, T., 1990, Foundations of Neural Networks: Addison-Wesley Publ. Co.

Klir, G., and Folger, T., 1988, Fuzzy Sets, Uncertainty, and Information: Prentice-Hall, Inc.

Klir, G. J., and Yuan, B., 1996, Fuzzy sets, fuzzy logic, and fuzzy systems: selected papers by Lofti A Zadeh: World Scientific.

Kohonen, T., 1989, Self Organisation and Associative Memory: Springer-Verlag.

Kohonen, T., 1997, Self-Organizing Maps: Springer Pub. Co., Inc.

Kosko, B., 1991, Fuzzy Logic and Neural Networks: Prentice-Hall, Inc.

Kosko, B., 1997, Fuzzy engineering: Prentice-Hall, Inc.

Kung, S. Y., 1995, Digital Neural Networks: Prentice-Hall, Inc.

Leondes, C. T., 1999, Fuzzy theory systems: techniques and applications: Academic Press Inc.

Lin, C. T., and Lee, C. S. G., 1996, Neural fuzzy systems: a neuro fuzzy synergism to intelligent systems: Prentice-Hall, Inc.

Lindblad, T., and Kinser, J. M., 1997, Image processing using pulse-coupled neural networks: Springer Pub. Co., Inc.

Looney, C. G., 1997, Pattern recognition using neural networks: theory and algorithms for engineers and scientists: Oxford Univ. Press, Inc.

Luo, F. L., and Unbehauen, R., 1999, Applied neural networks for signal processing: Cambridge Univ. Press.

Maass, W., and Bishop, C. M., 1999, Pulsed neural networks: M.I.T. Press.

MacIntyre, J., 1994, Neural computing and applications: Springer Pub. Co., Inc.

Maren, A., Hartston, C., and Pap, R., 1990, Handbook of Neural Computing Applications: Academic Press Inc.

Masters, T., 1993, Practical Neural Network Recipes in C++: Academic Press Inc.

Medsker, L. R., and Jain, L. C., 2000, Recurrent neural networks: design and applications: CRC Press.

Mehrotra, K., Mohan, C. K., and Ranka, S., 1996, Elements of artificial neural networks: M.I.T. Press.

Müller, B., Reinhardt, J., and Strickland, M., 1995, Neural Networks: An Introduction: Springer-Verlag.

Neelakanta, P. S., 1999, Information-theoretic aspects of neural networks: CRC Press.

Pandya, A. S., and Macy, R. B., 1996, Pattern recognition with neural networks in C++: CRC Press.

Pao, Y., 1989, Adaptive Pattern Recognition and Neural Networks: Addison-Wesley Publ. Co.

Pedrycz, W., and Gomide, F., 1998, An introduction to fuzzy sets: analysis and design: M.I.T. Press.

Perlovsky, L. I., 2000, Neural networks and intellect: using model-based concepts: Oxford Univ. Press, Inc.

Principe, J. C., Euliano, N. R., and Lefebvre, W. C., 1999, Neural and adaptive systems: fundamentals through simulation: John Wiley & Sons, Inc.

Reed, R. D., and Marks II, R. J., 1999, Neural smithing: supervised learning in feedforward artificial neural networks: M.I.T. Press.

Ripley, B. D., 1996, Pattern Recognition and Neural Networks: Cambridge Univ. Press.

Sanchez, E., Shibata, T., and Zadeh, L. A., 1996, Genetic algorithms and fuzzy logic systems: soft computing perspectives: World Scientific.

Sharkey, A. J. C., 1999, Combining artificial neural nets: ensemble and modular multi-net systems: Springer Pub. Co., Inc.

Shepherd, A. J., 1997, Second-order methods for neural networks: fast and reliable training methods for multi-layer perceptrons: Springer Pub. Co., Inc.

Simpson, P., 1990, Artificial Neural Systems: Foundations, Paradigms, Applications, and Implementations: Pergamon Press, Inc.

Simpson, P. K., 1995, IEEE Technology Update Series: Neural Networks: IEEE Press.

Tarassenko, L., 1998, A guide to neural computing applications: Arnold Publishers.

Veelenturf, L. P. J., 1995, Analysis and applications of artificial neural networks: Prentice-Hall, Inc.

Wang, P. Z., and Loe, K. F., 1994, Between mind and computer: fuzzy science and engineering: World Scientific.

Wasserman, P. D., 1989, Neuralcomputing: Theory and Practice: Van Nostrand Reinhold.

Wasserman, P. D., 1993, Advanced Methods in Neural Computing: Van Nostrand Reinhold.

White, H., 1992, Artificial Neural Networks: Approximation and Learning Theory: Blackwell Scientific Publications, Inc.

Yen, J., and Langari, R., 1998, Fuzzy logic: intelligence, control, and information: Prentice-Hall, Inc.

Zadeh, L. A., 1994, Advances in fuzzy systems: applications and theory: World Scientific.

Zimmermann, H. J., 1985, Fuzzy Set Theory and its Applications: Kluwer Academic Publ.

Zurada, J. M., 1992, Introduction to artificial neural systems: West Publ. Co.

2. REVIEW ARTICLES

Lippmann, R. P., 1987, An introduction to computing with neural nets: IEEE Acoust., Speech, and Sig. Proc. Mag., 4, 4-22.

Lippmann, R. P., 1989, Pattern classification using neural networks: IEEE Comms. Mag., 27, 47-64.

B. GEOPHYSICAL APPLICATIONS OF ARTIFICIAL NEURAL NETWORKS AND FUZZY LOGIC: JOURNAL, CONFERENCE PAPERS; BOOK CHAPTERS; THESES AND REPORTS

1. OVERVIEWS

Braunschweig, B., 1992, Artificial intelligence and the oil and gas industry: a variety of methods for a variety of applications: Rev. Inst. Fran. Petr., **47**, 299-304.

Calderon, C., Sen, M., and Stoffa, P., 2000, Artificial neural networks for parameter estimation in geophysics: Geophys. Prosp., **48**, 21-48.

Cary, P., and Upham, W., 1992, An evaluation of neural networks: The Leading Edge, **11**, 45-47.

Clare, A., Bolt, G., and Daneel, G., 1997, The use of neural analysis as an exploration tool: Proc. 12th Internat. Conf. Appl. Geologic Sensing, **2**, 199-205.

Day, R., and McCormack, M. D., 1993, Evolution of artificial intelligence in the petroleum industry: 63rd Ann. Internat. Mtg., Soc. Expl. Geophys., Expanded Abstracts, WS7.1, 1394.

Fogelman-Souli, F., 1992, Neural networks and their applications for the oil industry: Rev. Inst. Fran. Petr., **47**, 383-392.

McCormack, M. D., 1991, Neural computing in geophysics: The Leading Edge, **10**, 11-15.

McCormack, M. D., 1991, Neural networks in the petroleum industry: 61st Ann. Internat. Mtg., Soc. Expl. Geophys., Expanded Abstracts, RE2.1, 728-731.

Paola, J. D., and Schowengerdt, R. A., 1995, A review and analysis of back propagation neural networks for classification of remotely-sensed multi-spectral imagery: Internat. J. Remote Sensing, **16**, 3033-3058.

Raiche, A., 1991, A pattern recognition approach to geophysical inversion using neural nets: Geophys. J. Internat., **105**, 629-648.

Sandham, W. A., and Leggett, M., 1997, Geophysical applications of artificial neural networks and fuzzy logic: Overview and survey, 59th Ann. Internat. Mtg., Eur. Assn. Geosci. Eng., Pre-Conference Workshop on Geophysical Applications of Artificial Neural Networks and Fuzzy Logic, Paper 0.1.

Sandham, W. A., and Leggett, M., 1998, Geophysical applications of artificial neural networks and fuzzy logic: First Break, **16**, 58-60.

Tyapkin, O. K., and Troyan, J. G., 1998, The application of fuzzy sets in rock physics and exploration geophysics: 60th Ann. Internat. Mtg., Eur. Assn. Geosci. Eng., Extended Abstracts, R001.

Van der Baan, M., and Jutten, C., 2000, Neural networks in geophysical applications: Geophysics, **65**, 1032-1047.

2. PATTERN RECOGNITION

Bolt, G., Daneel, G., and Clare, A., 1997, Prospect explorer: an exploration neural analysis tool: 67th Ann. Internat. Mtg., Soc. Expl. Geophys., Expanded Abstracts, WS 1.2, 2041-2044.

Dai, H. C., and MacBeth, C., 1995, Automatic picking of seismic arrivals in local earthquake data using an artificial neural network: Geophys. J. Internat., **120**, 758-774.

FitzGerald, E. M., and Bean, G. J., 2000, Application of artificial neural networks (ANNs) to sub-basalt imaging: 70th Ann. Internat. Mtg., Soc. Expl. Geophys., Expanded Abstracts, ST P1.10, 2452-2455.

Hansen, K. V., 1993, Neural networks for primary reflection identification: 63rd Ann. Internat. Mtg., Soc. Expl. Geophys., Expanded Abstracts, CT4.4, 242-245.

Huang, K. Y., 1999, Neural network for robust pattern recognition in real seismic data: 69th Ann. Internat. Mtg., Soc. Expl. Geophys., Expanded Abstracts, SINT 2.7, 946-949.

Huang, K. Y., Liu, W. H., and Chang, I. C., 1989, Hopfield model of neural networks for detection of bright spots: 59th Ann. Internat. Mtg., Soc. Expl. Geophys., Expanded Abstracts, 444-446.

Huang, K. Y., and Liaw, J. Y., 1992, Neocognitron of a neural network for seismic pattern recognition: 62nd Ann. Internat. Mtg., Soc. Expl. Geophys., Expanded Abstracts, AI1.8, 26-29.

Huang, Z., and Williamson, M. A., 1994, Geological pattern recognition and modelling with a general regression neural network: J. Can. Soc. Expl. Geophys., **30**, 60-68.

Lashgari, B., and Estill, R. E., 1992, High resolution seismic analysis via fuzzy cluster partitioning: 54th Ann. Internat. Mtg., Eur. Assn. Expl. Geophys., Abstracts, A045, 72-73.

Wang, J. L., and Huang, K. Y., 1993, Neural networks for robust recognition of seismic reflection patterns: 63rd Ann. Internat. Mtg., Soc. Expl. Geophys., Expanded Abstracts, CT4.5, 246-249.

Yang, H. Z., and Huang, K. Y., 1991, Hybrid neural network for seismic pattern recognition: 61st Ann. Internat. Mtg., Soc. Expl. Geophys., Expanded Abstracts, CM1.4, 295-298.

Zhang, X., 1993, Neural network approaches for pattern recognition with application to petroleum exploration: Ph.D., Tsinghua University, Beijing, China.

3. EXPLORATION SEISMOLOGY

3.1. Data Processing

3.1.1. Trace Editing

Huang, K. Y., and Hu, Y. C., 1994, Fuzzy functional-link net for seismic trace editing: 64th Ann. Internat. Mtg., Soc. Expl. Geophys., Expanded Abstracts, CT1.5, 200-201.

Kusuma, T., and Brown, M. M., 1992, Cascade-correlation learning architecture for first-break picking and automated trace editing: 62nd Ann. Internat. Mtg., Soc. Expl. Geophys., Expanded Abstracts, SP2.8, 1136-1139.

Kusuma, T., and Brown, M. M., 1992, First break picking and trace editing using cascade-correlation learning architecture: Proc. Internat. Conf. Petroleum Expl. & Prod.

McCormack, M. D., 1990, Automated trace editing and refraction event picking using neural networks: Proc. Conf. Artif. Intell. for Petr. Expl. & Prod., **1**, 125-132.

McCormack, M. D., 1990, Seismic trace editing and first break picking using neural networks: 60th Ann. Internat. Mtg., Soc. Expl. Geophys., Expanded Abstracts, 321-324.

McCormack, M. D., and Zaucha, D. E., 1992, Automated trace editing and refraction event picking using neural networks: Rev. Inst. Fran. Petr., **47**, 393-405.

McCormack, M. D., Zaucha, D. E., and Dushek, D. W., 1993, First-break refraction event picking and seismic data trace editing using neural networks: Geophysics, **58**, 67-78 (Erratum in Geophysics **58**, 315-318).

Upham, W., and Cary, P., 1991, Comparison of linear and nonlinear neural networks for seismic trace editing: 61st Ann. Internat. Mtg., Soc. Expl. Geophys., Expanded Abstracts, CM1.3, 293-294.

3.1.2. First Break Picking

An, G., and Epping, W. J. M., 1993, Seismic first-arrival picking using neural networks: Proc. World Congr. Neural. Nets., **1**, 174-177.

Chu, C. K. P., and Mendel, J. M., 1994, First break refraction event picking using fuzzy logic systems: IEEE Trans. Fuzzy Systems, **2**, 255-266.

Dimitropoulos, C. H., and Boyce, J. F., 1993, Applications of neural nets to seismic signal analysis: Proc. IEE Int. Conf. Acoust. Sens. Imag. (ASI'93), IEE Conf. Publ. No. 369, 62-67.

Dimitropoulos, C. H., and Boyce, J. F., 1993, Neural nets for first break detection in seismic reflection data: Proc. 3rd IEE Int. Conf. Art. Neural Networks. (ANN'93), IEE Conf. Publ. No. 372, 153-157.

Dimitropoulos, C. D., and Boyce, J. F., 1994, First break detection in seismic reflection data with fuzzy ARTMAP neural networks, *in* Holt, M. J. J., Cowan, C. F. N., Grant, P. M., and Sandham, W. A., Eds., Signal Processing VII: Theories and Applications: European Association for Signal Processing, 221-224.

Fuller, B. N., and Kusuma, T., 1993, A method for picking first breaks in 3-component seismic data: 63rd Ann. Internat. Mtg., Soc. Expl. Geophys., Expanded Abstracts, DP2.8, 325-328.

Hart, D. I., 1996, Improving the reliability of first-break picking with neural networks: 66th Ann. Internat. Mtg., Soc. Expl. Geophys., Expanded Abstracts, PR 8.5, 1662-1665.

Hart, D. I., 1996, Reliably picking first breaks with neural networks: 58th Ann. Internat. Mtg., Eur. Assn. Geosci. Eng., Extended Abstracts, **1**, B033.

Huang, K. Y., and Yuan, Y. W., 1995, Neural network of fuzzy K-nearest neighbour classification rule for seismic first-arrival picking: 65th Ann.Internat. Mtg., Soc. Expl. Geophys., Expanded Abstracts, CT1.3, 145-148.

Kusuma, T., and Brown, M. M., 1992, Cascade-correlation learning architecture for first-break picking and automated trace editing: 62nd Ann. Internat. Mtg., Soc. Expl. Geophys., Expanded Abstracts, SP2.8, 1136-1139.

Kusuma, T., and Brown, M. M., 1992, First break picking and trace editing using cascade-correlation learning architecture: Proc. Internat. Conf. Petroleum Expl. & Prod.

Kusuma, T., and Fish, B. C., 1993, Toward more robust neural-network first break and horizon pickers: 63rd Ann. Internat. Mtg., Soc. Expl. Geophys., Expanded Abstracts, CT4.3, 238-241.

McCormack, M. D., 1990, Seismic trace editing and first break picking using neural networks: 60th Ann. Internat. Mtg., Soc. Expl. Geophys., Expanded Abstracts, 321-324.

Mendel, J., and Chu, C. K., 1992, First break refraction event picking using fuzzy logic systems: USC-SIPI Rep. 221, Dept. Elec. Eng. Syst., Univ. Southern California, Los Angeles, California, USA.

Murat, M. E., and Rudman, A. J., 1992, Automated first arrival picking: a neural network approach: Geophys. Prosp., 40, 587-604.
Taner, T., 1988, The use of supervised learning in first break picking, *in* Bielanski, E., Ed., Proc. Symp. Geophysical Soc. Tulsa.
Veezhinathan, J., and Wagner, D., 1990, A neural network approach to first break picking: Proc. Internat. Joint Conf. Neural Networks, San Diego, California, USA, 1, 235-240.
Veezhinathan, J., Wagner, D., and Ehlers, J., 1991, First break picking using a neural network, *in* Aminzadeh, F., and Simaan, M., Eds., Expert Systems in Exploration: Soc. Expl. Geophys., 179-202.
Wagner, D. E., Ehlers, J. W., and Veezhinathan, J., 1990, First break picking using neural networks: 60th Ann. Internat. Mtg., Soc. Expl. Geophys., Expanded Abstracts, 370-373.

3.1.3. Deconvolution and Multiple Suppression

Calderón-Macías, C., Sen, M. K., and Stoffa, P. L., 1997, Hopfield neural networks, and mean field annealing for seismic deconvolution and multiple attenuation: Geophysics, 62, 992-1002.
Essenreiter, R., 1999, Identification and attenuation of multiple reflections with neural networks: Ph.D., University of Karlsruhe, Germany.
Essenreiter, R., and Karrenbach, M., 1996, Deconvolution using neural networks: 66th Ann. Internat. Mtg., Soc. Expl. Geophys., Expanded Abstracts, PR 6.8, 1611-1614.
Essenreiter, R., Treitel, S., and Karrenbach, M., 1996, Elimination of multiples with neural networks: 58th Ann. Internat. Mtg., Eur. Assn. Geosci. Eng., Extended Abstracts, 1, B034.
Essenreiter, R., Karrenbach, M., and Treitel, S., 1997, Discrimination of primary against multiple reflections in marine seismograms using backpropagation neural networks: 59th Ann. Internat. Mtg., Eur. Assn. Geosci. Eng., Pre-Conference Workshop on Geophysical Applications of Artificial Neural Networks and Fuzzy Logic, Proceedings, Paper 1.1, 1-5.
Essenreiter, R., and Karrenbach, M., 1998, Constrained deconvolution with neural networks: 60th Ann. Internat. Mtg., Eur. Assn. Geosci. Eng., Extended Abstracts, 1, 1-31.
Essenreiter, R., Karrenbach, M., and Treitel, S., 1998, Multiple reflection attenuation in seismic data using backpropagation: IEEE Trans. Sig. Proc., 46, 2001-2011.
Essenreiter, R., Karrenbach, M., and Treitel, S., 2001, Identification and classification of multiple reflections with self-organizing maps: Geophys. Prosp., 49, 341-352.
Karrenbach, M., Essenreiter, R., and Treitel, S., 2000, Multiple attenuation with attribute-based neural networks: 70th Ann. Internat. Mtg., Soc. Expl. Geophys., Expanded Abstracts, SP 3.3, 1989-1992.
Treitel, S., and Essenreiter, R., 1995, Predictive deconvolution revisited with neural nets: 57th Ann. Internat. Mtg., Eur. Assn. Expl. Geophys. & Eur. Assn. Petr. Geol., Extended Abstracts, 1, P065.
Treitel, S., and Essenreiter, R., 1996, Adaptive minimum prediction-error deconvolution and source wavelet estimation using Hopfield neural networks, *in* Osman, O. M., and Robinson, E. A., Eds., Seismic source signature estimation and measurement: SEG, 234-243.
Wang, L. X., and Mendel, J. M., 1992, Adaptive minimum prediction-error deconvolution and source wavelet estimation using Hopfield neural networks: Geophysics, 57, 670-679.
Zhao, X., and Mendel, J. M., 1988, Minimum-variance deconvolution using artificial neural networks: 58th Ann. Internat. Mtg., Soc. Expl. Geophys., Expanded Abstracts, S3.8.

3.1.4. Wavelet Estimation

Huang, K. Y., and Wang, S. P., 2000, Neural networks for seismic wavelet extraction and clustering: 70th Ann. Internat. Mtg., Soc. Expl. Geophys., Expanded Abstracts, INT P1.11, 741-744.
Liu, Z., and Castagna, J. P., 2000, Seismic wavelet estimation constrained by well data using artificial neural networks: 70th Ann. Internat. Mtg., Soc. Expl. Geophys., Expanded Abstracts, ST P1.12, 2460-2463.
Palaz, I., and Weger, R. C., 1990, Waveform recognition using neural networks: The Leading Edge, 9, 28-31.
Wang, L. X., and Mendel, J. M., 1996, Adaptive minimum prediction-error deconvolution and source wavelet estimation using Hopfield neural networks, *in* Osman, O. M., and Robinson, E. A., Eds., Seismic source signature estimation and measurement: SEG, 234-243.
Yang, F. M., and Huang, K. Y., 1991, Multilayer perceptron for detection of seismic anomalies: 61st Ann. Internat. Mtg., Soc. Expl. Geophys., Expanded Abstracts, CM1.8, 309-312.

3.1.5. Velocity Analysis

Calderón-Macías, C., Sen, M. K., and Stoffa, P. L., 1996, A neural network optimization approach for automatic NMO correction and velocity estimation: 66th Ann. Internat. Mtg., Soc. Expl. Geophys., Expanded Abstracts, TG 2.4, 1979-1982.

Calderón-Macías, C., Sen, M. K., and Stoffa, P. L., 1998, Automatic NMO correction and velocity estimation by a feedforward neural network: Geophysics, 63, 1696-1707.

Calderón-Macías, C., Sen, M. K., and Stoffa, P. L., 2000, Artificial neural networks for parameter estimation in geophysics: Geophys. Prosp., 48, 21-47.

Fish, B. C., and Kusuma, T., 1994, A neural network approach to automate velocity picking: 64th Ann. Internat. Mtg., Soc. Expl. Geophys., Expanded Abstracts, CT1.1, 185-188.

Schmidt, J., 1992, Neural network velocity picking: Ph.D., Colorado School of Mines, Golden, Colorado, USA.

Schmidt, J., and Hadsell, F. A., 1992, Neural network stacking velocity picking: 62nd Ann. Internat. Mtg., Soc. Expl. Geophys., Expanded Abstracts, AI1.6, 18-21.

Yao, Y., 1992, Fuzzy velocity inversion: 54th Ann. Internat. Mtg., Eur. Assn. Expl. Geophys., Abstracts, P163, 754-755.

3.1.6. Noise Identification/Reduction

Buffenmyer, V., Poulton, M., and Johnson, R., 1999, Neural network approach to seismic crew noise identification in marine surveys: 69th Ann. Internat. Mtg., Soc. Expl. Geophys., Expanded Abstracts, SPRO 6.6, 1224-1227.

Buffenmyer, V., Poulton, M., and Johnson, R., 2000, Identification of seismic crew noise in marine surveys by neural networks: The Leading Edge, 19, 370-376.

Mousset, E., 1992, Spikes filtering with neural networks: a two-stage detection system: Rev. Inst. Fran. Petr., 47, 407-421.

3.1.7. Statics Analysis

Chen, K. C., Chen, T. C., and Hu, T., 1991, Fuzzy residual statics computation: 61st Ann. Internat. Mtg., Soc. Expl. Geophys., Expanded Abstracts, SP4.4, 1409-1412.

3.1.8. Dataset Matching/Prediction

Al-Ismaili, I. Y., and Warner, M. R., 2002, Non-linear cross-equalization of time-lapse seismic surveys using artificial neural networks: 64th Ann. Internat. Mtg., Eur. Assn. Geosci. Eng., Extended Abstracts, P035.

Davies, P., and Warner, M., 1999, Non-linear cross-equalisation of seismic surveys acquired with different bandwidths: 69th Ann. Internat. Mtg., Soc. Expl. Geophys., Expanded Abstracts, SRC 5.3, 1687-1690.

Davies, P., and Warner, M., 1999, Prediction of multi-channel seismic data using parallel arrays of neural networks: 61st Ann. Internat. Mtg., Eur. Assn. Geosci. Eng., Extended Abstracts, 1, P012.

Davies, P., and Warner, M., 2000, Prediction of near-offset seismic data using parallel arrays of neural networks: 62nd Ann. Internat. Mtg., Eur. Assn. Geosci. Eng., Extended Abstracts, 1, P-124.

3.1.9. Attenuation

Boadu, F. K., 1997, Rock properties and seismic attenuation: Neural network analysis: Pure Appl. Geophys., 149, 507-524.

3.2. AVO Analysis

Dai, H. C., MacBeth, C., and Li, X. Y., 1998, Using the BPNN as a general feature detector: 60th Ann. Internat. Mtg., Eur. Assn. Geosci. Eng., Extended Abstracts, 1, P003.

Eastwood, J., and West, B., 2001, Interactive AVO time-allignment and neural network classification: 71st Ann. Internat. Mtg., Soc. Expl. Geophys., Expanded Abstracts, 257-260.

Mogensen, S., and Link, C., 2001, Artificial neural network solutions to AVO inversion problems: 71st Ann. Internat. Mtg., Soc. Expl. Geophys., Expanded Abstracts, 316-319.

Russell, B., Ross, C., and Lines, L., 2002, Neural networks and AVO: The Leading Edge, 21, 268-276,314.

Sun, Q., Castagna, J., and Liu, Z., 2000, AVO inversion by artificial neural networks (ANN): 70th Ann. Internat. Mtg., Soc. Expl. Geophys., Expanded Abstracts, AVO 2.1, 130-133.

West, B., May, S., Eastwood, J., and Rossen, C., 2001, Interactive seismic facies classification of stack and AVO data using textural attributes and neural networks: 71st Ann. Internat. Mtg., Soc. Expl. Geophys., Expanded Abstracts, 576-579.

3.3. Seismic Chimneys

Heggland, R., Meldahl, P., Bert, B., and De Groot, P., 1999, The chimney cube, an example of semi-automated detection of seismic objects by directive attributes and neural networks, Part II: Interpretation: 69th Ann. Internat. Mtg., Soc. Expl. Geophys., Expanded Abstracts, 935-937.

Heggland, R., Meldahl, P., De Groot, P., and Bril, B., 2000, Detection of seismic chimneys by neural networks - a new prospect evaluation tool: 62nd Ann. Internat. Mtg., Eur. Assn. Geosci. Eng., Extended Abstracts, A-06.

Meldahl, P., Heggland, R., and De Groot, P., 1999, The chimney cube, an example of semi-automated detection of seismic objects by directive attributes and neural networks, Part I: Methodology: 69th Ann. Internat. Mtg., Soc. Expl. Geophys., Expanded Abstracts, 931-934.

Meldahl, P., Heggland, R., Bril, B., and De Groot, P., 2001, Identifying faults and gas chimneys using multiattributes and neural networks: The Leading Edge, **20**, 474-478.

3.4. Compression / Dimensionality Reduction

Cheng-Dang, Z., Yong-Zhong, G., Zhen-Wu, J., and Quan, G. S., 1992, Bi-directional optimal reduction of dimensionality in multi-channel seismic signal processing using neural networks: 9th Ann. Internat. Mtg., Austral. Soc. Expl. Geophys., Abstracts, **23**, 57-60.

Epping, W. J. M., and L'Istelle, A. R., 1992, Data compression of seismic images by neural networks: Rev. Inst. Fran. Petr., **47**, 423-430.

Huang, K. Y., 1996, Seismic principal components analysis using neural networks: 66th Ann. Internat. Mtg., Soc. Expl. Geophys., Expanded Abstracts, PP 9.3, 1259-1262.

Zhang, Z., Zhou, Z., Frenkel, M. A., Chunduru, R., and Mezzatesta, A., 2000, Real time inversion of array resistivity logging data using dimensional reduction and neural network simulation: 70th Ann. Internat. Mtg., Soc. Expl. Geophys., Expanded Abstracts, RPB 5.3, 1802-1805.

Zhou, C. D., Gan, Y. Z., Jin, Z. W., and Guo, S. Q., 1992, Bi-directional optimal reduction of dimensionality in multi-channel seismic signal processing using neural networks: Expl. Geophys., **23**, 57-60.

3.5. Shear Wave Analysis

Dai, H. C., and MacBeth, C., 1993, Analysis of split shear-waves from near-offset VSP data using a neural network: 55th Ann. Internat. Mtg., Eur. Assn. Expl. Geophys., Extended Abstracts, C010.

Dai, H. C., and MacBeth, C., 1994, A neural network picker for VSP data: 56th Ann. Internat. Mtg., Eur. Assn. Expl. Geophys., Extended Abstracts, G032.

Dai, H. C., and MacBeth, C., 1994, Split shear-wave analysis using an artificial neural network: First Break, **12**, 605-613.

Fuller, B. N., and Kusuma, T., 1993, A method for picking first breaks in 3-component seismic data: 63rd Ann. Internat. Mtg., Soc. Expl. Geophys., Expanded Abstracts, DP2.8, 325-328.

Schmidt, K., and James, C., 1995, Application of Euler deconvolution and neural network system as interpretation aids for three component downhole TEM data: 11th Ann. Internat. Mtg., Austral. Soc. Expl. Geophys., Abstracts, **26**, 154-157.

3.6. Interpretation

3.6.1. Interpretation

Addy, S. K., 1998, Neural network classification method helps seismic interval interpretation: Oil and Gas J., **38**, 47-59.

Calderón-Macías, C., 1997, Artifical neural systems for interpretation and inversion of seismic data: Ph.D., University of Texas at Austin, Austin, Texas, USA.

Huang, K. Y., Yang, H. Z., and Lin, C. M., 1990, Connectionist expert system for seismic interpretations: Expanded Abstracts of the 60th SEG Annual International Meeting & Exposition, CM1.1, 310-312.

Leggett, M., Smyth, M., Manning, A., Prescott, C. N., and Edwards, H., 1995, Neural networks and paper seismic interpretation: 65th Ann.Internat. Mtg., Soc. Expl. Geophys., Expanded Abstracts, CT1.2, 142-144.

Meldahl, P., Heggland, R., De Groot, P., and Bril, B., 2000, Semi-automated detection of seismic objects by directive attributes and neural networks, method and applications: 62nd Ann. Internat. Mtg., Eur. Assn. Geosci. Eng., Extended Abstracts, Vol 1, A-04.

Pearson, W. C., Wiener, J. M., and Moll, R. F., 1990, Aeromagnetic structural interpretation using neural networks: a case study from the northern Denver-Julesburg Basin: 60th Ann. Internat. Mtg., Soc. Expl. Geophys., Expanded Abstracts, 587-590.

Schmidt, K., and James, C., 1995, Application of Euler deconvolution and neural network system as interpretation aids for three component downhole TEM data: 11th Ann. Internat. Mtg., Austral. Soc. Expl. Geophys., Abstracts, **26**, 154-157.

3.6.2. Event Tracking

Alberts, P., and Moorfeld, K., 1999, Artificial neural networks for seismic horizon tracking across discontinuities: 61st Ann. Internat. Mtg., Eur. Assn. Geosci. Eng., Extended Abstracts, 6-31.

Alberts, P., Warner, M., and Lister, D., 2000, Artificial neural networks for simultaneous multihorizon tracking across discontinuities: 70th Ann. Internat. Mtg., Soc. Expl. Geophys., Expanded Abstracts, INT 5.4, 651-653.

Cheng, P. Z., 1999, Seismic event picking for velocity estimation using neural networks: Ph.D., University of California, USA.

Clark, G. A., Glinsky, M. E., Devi, K. R. S., Robinson, J. H., Cheng, P. K. Z., and Ford, G. E., 1996, Automatic event picking in pre-stack migrated gathers using a probabilistic neural network: 66th Ann. Internat. Mtg., Soc. Expl. Geophys., Expanded Abstracts, MOD 3.8, 735-738.

Glinsky, M. E., Clark, G. A., Cheng, P. K. Z., Devi, K. R. S., Robinson, J. H., and Ford, G. E., 2001, Automatic event picking in prestack migrated gathers using a probabilistic neural network: Geophysics, **66**, 1488-1496.

Harrigan, E., 1992, Seismic event tracking: Ph.D., University of Strathclyde, Glasgow, Scotland.

Harrigan, E., Kroh, J. R., Sandham, W. A., and Durrani, T. S., 1991, An artificial neural network for horizon picking: Proc. IEEE Internat. Workshop on Parallel Architectures for Seismic Data Processing, 2-3 Sept., 1991, Glasgow, Scotland, 206-213.

Harrigan, E., Kroh, J. R., Sandham, W. A., and Durrani, T. S., 1991, Seismic wavelet extraction using artificial neural networks: Proc. 2nd IEE Internat. Conf. Artificial Neural Networks, 18-20 Nov., 1991, Bournemouth, UK, IEE Conf. Publ. No. 349, 214-218.

Harrigan, E., Kroh, J. R., Sandham, W. A., and Durrani, T. S., 1992, Seismic horizon picking using an artificial neural network: Proc. IEEE Internat. Conf. Acoustics, Speech and Signal Processing, 23-26 March, 1992, San Francisco, USA, 3, 105-108.

Huang, K. Y., 1997, Hopfield neural network for seismic horizon picking: 67th Ann. Internat. Mtg., Soc. Expl. Geophys., Expanded Abstracts, INT 1.02, 562-565.

Huang, K. Y., Chang, W. R. I., and Yen, H. T., 1990, Self-organizing neural network for picking seismic horizons: 60th Ann. Internat. Mtg., Soc. Expl. Geophys., Expanded Abstracts, CM1.2, 313-316.

Kemp, F., Threet, J. R., and Veezhinathan, J., 1992, A neural net branch and bound seismic horizon tracker: 62nd Ann. Internat. Mtg., Soc. Expl. Geophys., Expanded Abstracts, AI 1.3, 10-13.

Kusuma, T., and Fish, B. C., 1993, Toward more robust neural-network first break and horizon pickers: 63rd Ann. Internat. Mtg., Soc. Expl. Geophys., Expanded Abstracts, CT4.3, 238-241.

Lashgari, B., and Estill, R. E., 1991, Application of fuzzy pattern recognition to seismic stratigraphy: 53rd Ann. Internat. Mtg., Eur. Assoc. Expl. Geophys., Abstracts, A040, 78-79.

Leggett, M., Sandham, W. A., and Durrani, T. S., 1993, Seismic event classification using a self-organizing Kohonen network: Proc. IEE/IEEE Internat. Workshop on Natural Algorithms in Sig. Proc., 14-16 Nov., 1993, Chelmsford, UK.

Leggett, M., Sandham, W. A., and Durrani, T. S., 1994, 3D seismic horizon tracking using an artificial neural network: 56th Ann. Internat. Mtg., Eur. Assn. Expl. Geophys., Extended Abstracts, B049.

Leggett, M., Smyth, M., Manning, A., Prescott, C. N., and Edwards, H., 1995, Neural networks and paper seismic interpretation: 65th Ann.Internat. Mtg., Soc. Expl. Geophys., Expanded Abstracts, CT1.2, 142-144.

Leggett, M., Sandham, W. A., and Durrani, T. S., 1996, 3D horizon tracking using artificial neural networks: First Break, **14**, 413-418.

Leggett, M., Sandham, W. A., and Durrani, T. S., 1997, 3D event tracking using artificial neural networks: 59th Ann. Internat. Mtg., Eur. Assn. Geosci. Eng., Pre-Conference Workshop on Geophysical Applications of Artificial Neural Networks and Fuzzy Logic, Proceedings, Paper 1.2, 6-11.

Liu, X., Xue, P., and Li, P., 1989, Neural network method for tracing seismic events: 59th Ann. Internat. Mtg., Soc. Expl. Geophys., Expanded Abstracts, 716-718.

McCormack, M. D., 1990, Automated trace editing and refraction event picking using neural networks: Proc. Conf. Artif. Intell. for Petr. Expl. & Prod., 1, 125-132.

McCormack, M. D., and Zaucha, D. E., 1992, Automated trace editing and refraction event picking using neural networks: Rev. Inst. Fran. Petr., **47**, 393-405.

McCormack, M. D., Zaucha, D. E., and Dushek, D. W., 1993, First-break refraction event picking and seismic data trace editing using neural networks: Geophysics, **58**, 67-78 (Erratum in Geophysics **58**, 315-318).

Prochnow, U., 1999, Horizon tracking in 3D seismic datasets using neural networks: 61st Ann. Internat. Mtg., Eur. Assn. Geosci. Eng., Extended Abstracts, 6-32.

Veezhinathan, J., and Threet, J., 1993, Multiple horizons tracking: Proc. World Congr. Neural. Nets., **1**, 379-382.

3.6.3. *Lithology Prediction and Well Log Analysis*

Addy, S. K., 1997, Attribute analysis in Edwards limestone in Lava County, Texas: 67th Ann. Internat. Mtg., Soc. Expl. Geophys., Expanded Abstracts, INT7.7, 737-740.

Baldwin, J. L., and Otte, D. N., 1989, Computer emulation of human mental processes: Application of neural network simulators to problems in well log interpretation: Proc. 64th Ann. Tech. Conf. Exhibition Soc. Petroleum Eng.

Baldwin, J. L., Bateman, R. M., and Wheatley, C. L., 1990, Application of a neural network to the problem of mineral identification from well logs: The Log Analyst, **31**, 279-293.

Banchs, R. E., and Michelena, R. J., 2000, Well log estimates and confidence intervals by using artificial neural networks: 70th Ann. Internat. Mtg., Soc. Expl. Geophys., Expanded Abstracts, RC 2.4, 1430-1432.

Bhatt, A., and Helle, H. B., 1999, Porosity, permeability and TOC prediction from well logs using a neural network approach: 61st Ann. Internat. Mtg., Eur. Assn. Geosci. Eng., Extended Abstracts, 2-52.

Bhatt, A., Helle, H. B., and Ursin, B., 2002, Determination of facies from well logs using modular neural networks: 64th Ann. Internat. Mtg., Eur. Assn. Geosci. Eng., Extended Abstracts, P091.

Chakravarthy, S. V., Chunduru, R. K., Mezzatesta, A. G., and Fanini, O., 1999, Detection of layer boundaries from array induction tool responses using neural networks: 69th Ann. Internat. Mtg., Soc. Expl. Geophys., Expanded Abstracts, BH/RP 5.6, 140-143.

De Groot, P. F. M., 1999, Volume transformation by way of neural network mapping: 61st Ann. Internat. Mtg., Eur. Assn. Geosci. Eng., Extended Abstracts, 3-37.

Derek, H., Johns, R., and Pasternack, E., 1990, Comparative study of backpropagation neural network and statistical pattern recognition techniques in identifying sandstone lithofacies: Proc. Conf. Artif. Intell. for Petr. Expl. & Prod., **1**, 41-49.

Ecoublet, P., Symes, W. W., and Levin, S., 1997, Seismic inversion for porosity using a backpropagation neural network: TRIP Annual Report 97, Rice University.

Ecoublet, P. O., Symes, W. W., and Levin, S. A., 1998, Porosity prediction from seismic inversion using artificial neural networks: 60th Ann. Internat. Mtg., Eur. Assn. Geosci. Eng., Extended Abstracts, **1**, 3-52.

Fitzgerald, E. M., Bean, C. J., and Reilly, R. J., 1999, Application of neural networks to fracture frequency prediction in boreholes: 61st Ann. Internat. Mtg., Eur. Assn. Geosci. Eng., Extended Abstracts, P051.

Fitzgerald, E. M., Bean, C. J., and Reilly, R., 1999, Fracture-frequency prediction from borehole wireline logs using artificial neural networks: Geophys. Prosp., 47, 1031-1044.

Ford, D., and Kelly, M., 2001, Using neural networks to predict lithology from well logs: 71st Ann. Internat. Mtg., Soc. Expl. Geophys., Expanded Abstracts, 666-667.

Fu, L. Y., 1999, Looking for links between deterministic and statistical methods for porosity and clay-content estimation: 69th Ann. Internat. Mtg., Soc. Expl. Geophys., Expanded Abstracts, BH/RP 3.6, 81-84.

Helle, H. B., Bhatt, A., and Ursin, B., 2001, Porosity and permeability prediction from wireline logs using artificial neural networks: a North Sea case study: Geophys. Prosp., 49, 431-444.

Helle, H. B., Bhatt, A., and Ursin, B., 2002, Fluid saturation from well logs using modular neural networks: 64th Ann. Internat. Mtg., Eur. Assn. Geosci. Eng., Extended Abstracts, B045.

Huang, Y. T., Wong, P., and Gedeon, T., 1997, Spatial interpolation in log analysis using neural-fuzzy technique: 59th Ann. Internat. Mtg., Eur. Assn. Geosci. Eng., Extended Abstracts, Vol 1, P174.

Huang, Y., Wong, P. M., and Gedeon, T. D., 1998, Neural-fuzzy-genetic-algorithm interpolator in log analysis: 60th Ann. Internat. Mtg., Eur. Assn. Geosci. Eng., Extended Abstracts, **1**, P106.

Huang, Z., Shimeld, J., Williamson, M., and Katsube, J., 1996, Permeability prediction with artificial neural network modeling in the Venture gas field, offshore eastern Canada: Geophysics, 61, 422-436.

Imamura, S., 1994, Integrated interpretation of exploration data in geotechnical engineering: an approach using fuzzy theory: 64th Ann. Internat. Mtg., Soc. Expl. Geophys., Expanded Abstracts, CT1.6, 202-205.

Johnston, D. H., 1993, Seismic attribute calibration using neural networks: 63rd Ann. Internat. Mtg., Soc. Expl. Geophys., Expanded Abstracts, CT4.6, 250-253.

Kalantzis, F., and An, P., 1996, Porosity mapping in a Devonian reef using 3D seismic waveforms and neural networks: 58th Ann. Internat. Mtg., Eur. Assn. Geosci. Eng., Extended Abstracts, X004.

Kovalevsky, E. V., and Kharchenko, V. I., 1993, Testing an engineering geological expert system for marine soil identification: 63rd Ann. Internat. Mtg., Soc. Expl. Geophys., Expanded Abstracts, SS2.16, 1277-1280.

Lau, L., Gonzalez-Serrano, A., Mallick, S., and Gillespie, D., 2002, Prestack waveform inversion and neural network interpolation - a two-step approach to log prediction: 64th Ann. Internat. Mtg., Eur. Assn. Geosci. Eng., Extended Abstracts, G016.

Le Nir, I., and Ruiz, G., 2000, Fluvial-marine successions facies evaluation and neural network distribution - methods and limitations: 62nd Ann. Internat. Mtg., Eur. Assn. Geosci. Eng., Extended Abstracts, 1, P-26.

Lindsay, R., and Bocanegra, D., 2001, Sand thickness prediction of the Basal Tena/M1 Sands from seismic attributes using neural networks: Oriente Basin,Ecuador: 71st Ann. Internat. Mtg., Soc. Expl. Geophys., Expanded Abstracts, 547-550.

Link, C. A., 1995, Artificial neural networks for lithology prediction and reservoir characterization: Proceedings of the SPIE: Mathematical Methods in Geophysical Imaging III, 2571, 163-174.

Liu, Z., and Castagna, J. P., 2000, Seismic wavelet estimation constrained by well data using artificial neural networks: 70th Ann. Internat. Mtg., Soc. Expl. Geophys., Expanded Abstracts, ST P1.12, 2460-2463.

Liu, Z. P., and Liu, J. Q., 1998, Seismic-controlled nonlinear extrapolation of well parameters using neural networks: Geophysics, 63, 2035-2041.

Logel, J. D., 1997, Lithologic prediction from density logs using seismic attributes and neural networks: 59th Ann. Internat. Mtg., Eur. Assn. Geosci. Eng., Pre-Conference Workshop on Geophysical Applications of Artificial Neural Networks and Fuzzy Logic, Proceedings, Paper 3.2, 31-33.

Lorenzetti, E. A., 1992, Predicting lithology from Vp and Vs using neural networks: 62nd Ann. Internat. Mtg., Soc. Expl. Geophys., Expanded Abstracts, AI1.5, 14-17.

Monson, G., and Pita, J., 1997, Neural network prediction of pseudo-logs for net pay and reservoir property interpretation: Greater Zafiro field area,Equatorial Guinea: 67th Ann. Internat. Mtg., Soc. Expl. Geophys., Expanded Abstracts, 627.

Oldenziel, T., De Groot, P., and Kvamme, L. B., 2000, Statfjord study demonstrates use of neural network to predict porosity and water saturation from time-lapse seismic: First Break, 18, 65-69.

Poupon, M., Azbel, K., and Palmer, G., 1999, A new methodology based on seismic facies analysis and litho-seismic modeling. The Elkhorn Slough field project, Solano County, California: 69th Ann. Internat. Mtg., Soc. Expl. Geophys., Expanded Abstracts, SINT 2.2, 927-930.

Qu, S., Zhang, F., and Yin, X., 1998, A simulated annealing neural network and Lanczos inverse approach for impedance inversion: 68th Ann. Internat. Mtg., Soc. Expl. Geophys., Expanded Abstracts, SP 1.8, 1100-1103.

Rogers, S. J., Fang, J. H., Karr, C. L., and Stanley, D. A., 1992, Determination of lithology from well logs using a neural network: AAPG Bull., 76, 731-739.

Rogers, S. J., Chen, H. C., Kopaska-Merkel, D. C., and Fang, J. H., 1995, Predicting permeability from porosity using artificial neural networks: AAPG Bull., 79, 1786-1797.

Ronen, S., Schultz, P. S., Hattori, M., and Corbett, C., 1994, Seismic-guided estimation of log properties; part 2: using artificial neural networks for nonlinear attribute calibration: The Leading Edge, 13, 674-678 (Erratum in TLE 13, 822).

Sanchez Arandia, M., Uzcategui, O., and Taheri, M., 2001, Resolution analysis in well log estimation by using neural networks: 71st Ann. Internat. Mtg., Soc. Expl. Geophys., Expanded Abstracts, 642-645.

Schmidt, K., and James, C., 1995, Application of Euler deconvolution and neural network system as interpretation aids for three component downhole TEM data: 11th Ann. Internat. Mtg., Austral. Soc. Expl. Geophys., Abstracts, 26, 154-157.

Shuki, R., Schultz, P., Hattori, M., and Corbett, C., 1994, Seismic-guided estimation of log properties. Part II: Using artificial neural networks for nonlinear attribute calibration: The Leading Edge, 13, 674-678.

Suppala, I., and Virransalo, P., 1992, Use of Kohonen's self-organising map to combine images of airborne geophysical data and to classify lithological units: 54th Ann. Internat. Mtg., Eur. Assn. Expl. Geophys., Abstracts, P003.

Todorov, T., Stewart, R., Hampson, D., and Russell, B., 1998, Well log prediction using attributes from 3C-3D seismic data: 68th Ann. Internat. Mtg., Soc. Expl. Geophys., Expanded Abstracts, ST 5.4, 1574-1576.

Toumani, A., Schmitz, D., and Schepers, R., 1994, Automatic determination of lithology from well logs using fuzzy classification: 56th Ann. Internat. Mtg., Eur. Assn. Expl. Geophys., Extended Abstracts, H041.

Toumani, A., 1997, Fuzzy hybrid classification for lithology prediction from well logs: 59th Ann. Internat. Mtg., Eur. Assn. Geosci. Eng., Pre-Conference Workshop on Geophysical Applications of Artificial Neural Networks and Fuzzy Logic, Proceedings, Paper 3.1, 25-30.

Vallas, J., HA, Messatesta, A., and Strack, K. M., 1993, Resistivity-log inversion using a heterogeneous artificial neural network: 63rd Ann. Internat. Mtg., Soc. Expl. Geophys., Expanded Abstracts, WS7.2, 1395 (Neural Networks, Expert Systems, Fuzzy Logic & Unconventional Mathematical Techniques in Geophysics; Post-Convention Workshop).

Wang, B., Pann, K., Schuelke, J., Shirley, T., and Ferguson, B., 1997, View of neural network training as constrained optimisation and applications to rock porosity prediction: 67th Ann. Internat. Mtg., Soc. Expl. Geophys., Expanded Abstracts, RC 2.3, 838-841.

Wanstedt, S., Huang, Y., and Malmstram, L., 1996, Using neural networks to interpret geophysical logs in the Zinkgruvan Mine; a case study: 66th Ann. Internat. Mtg., Soc. Expl. Geophys., Expanded Abstracts, 2049-2050.

Wawrzyniak, K., 2002, Improving of acoustic full waveforms using artificial neural network: 64th Ann. Internat. Mtg., Eur. Assn. Geosci. Eng., Extended Abstracts, P144.

Wawrzyniak, K., and Olesinski, K., 2001, Processing and interpretation of acoustic full waveforms using artificial neural network: M.Sc. Thesis, University of Mining and Metallurgy, Cracow, Poland.

Wiener, J. M., Rogers, J. A., Rogers, J. R., and Moll, R. F., 1991, Predicting carbonate permeabilities from wireline logs using a back-propagation neural network: 61st Ann. Internat. Mtg., Soc. Expl. Geophys., Expanded Abstracts, CM1.1, 285-288.

Wong, P. M., and Shibli, S. A. R., 1998, Use of interpolation neural networks for permeability estimation from well logs: The Log Analyst, **41**, 18-26.

Zhang, L., and Poulton, M., 1997, Neural network interpretation of EM39 well log data: Proc. Symp. Appl. Geophys. Eng. Environ. Problems, 23-26 March 1997, Reno, Nevada.

Zhang, L., Poulton, M., and Mezzatesta, A., 1999, Neural network based layer picking for unfocused resistivity log parameterization: 69th Ann. Internat. Mtg., Soc. Expl. Geophys., Expanded Abstracts, BH/RP 5.5, 136-139.

Zhang, L., Poulton, M., Zhang, Z., Mezzatesta, A., and Chakravarthy, S., 1999, Fast forward modelling simulation of resistivity well logs using neural networks: 69th Ann. Internat. Mtg., Soc. Expl. Geophys., Expanded Abstracts, BH/RP 5.2, 124-127.

Zhang, Y., Salisch, H. A., and McPherson, J. G., 1999, Application of neural networks to identify lithofacies from well-logs: Expl. Geophys., **30**, 45-49.

Zhang, Z., and Zhou, Z., 2002, Real-time quasi-2-D inversion of array resistivity logging data using neural network: Geophysics, **67**, 517-524.

3.6.4. *Prospect Appraisal*

Bois, P., 1987, Improvement of seismic prospecting by using fuzzy subsets theory, *in* Aminzadeh, F., Ed., Pattern recognition and image processing: Geophysical Press Limited, 473-503.

Chung, P. W. H., and Inder, R., 1992, Handling uncertainty in accessing petroleum exploration data: Rev. Inst. Fran. Petr., **47**, 305-314.

Eddy, B. G., Bonham-Carter, G. F., and Jefferson, C. W., 1995, Mineral resource assessment of the Parry Islands, high Arctic, Canada: a GIS-based fuzzy logic model: Proc. Canadian Conf. GIS, C3, 4.

Guberman, S., Pikovski, Y., and Rantsman, E., 1998, Risk reduction in oil/gas exploration by combining geomorphological analysis and artificial intelligence: 68th Ann. Internat. Mtg., Soc. Expl. Geophys., Expanded Abstracts, INT 3.5, 640-644.

Harris, D., and Pan, G., 1999, Mineral favourability mapping: a comparison of artificial neural networks, logistic regression, and discriminant analysis: Natural Resources Research, **8**, 93-109.

Li, L. H., Chen, H. C., and Fang, J. H., 1992, XPROS: a fuzzy expert system for prospect appraisal: Rev. Inst. Fran. Petr., **47**, 315-323.

3.6.5. *Hydrocarbon Prediction*

Cai, Y., 1994, The artificial neural network approach for hydrocarbon prediction by synthesising multiple seismic information: 56th Ann. Internat. Mtg., Eur. Assn. Expl. Geophys., Extended Abstracts, P153.

Cai, Y., 1994, The artificial neural network for research of the recovery ratio of oil fields: 64th Ann. Internat. Mtg., Soc. Expl. Geophys., Expanded Abstracts, PP5.7, 791-793.

Chen, Z., and Zhu, G., 1995, Application of Kohonen network to the lateral prediction of oil and gas: Geophys. Prosp. Ptr., **34**, 53-56.

Cook, J., Bøe, Ø., and Møller-Holst, J., 1993, Prediction of oil reservoir output by neural networks: Proc. World Congr. Neural. Nets., **1**, 130-133.

De Groot, P. F. M., 1999, Volume transformation by way of neural network mapping: 61st Ann. Internat. Mtg., Eur. Assn. Geosci. Eng., Extended Abstracts, 3-37.

Kalkomey, C. T., 1996, Use of seismic attributes as predictors of reservoir properties - potential risks: 66th Ann. Internat. Mtg., Soc. Expl. Geophys., Expanded Abstracts, SL 3.5, 1789-1792.

Moon, W. M., and An, P., 1991, Integration of mineral exploration data using fuzzy set theory: 53rd Ann. Internat. Mtg., Eur. Assoc. Expl. Geophys., Abstracts, P126, 460.

Zhang, X., Li, Y., and Liu, H., 2000, Hydrocarbon prediction using dual neural network: 70th Ann. Internat. Mtg., Soc. Expl. Geophys., Expanded Abstracts, RC 2.7, 1440-1443.

3.6.6. Inversion

Boadu, F. K., 1998, Inversion of fracture density from field seismic velocities using artificial neural networks: Geophysics, **63**, 534-545.

Calderón-Macías, C., 1997, Artificial neural systems for interpretation and inversion of seismic data: Ph.D., University of Texas at Austin, Austin, Texas, USA.

Calderón-Macías, C., and Sen, M. K., 1993, Geophysical interpretation by artificial neural systems: a feasibility study: 63rd Ann. Internat. Mtg., Soc. Expl. Geophys., Expanded Abstracts, CT4.7, 254-257.

Calderón-Macías, C., Sen, M. K., and Stoffa, P. L., 2000, Artificial neural networks for parameter estimation in geophysics: Geophys. Prosp., **48**, 21-47.

Ecoublet, P., Symes, W. W., and Levin, S., 1997, Seismic inversion for porosity using a backpropagation neural network: TRIP Annual Report 97, Rice University.

Ecoublet, P. O., Symes, W. W., and Levin, S. A., 1998, Porosity prediction from seismic inversion using artificial neural networks: 60th Ann. Internat. Mtg., Eur. Assn. Geosci. Eng., Extended Abstracts, 1, 3-52.

Francis, A., 1997, Acoustic impedance inversion pitfalls and some fuzzy analysis: The Leading Edge, **16**, 275-278.

Fu, L. Y., 1997, Application of the Caianiello neuron-based network to joint inversion: 67th Ann. Internat. Mtg., Soc. Expl. Geophys., Expanded Abstracts, ST 7.3, 1624-1627.

Fu, L. Y., 1999, Looking for links between deterministic and statistical methods for porosity and clay-content estimation: 69th Ann. Internat. Mtg., Soc. Expl. Geophys., Expanded Abstracts, BH/RP 3.6, 81-84.

Fu, L. Y., Chen, S., and Duan, Y., 1997, ANNLOG technique for seismic wave impedance inversion and its application effect: Oil Geophys. Prosp., **32**, 34-44.

Lau, L., Gonzalez-Serrano, A., Mallick, S., and Gillespie, D., 2002, Prestack waveform inversion and neural network interpolation - a two-step approach to log prediction: Extended Abstracts of the 64th EAGE Conference & Technical Exhibition, G016.

Michaels, P., and Smith, R. B., 1992, Recurrent neural network representation of the inelastic wave equation and full waveform inversion without local minima: 62nd Ann. Internat. Mtg., Soc. Expl. Geophys., Expanded Abstracts, AI1.7, 22-25.

Mogensen, S., and Link, C., 2001, Artificial neural network solutions to AVO inversion problems: 71st Ann. Internat. Mtg., Soc. Expl. Geophys., Expanded Abstracts, 316-319.

Qu, S., Zhang, F., and Yin, X., 1998, A simulated annealing neural network and Lanczos inverse approach for impedance inversion: 68th Ann. Internat. Mtg., Soc. Expl. Geophys., Expanded Abstracts, SP 1.8, 1100-1103.

Raiche, A., 1991, A pattern recognition approach to geophysical inversion using neural nets: Geophys. J. Internat., **105**, 629-648.

Roethe, G., 1993, Neural networks and seismic inverse problems: Ph.D., University of Paris, Paris, France.

Roethe, G., and Tarantola, A., 1991, Use of neural networks for inversion of seismic data: 61st Ann. Internat. Mtg., Soc. Expl. Geophys., Expanded Abstracts, CM1.6, 302-305.

Röth, G., and Tarantola, A., 1992, Inversion of seismic waveforms using neural networks: 62nd Ann. Internat. Mtg., Soc. Expl. Geophys., Expanded Abstracts, SI2.7, 788-791.

Röth, G., and Tarantola, A., 1994, Neural networks and inversion of seismic data: J. Geophys. Res., 99(B4), 6753-6768.

Roy-Chowdhury, K., Devilee, R., and Curtis, A., 1997, Use of neural networks for non-linear non-unique inversion in geophysics: 67th Ann. Internat. Mtg., Soc. Expl. Geophys., Expanded Abstracts, 1620-1623.

Sun, Q., Castagna, J., and Liu, Z., 2000, AVO inversion by artificial neural networks (ANN): 70th Ann. Internat. Mtg., Soc. Expl. Geophys., Expanded Abstracts, AVO 2.1, 130-133.

Sun, Q., Eissa, M., Castagna, J., Cersosimo, D., Sun, S., and Decker, C., 2001, Porosity from artificial neural network inversion for Bermejo Field, Ecuador: 71st Ann. Internat. Mtg., Soc. Expl. Geophys., Expanded Abstracts, 734-737.

Wang, B., Pann, K., Schuelke, J., Shirley, T., and Ferguson, B., 1997, View of neural network training as constrained optimisation and applications to rock porosity prediction: 67th Ann. Internat. Mtg., Soc. Expl. Geophys., Expanded Abstracts, RC 2.3, 838-841.

3.6.7. Reservoir Characterisation

Addy, S. K., 1998, Neural network classification method helps seismic interval interpretation: Oil and Gas J., **38**, 47-59.

Aminzadeh, F., 1997, Model-based neural network and data mining applied to the reservoir parameter estimation problem: 59th Ann. Internat. Mtg., Eur. Assn. Geosci. Eng., Pre-Conference Workshop on Geophysical Applications of Artificial Neural Networks and Fuzzy Logic, Proceedings, Paper 3.4, 39.

An, P., 1994, The effect of random noise in lateral reservoir characterisation using feed-forward neural networks: 64th Ann. Internat. Mtg., Soc. Expl. Geophys., Expanded Abstracts, PP5.6, 787-790.

An, P., and Moon, W. M., 1993, Reservoir characterisation using feedforward neural networks: 63rd Ann. Internat. Mtg., Soc. Expl. Geophys., Expanded Abstracts, CT4.8, 258-262.

An, P., Moon, W. M., and Kalantzis, F., 2001, Reservoir characterization using seismic waveform and feedforword neural networks: Geophysics, 66, 1450-1456.

Balz, O., Pivot, F., and Veeken, P., 1999, Reservoir characterisation using neural networks controlled by petrophysical and seismic modelling: 61st Ann. Internat. Mtg., Eur. Assn. Geosci. Eng., Extended Abstracts, 5-15.

Bhatt, A., Helle, H. B., and Ursin, B., 2001, Application of parallel neural networks for reservoir characterisation while drilling: Extended Abstracts of the 63rd EAGE Conference & Technical Exhibition, 11 - 15 June 2001, RAI, Amsterdam, The Netherlands, L018.

Boadu, F. K., 2000, Neural inversion for oil saturation from seismic velocities: 70th Ann. Internat. Mtg., Soc. Expl. Geophys., Expanded Abstracts, RC 2.1, 1417-1421.

Cheng, D. Z., Wu, X. L., and Cheng, J. A., 1993, Determining reservoir properties in reservoir studies using a fuzzy neural network: Proc. 68th Ann. Tech. Conf. Exhibition Soc. Petroleum Eng.

De Groot, P. F. M., Campbell, A. E., Kavli, T., and Melnyk, D., 1993, Reservoir characterisation from 3D seismic data using artificial neural networks and stochastic modelling techniques: 55th Ann. Internat. Mtg., Eur. Assn. Expl. Geophys., Extended Abstracts, B047.

De Groot, P., Krajewski, P., and Bischoff, R., 1998, Evaluation of remaining oil potential with 3D seismic using neural networks: 60th Ann. Internat. Mtg., Eur. Assn. Geosci. Eng., Extended Abstracts, 1, 2-47.

Durrani, J. A., and Baumel, R. T., 2002, Reservoir characterization by prestack inversion and neural net calibration: Extended Abstracts of the 64th EAGE Conference & Technical Exhibition, G020.

Fogg, A. N., 2000, Petro-seismic classification using neural networks: UK onshore: 70th Ann. Internat. Mtg., Soc. Expl. Geophys., Expanded Abstracts, RC 2.3, 1426-1429.

Himmer, P., and Link, C., 1997, Reservoir porosity prediction from 3-D seismic data using neural networks: 67th Ann. Internat. Mtg., Soc. Expl. Geophys., Expanded Abstracts, RC 2.4, 842-845.

Kalkomey, C. T., 1996, Use of seismic attributes as predictors of reservoir properties - potential risks: 66th Ann. Internat. Mtg., Soc. Expl. Geophys., Expanded Abstracts, SL 3.5, 1789-1792.

Key, S. C., Nielsen, H. H., Signer, C., Sønneland, L., Waagbø, K., and Veire, H. H., 1997, Fault and fracture classification using artificial neural networks - case study from the Ekofisk Field: 67th Ann. Internat. Mtg., Soc. Expl. Geophys., Expanded Abstracts, INT 3.2, 623-626.

Kovalevsky, E. V., and Denisov, S. B., 1997, Fuzzy geomodel expert system - application results: 59th Ann. Internat. Mtg., Eur. Assn. Geosci. Eng., Extended Abstracts, 1, 180.

Kovalevsky, E. V., and Kharchenko, V. I., 1995, "Fuzzy Geomodel" expert system: 57th Ann. Internat. Mtg., Eur. Assn. Expl. Geophys. & Eur. Assn. Petr. Geol., Extended Abstracts, 1, 157.

Leiphart, D. J., and Hart, B. S., 2001, Comparison of linear regression and a probabilistic neural network to predict porosity from 3-D seismic attributes in Lower Brushy Canyon channeled sandstones, southeast New Mexico: Geophysics, 66, 1349-1358.

Link, C. A., 1995, Artificial neural networks for lithology prediction and reservoir characterization: Proceedings of the SPIE: Mathematical Methods in Geophysical Imaging III, 2571, 163-174.

Monson, G., and Pita, J., 1997, Neural network prediction of pseudo-logs for net pay and reservoir property interpretation: Greater Zafiro field area,Equatorial Guinea: 67th Ann. Internat. Mtg., Soc. Expl. Geophys., Expanded Abstracts, 627.

Osborne, D. A., 1992, Permeability estimation using a neural network: a case study from the Roberts Unit, Wasson field, Yoakum County, Texas: Proc. Conf. Artif. Intell. for Petr. Expl. & Prod., 1, 89-96.

Saggaf, M. M., Toksöz, M. N., and Mustafa, H. M., 2000, Estimation of reservoir properties from seismic data by regularized neural networks: 70th Ann. Internat. Mtg., Soc. Expl. Geophys., Expanded Abstracts, RC 2.2, 1422-1425.

Schuelke, J. S., Quirein, J. A., Sarg, J. F., Altany, D. A., and Hunt, P. E., 1997, Reservoir architecture and porosity distribution, Pegasus Field, West Texas - an integrated sequence stratigraphic-seismic attribute study using neural networks: 67th Ann. Internat. Mtg., Soc. Expl. Geophys., Expanded Abstracts, INT 6.2, 868-871.

Shimada, N., Nishikawa, N., and Tani, A., 1997, Reservoir characterisation using artificial neural networks: 59th Ann. Internat. Mtg., Eur. Assn. Geosci. Eng., Pre-Conference Workshop on Geophysical Applications of Artificial Neural Networks and Fuzzy Logic, Proceedings, Paper 3.3, 34-38.

Signer, C., Key, S. C., Nielsen, H. H., Sønneland, L., Waagbø, K., and Veire, H. H., 1997, Fault and fracture classification using artificial neural networks - case study from the Ekofisk field: 59th Ann. Internat. Mtg., Eur. Assn. Geosci. Eng., Extended Abstracts, **2**, C038.

Thadani, S. G., 1994, Stochastic imaging of reservoir properties using probabilistic neural networks: 56th Ann. Internat. Mtg., Eur. Assn. Expl. Geophys., Extended Abstracts, H047.

Thadani, S. G., 1994, Stochastic imaging of reservoir properties using probabilistic neural networks: 64th Ann. Internat. Mtg., Soc. Expl. Geophys., Expanded Abstracts, CT1.2, 189-191 (Paper withdrawn).

Tonn, R., 2002, Neural network seismic reservoir characterization in a heavy oil reservoir: The Leading Edge, **21**, 309-312.

Trappe, H., and Hellmich, C., 1997, Areal prediction of porosity thickness from 3D seismic data by means of neural networks: 59th Ann. Internat. Mtg., Eur. Assn. Geosci. Eng., Extended Abstracts, **2**, C007.

Trappe, H., and Hellmich, C., 2000, Using neural networks to predict porosity thickness from 3D seismic: First Break, **18**, 377-384.

Trappe, H., Hellmich, C., Knudsen, J., and Baartman, H., 2000, Mapping the gas column in an aquifer gas storage with neural networks: 62nd Ann. Internat. Mtg., Eur. Assn. Geosci. Eng., Extended Abstracts, X-08.

Trappe, H., Schubart-Engelschall, J., and Laggiard, E., 2002, Seismic facies classification of deep Rotliegend sandstones by neural network techniques: 64th Ann. Internat. Mtg., Eur. Assn. Geosci. Eng., Extended Abstracts, H039.

Walls, J. D., Taner, M. T., Guidish, T., Taylor, G., Dumas, D., and Derzhi, N., 1999, North Sea reservoir characterisation using rock physics, seismic attributes, and neural networks; a case history: 69th Ann. Internat. Mtg., Soc. Expl. Geophys., Expanded Abstracts, SRC 1.4, 1572-1575.

Walls, J. D., Taner, M. T., Taylor, G., Smith, M., Carr, M., Derzhi, N., Drummond, J., McGuire, D., Morris, S., Bregar, J., and Lakings, J., 2000, Seismic reservoir characterization of a mid-continent fluvial system using rock physics, poststack seismic attributes and neural networks: a case history: 70th Ann. Internat. Mtg., Soc. Expl. Geophys., Expanded Abstracts, RC 2.6, 1437-1439.

Wong, P. M., Gedeon, T. D., and Taggart, I. J., 1995, An improved technique in porosity prediction: a neural network approach: IEEE Trans. Geosci. Remote Sensing, **33**, 971-980.

Zhang, X., Li, Y., Hu, Q., and Feng, D., 1995, Early-stage reservoir analysis with SOMA: a neural network approach: 65th Ann.Internat. Mtg., Soc. Expl. Geophys., Expanded Abstracts, CT1.1, 138-141.

Zhang, X., Li, Y., Liu, F., and Wang, L., 1995, Estimating reservoir's lithological parameters from seismic data using neural network: 65th Ann.Internat. Mtg., Soc. Expl. Geophys., Expanded Abstracts, PP4.2, 606-608.

3.6.8. Quality Assessment

Jakubowicz, H., and Holland Smith, D. J., 1997, Automated quality assessment of seismic data: 67th Ann. Internat. Mtg., Soc. Expl. Geophys., Expanded Abstracts, ACQ 3.7.

3.6.9. Tomography

Lo, T. W., 1991, Fuzzy logic approach for solving a traveltime tomography problem: 61st Ann. Internat. Mtg., Soc. Expl. Geophys., Expanded Abstracts, RE2.2, 732-736.

Lo, T. W., 1991, Fuzzy logic traveltime tomography: synthetic data test: 61st Ann. Internat. Mtg., Soc. Expl. Geophys., Expanded Abstracts, SI2.6, 925-928.

4. EARTHQUAKE SEISMOLOGY AND SUBTERRANEAN NUCLEAR EXPLOSIONS

Dai, H. C., 1995, Analysis of local earthquake data using artificial neural networks: Ph.D., University of Edinburgh, Edinburgh, Scotland.

Dai, H. C., and MacBeth, C., 1995, Automatic picking of seismic arrivals in local earthquake data using an artificial neural network: Geophys. J. Internat., **120**, 758-774.

Dai, H. C., and MacBeth, C. D., 1995, Identifying P- and S-waves using artificial neural network: 57th Ann. Internat. Mtg., Eur. Assn. Expl. Geophys. & Eur. Assn. Petr. Geol., Extended Abstracts, **1**, C039.

Dai, H. C., and MacBeth, C., 1997, The application of back-propagation neural network to automatic picking seismic arrivals from single-component recordings: J. Geophys. Res., **102**, 15105-15113.

Dai, H. C., and MacBeth, C., 1997, Application of back-propagation neural networks to identification of seismic arrival types: Phys. Earth Plan. Int., **101**, 177-188.

Dowla, F. U., De Groot, A. J., Parker, S. R., and Vermuri, V. R., 1989, Back-propagation neural networks: systolic implementation for seismic signal filtering: Neural Networks, **1**, 138-153.

Dowla, F. U., Taylor, S. R., and Anderson, R. W., 1990, Seismic discrimination with artificial neural networks: preliminary results with regional spectral data: Bull. Seis. Soc. Am., **80**, 1346-1373.

Dysart, P. S., and Pulli, J. J., 1990, Regional seismic event classification at the NORESS array: seismological measurements and the use of trained neural networks: Bull. Seis. Soc. Am., **80**, 1910-1933.

Falsaperla, S., Fortuna, L., Graziani, S., and Nunnari, G., 1992, Automatic classification of seismic events by neural networks: Proc. IEEE IGARSS'92, 1, 224-226.

Fortuna, L., Graziani, S., Muscato, G., and Nunnari, G., 1992, Auto-associative multi-layered neural networks for the classification of seismic signals: Proc. Int. Joint Conf. Neural Networks, **3**, 769-774.

Ifantis, A., Fotinos, A., Economou, G., and Fotopoulos, S., 1999, Fuzzy identification of seismic electric signals: Appl. Sig. Proc., **6**, 33-41.

5. MINERAL EXPLORATION

An, P., and Moon, W. M., 1990, Application of fuzzy set theory for integrating geological and geophysical information: 60th Ann. Internat. Mtg., Soc. Expl. Geophys., Expanded Abstracts, 366-369.

An, P., Moon, W. M., and Rencz, A., 1991, Application of fuzzy set theory to integrated mineral exploration: J. Can. Soc. Expl. Geophys., **27**, 1-11.

Bolt, G., Daneel, G., and Clare, A., 1997, Prospect explorer: an exploration neural analysis tool: 67th Ann. Internat. Mtg., Soc. Expl. Geophys., Expanded Abstracts, WS 1.2, 2041-2044.

Clare, A., Bolt, G., and Daneel, G., 1997, The use of neural analysis as an exploration tool: Proc. 12th Internat. Conf. Appl. Geologic Sensing, **2**, 199-205.

Fang, J. H., Chen, H. C., and Wright, D., 1991, A fuzzy expert system for thin-section mineral identification, *in* Aminzadeh, F., and Simaan, M., Eds., Expert Systems in Exploration: Soc. Expl. Geophys., 203-220.

Singer, D. A., and Kouda, R., 1996, Application of a feedforward neural network in the search for Kuroko deposits in the Hokuroku District, Japan: Math. Geol., **28**, 1017-1023.

6. ELECTROMAGNETIC / POTENTIAL FIELD EXPLORATION

6.1. Electromagnetic Methods

Ahl, A., 1997, Classification of airborne electromagnetic data with neural networks: 59th Ann. Internat. Mtg., Eur. Assn. Geosci. Eng., Pre-Conference Workshop on Geophysical Applications of Artificial Neural Networks and Fuzzy Logic, Proceedings, Paper 2.3, 23-24.

Ahl, A., Winkler, E., and Seiberl, W., 1999, Case study of neural network interpretation of airborne electromagnetic measurements: 61st Ann. Internat. Mtg., Eur. Assn. Geosci. Eng., Extended Abstracts, 1, 2-26.

Ashley, D., and Poulton, M., 1993, Geophysical target identification in environmental investigations, *in*, Eds., Intelligent Engineering Systems through Artificial Neural Networks:, 903-908.

Birken, R., 1997, Neural network interpretation of electromagnetic ellipticity data in a frequency range from 1kHz to 32MHz: Ph.D., University of Arizona, Tucson, USA.

Birken, R. A., and Poulton, M. M., 1995, Neural network interpretation scheme for a frequency-domain electromagnetic ellipticity surveys: Proc. Symp. Appl. Geophys. Eng. Environ. Problems, 1, 349-357.

Birken, R., and Poulton, M., 1997, Neural network interpretation of high-frequency electromagnetic ellipticity data: Proc. Symp. Appl. Geophys. Eng. Environ. Problems.

Brown, M. P., and Poulton, M. M., 1995, Neural network interpretation with electromagnetic and magnetic data for environmental site investigations: Proc. Symp. Appl. Geophys. Eng. Environ. Problems, 1, 689-698.

Brown, M., and Poulton, M., 1996, Locating buried objects for environmental site investigations using neural networks: J. Environ. and Eng. Geophys., **1**, 179-188.

Cisar, D., Novotny, T. J., and Dickerson, J., 1992, Electromagnetic data evaluation using a neural network: initial investigation - underground storage tanks: 62nd Ann. Internat. Mtg., Soc. Expl. Geophys., Expanded Abstracts, EM1.6, 409-411.

Cisar, D., Dickerson, J., and Novotny, T., 1993, Electromagnetic data evaluation using a neural network: initial investigation - underground storage tanks: Proc. Symp. Appl. Geophys. Eng. Environ. Problems, 1, 599-612.

El-Kaliouby, H. M., Poulton, M. M., and El-Diwany, E. A., 1999, Inversion of coincident loop TEM data for layered polarizable ground using neural networks: 69th Ann. Internat. Mtg., Soc. Expl. Geophys., Expanded Abstracts, EM 1.5, 259-262.

Fei, D., Teng, Y. C., and Kuo, J. T., 1994, Detection of conductive thick plate based on finite-element method and neural networks: 64th Ann. Internat. Mtg., Soc. Expl. Geophys., Expanded Abstracts, PP1.2, 636-639.
Poulton, M. M., 1990, Neural network pattern recognition of electromagnetic ellipticity images: Ph.D., University of Arizona, Tucson, USA.
Poulton, M. M., 1991, Comparison of accuracy and sensitivity of neural network parameter-estimation paradigms for EM subsurface images: 61st Ann. Internat. Mtg., Soc. Expl. Geophys., Expanded Abstracts, CM1.5, 299-301.
Poulton, M. M., Glass, C. E., and Sternberg, B. K., 1989, Recognizing EM ellipticity patterns with neural networks: 59th Ann. Internat. Mtg., Soc. Expl. Geophys., Expanded Abstracts, 208-212.
Poulton, M. M., Sternberg, B. K., and Glass, C. E., 1992, Location of subsurface targets in geophysical data using neural networks: Geophysics, **57**, 1534-1544.
Poulton, M. M., Sternberg, B. K., and Glass, C. E., 1992, Neural network pattern recognition of subsurface EM images: J. Appl. Geophys., **29**, 21-36.
Poulton, M. M., Brown, M., Larson, T., and Callow, R., 1995, Hazardous waste site characterization using virtual environments and neural networks, *in* Wang, F., and Lever, P., Eds., Advances in Robotics and Automatic Systems for Hazardous Environments: World Scientific Publishers.
Poulton, M. M., Birken, R. A., and Zhang, L., 1997, Integrating neural networks in field interpretations for environmental geophysics surveys: 59th Ann. Internat. Mtg., Eur. Assn. Geosci. Eng., Pre-Conference Workshop on Geophysical Applications of Artificial Neural Networks and Fuzzy Logic, Proceedings, Paper 2.1, 12-17.
Poulton, M. M., and Birken, 1998, Estimating one-dimensional models from frequency-domain electromagnetic data using modular neural networks: IEEE Trans. Geosci. Remote Sensing, **36**, 547-555.
Seiberl, W., Ahl, A., and Winkler, E., 1998, Interpretation of airborne electromagnetic data with neural networks: Expl. Geophys., **29**, 152-156.
Schneiderbauer, K., 1994, Application of a neural network in the interpretation of electromagnetic data: 56th Ann. Internat. Mtg., Eur. Assn. Expl. Geophys., Extended Abstracts, P039.
Sternberg, B. K., and Poulton, M. M., 1994, High-resolution subsurface imaging and neural network recognition: Proc. Symp. Appl. Geophys. Eng. Environ. Problems, 847-855.
Suppala, I., and Virransalo, P., 1992, Use of Kohonen's self-organising map to combine images of airborne geophysical data and to classify lithological units: 54th Ann. Internat. Mtg., Eur. Assn. Expl. Geophys., Abstracts, P003.
Winkler, E., 1994, Inversion of electromagnetic data using neural networks: 56th Ann. Internat. Mtg., Eur. Assn. Expl. Geophys., Extended Abstracts, P124.
Winkler, E., and Seiberl, W., 1997, Inversion of airborne electromagnetic data using neural networks: 59th Ann. Internat. Mtg., Eur. Assn. Geosci. Eng., Pre-Conference Workshop on Geophysical Applications of Artificial Neural Networks and Fuzzy Logic, Proceedings, Paper 2.2, 18-22.
Zhang, L., and Poulton, M. M., 1997, Neural network interpretation of EM39 well log data: Proc. Symp. Appl. Geophys. Eng. Environ. Problems.

6.2. Potential Field Methods

Cartabia, G., Zerilli, A., and Apolloni, B., 1994, Lineaments recognition for potential fields images using a learning algorithm for Boltzmann machines: 64th Ann. Internat. Mtg., Soc. Expl. Geophys., Expanded Abstracts, GM1.5, 432-435.
Fossati, M., Zerilli, A., Ronchini, G., and Apolloni, B., 1992, Lineaments analysis for potential-fields using neural networks: 62nd Ann. Internat. Mtg., Soc. Expl. Geophys., Expanded Abstracts, AI1.2, 6-9.
Guo, Y., Hansen, R. O., and Harthill, N., 1992, Feature recognition from potential fields using neural networks: 62nd Ann. Internat. Mtg., Soc. Expl. Geophys., Expanded Abstracts, AI1.1, 1-5.

6.3. Ground Penetrating Radar

Minior, D. V., and Smith, S., 1993, Neural networks for highway maintenance investigations using ground penetrating radar: Proc. Symp. Appl. Geophys. Eng. Environ. Problems, **2**, 449-462.
Minior, D. V., and Smith, S., 1993, The use of neural networks in the interpretation of ground penetrating radar: 55th Ann. Internat. Mtg., Eur. Assn. Expl. Geophys., Extended Abstracts, D022, (Paper withdrawn).
Poulton, M. M., and El-Fouly, A., 1991, Preprocessing GPR signatures for cascading neural network classification: 61st Ann. Internat. Mtg., Soc. Expl. Geophys., Expanded Abstracts, E/G1.8, 507-509.

6.4. Remote Sensing

Alexander, J. A., Lloyd, D. E., Logar, A. M., Corwin, E. M., and Welch, R. M., 1996, Cloud classification in polar and desert regions and smoke classification from biomass burning using a hierarchical neural network: Proc. 8th Conf. Satellite Meteorol. Ocean., 303-307.

An, P., Moon, W. M., and Rencz, A., 1991, Integration of geological, geophysical and remote sensing data using fuzzy logic: J. Can. Soc. Expl. Geophys., 27, 1-11.

An, P., Chung, C. J. F., and Rencz, A. N., 1994, A preliminary study of alteration mapping from airborne geophysical and remote sensing data using feed-forward neural networks: 64th Ann. Internat. Mtg., Soc. Expl. Geophys., Expanded Abstracts, PP7.7, 837-840.

Baraldi, A., and Parmiggiani, F., 1995, A neural network for unsupervised categorization of multivalued input patterns: an application to satellite image clustering: IEEE Trans. Geosci. Remote Sensing, **33**, 305-316.

Benediktsson, J. A., Philip, H. S., and Erosy, O. K., 1990, Neural network approaches versus statistical method in classification of multi-source remote sensing data: IEEE Trans. Geosci. Remote Sensing, **28**, 185-192.

Benediktsson, J. A., Swain, P. H., and Ersoy, O. K., 1990, Neural network approaches versus statistical methods in classification of multisource remote sensing data: IEEE Trans. Geosci. Remote Sensing, **28**, 540-552.

Benediktsson, J. A., Swain, P. H., and Ersoy, O. K., 1993, Conjugate-gradient neural networks in classification of multisource and very high dimensional remote sensing data: Internat. J. Remote Sensing, **14**, 2883-2903.

Bischof, H., Schneider, W., and Pinz, A. J., 1992, Multispectral classification of landsat images using neural networks: IEEE Trans. Geosci. Remote Sensing, **30**, 482-490.

Bischof, H., and Leonardis, A., 1998, Finding optimal neural networks for land use classification: IEEE Trans. Geosci. Remote Sensing, **36**, 337-341.

Bruzzone, L., Conese, C., Maselli, F., and Roll, F., 1997, Multisource classification of complex rural areas by statistical and neural-network approaches: Photo. Eng. & Rem. Sens., **63**, 523-533.

Caillol, H., Hillion, A., and Pieczynski, W., 1993, Fuzzy random fields and unsupervised image segmentation: IEEE Trans. Geosci. Remote Sensing, **31**, 801-810.

Chen, K. S., Tzeng, Y. C., Chen, C. F., and Kao, W. L., 1995, Land-cover classification of multispectral imagery using a dynamic learning neural network: Photo. Eng. & Rem. Sens., **61**.

Chen, K. S., Huang, W. P., Tsay, D. H., and Amar, F., 1996, Classification of multifrequency polarimetric SAR imagery using a dynamic learning neural network: IEEE Trans. Geosci. Remote Sensing, **34**, 814-820.

Decateur, S. E., 1989, Application of neural networks to terrain classification: IEEE INNS Internat. Joint Conf. Neural Networks, **1**, 283-288.

Fardanesh, M. T., and Ersoy, O. K., 1998, Classification accuracy improvement of neural network classifiers by using unlabeled data: IEEE Trans. Geosci. Remote Sensing, **36**, 1020-1025.

Hara, Y., Atkins, R. G., Yueh, S. H., Shin, R. T., and Kong, J. A., 1994, Application of neural network to radar image classification: IEEE Trans. Geosci. Remote Sensing, **32**, 100-109.

Greenspan, H. K., and Goodman, R., 1993, Remote sensing image analysis via a texture classification neural network, *in* Hanson, S. J., Cowan, J. D., and Giles, C. L., Eds., Advances in Neural Information Processing Systems 5: Morgan Kaufmann, 425-432.

Heermann, P. D., and Khazene, N., 1992, Classification of multispectral remote sensing data using a back-propagation neural network: IEEE Trans. Geosci. Remote Sensing, **30**, 81-88.

Kanellopoulos, I., Varfis, A., Wilkinson, G. G., and Mégier, J., 1991, Neural network classification of multi-date satellite imagery: Proc. IEEE Internat. Geosci. Remote Sensing Symp. (IGARSS'91), 2215-2218.

Kriebel, S. K. T., 1997, Prediction of the upwelling phenomenon at the northwest African Atlantic coast - a connectionist approach: Proc. IEEE Internat. Geosci. Remote Sensing Symp. (IGARSS'97), 1035-1037.

Kriebel, S. K. T., Brauer, W., and Eifler, W., 1998, Coastal upwelling prediction with a mixture of neural networks: IEEE Trans. Geosci. Remote Sensing, **36**, 1508-1518.

Logar, A. M., Corwin, E. M., Alexander, J. A., Lloyd, D. E., Berendes, T. A., and Welch, R. M., 1997, A hybrid histogram/neural network classifier for creating global cloud masks: Internat. J. Remote Sensing, **18**, 847-869.

Logar, A. M., Lloyd, D. E., Corwin, E. M., Penaloza, M. L., Feind, R. E., Berendes, T. A., Kuo, K. S., and Welch, R. M., 1998, The ASTER polar cloud mask: IEEE Trans. Geosci. Remote Sensing, **36**, 1302-1312.

Paola, J. D., and Schowengerdt, R. A., 1995, A review and analysis of back propagation neural networks for classification of remotely-sensed multi-spectral imagery: Internat. J. Remote Sensing, **16**, 3033-3058.

Penaloza, M. A., and Welch, R. M., 1996, Feature selection for classification of polar regions using a fuzzy expert system: Remote Sens. of Environ., **58**, 81-100.
Ritter, N. D., and Hepner, G. F., 1990, Application of an artificial neural network to land-cover classification of thematic mapper imagery: Comput. and Geosci., **16**, 873-880.
Schöneburg, E., Straub, R., Van Camp, L., Schlittenhardt, P., and Nickolay, B., 1993, Analysis of the upwelling phenomenon at the northwest African coast using neural networks: Proc. World Congr. Neural. Nets., **1**, 153-156.
Spina, M. S., Schwartz, M. J., Staelin, D. H., and Gasiewski, A. J., 1998, Application of multilayer feedforward neural networks to precipitation cell-top altitude estimation: IEEE Trans. Geosci. Remote Sensing, **36**, 154-162.
Tzeng, Y. C., and Chen, K. S., 1998, A fuzzy neural network to SAR image classification: IEEE Trans. Geosci. Remote Sensing, **36**, 301-307.
Tovinkere, V. R., 1992, Fuzzy logic expert systems for classification of polar regions: M.Sc., South Dakota School of Mines & Technology, Rapid City, South Dakota, USA.
Tzeng, Y. C., Chen, K. S., Kao, W. L., and Fung, A. K., 1994, A dynamic learning neural network for remote sensing applications: IEEE Trans. Geosci. Remote Sensing, **32**, 1096-1102.
Wang, F., 1990, Fuzzy supervised classification of remote sensing images: IEEE Trans. Geosci. Remote Sensing, **28**, 194-201.
Watters, S. E., Logar, A. M., and Corwin, E. M., 1993, A comparison of neural network algorithms for polar scene identification: Proc. Conf. Intell. Eng. Syst. Through Artif. Neural Networks, **3**, 257-264.

6.5. Inversion

El-Kaliouby, H. M., Poulton, M. M., and El-Diwany, E. A., 1999, Inversion of coincident loop TEM data for layered polarizable ground using neural networks: 69th Ann. Internat. Mtg., Soc. Expl. Geophys., Expanded Abstracts, EM 1.5, 259-262.
El-Qady, G., and Ushijima, K., 2001, Inversion of DC resistivity data using neural networks: Geophys. Prosp., **49**, 417-430.
Poulton, M. M., Glass, C. E., and Sternberg, B. K., 1989, Recognizing EM ellipticity patterns with neural networks: 59th Ann. Internat. Mtg., Soc. Expl. Geophys., Expanded Abstracts, 208-212.
Sun, C., and Rabinowitz, P. D., 1993, Inversion of the magnetotelluric data of the inhomogeneous medium: 63rd Ann. Internat. Mtg., Soc. Expl. Geophys., Expanded Abstracts, SS2.9, 1249-1252.
Swiniarski, R., Hidalgo, H., and Gómez-Treviño, E., 1993, Neural network applied in the geophysical inversion problem: Proc. SPIE: Ground Sensing, 1941, 151-158.
Taylor, C. L., and Vasco, D. W., 1990, Inversion of gravity gradiometry data using a neural network: 60th Ann. Internat. Mtg., Soc. Expl. Geophys., Expanded Abstracts, 591-593.
Teles, T. N., and Do Carmo, E. A., 1999, DC resistivity data inversion using neural networks: 61st Ann. Internat. Mtg., Eur. Assn. Geosci. Eng., Extended Abstracts, P097.
Vallas, J., HA, Messatesta, A., and Strack, K. M., 1993, Resistivity-log inversion using a heterogeneous artificial neural network: 63rd Ann. Internat. Mtg., Soc. Expl. Geophys., Expanded Abstracts, WS7.2, 1395 (Neural Networks, Expert Systems, Fuzzy Logic & Unconventional Mathematical Techniques in Geophysics; Post-Convention Workshop).
Winkler, E., 1994, Inversion of electromagnetic data using neural networks: 56th Ann. Internat. Mtg., Eur. Assn. Expl. Geophys., Extended Abstracts, P124.
Zerilli, A., Botta, M., and Apolloni, B., 1997, Improving magnetotelluric data degraded by correlated noise with robust regression analysis and recurrent neural networks: 67th Ann. Internat. Mtg., Soc. Expl. Geophys., Expanded Abstracts, 366-369.
Zhang, L., Poulton, M., and Mezzatesta, A., 1999, Neural network based layer picking for unfocused resistivity log parameterization: 69th Ann. Internat. Mtg., Soc. Expl. Geophys., Expanded Abstracts, BH/RP 5.5, 136-139.
Zhang, M., Hou, J., and Liu, K., 1999, Improved BP method for 3D gravity inversion for multi-interfaces: 61st Ann. Internat. Mtg., Eur. Assn. Geosci. Eng., Extended Abstracts, 2-40.
Zhang, Y., and Paulson, K. V., 1997, Magnetotelluric inversion using regularized Hopfield neural networks: Geophys. Prosp., **45**, 725-743.
Zhang, Z., Zhou, Z., Frenkel, M. A., Chunduru, R., and Mezzatesta, A., 2000, Real time inversion of array resistivity logging data using dimensional reduction and neural network simulation: 70th Ann. Internat. Mtg., Soc. Expl. Geophys., Expanded Abstracts, RPB 5.3, 1802-1805.
Zhang, Z., and Zhou, Z., 2002, Real-time quasi-2-D inversion of array resistivity logging data using neural network: Geophysics, 67, 517-524.

1. E.I. Galperin: *Vertical Seismic Profiling and Its Exploration Potential.* 1985 ISBN 90-277-1450-9
2. E.I. Galperin, I.L. Nersesov and R.M. Galperina: *Borehole Seismology.* 1986 ISBN 90-277-1967-5
3. Jean-Pierre Cordier: *Velocities in Reflection Seismology.* 1985 ISBN 90-277-2024-X
4. Gregg Parkes and Les Hatton: *The Marine Seismic Source.* 1986 ISBN 90-277-2228-5
5. Guust Nolet (ed.): *Seismic Tomography.* 1987 Hb: ISBN 90-277-2521-7; Pb: ISBN 90-277-2583-7
6. N.J. Vlaar, G. Nolet, M.J.R. Wortel and S.A.P.L. Cloetingh (eds.): *Mathematical Geophysics.* 1988 ISBN 90-277-2620-5
7. J. Bonnin, M. Cara, A. Cisternas and R. Fantechi (eds.): *Seismic Hazard in Mediterranean Regions.* 1988 ISBN 90-277-2779-1
8. Paul L. Stoffa (ed.): *Tau-p: A Plane Wave Approach to the Analysis of Seismic Data.* 1989 ISBN 0-7923-0038-6
9. V.I. Keilis-Borok (ed.): *Seismic Surface Waves in a Laterally Inhomogeneous Earth.* 1989 ISBN 0-7923-0044-0
10. V. Babuska and M. Cara: *Seismic Anisotropy in the Earth.* 1991 ISBN 0-7923-1321-6
11. A.I. Shemenda: *Subduction.* Insights from Physical Modeling. 1994 ISBN 0-7923-3042-0
12. O. Diachok, A. Caiti, P. Gerstoft and H. Schmidt (eds.): *Full Field Inversion Methods in Ocean and Seismo-Acoustics.* 1995 ISBN 0-7923-3459-0
13. M. Kelbert and I. Sazonov: *Pulses and Other Wave Processes in Fluids.* An Asymptotical Approach to Initial Problems. 1996 ISBN 0-7923-3928-2
14. B.E. Khesin, V.V. Alexeyev and L.V. Eppelbaum: *Interpretation of Geophysical Fields in Complicated Environments.* 1996 ISBN 0-7923-3964-9
15. F. Scherbaum: *Of Poles and Zeros.* Fundamentals of Digital Seismology. 1996 Hb: ISBN 0-7923-4012-4; Pb: ISBN 0-7923-4013-2
    2nd Edition. Hb: ISBN 0-7923-6834-7; Pb: ISBN 0-7923-6835-5
16. J. Koyama: *The Complex Faulting Process of Earthquakes.* 1997 ISBN 0-7923-4499-5
17. L. Tauxe: *Paleomagnetic Principles and Practice.* 1998 ISBN 0-7923-5258-0
18. C.H. Thurber and N. Rabinowitz (eds.): *Advances in Seismic Event Location.* 2000 ISBN 0-7923-6392-2
19. A.P. Nickolaenko and M. Hayakawa: *Resonances in the Earth-Ionosphere Cavity.* 2002 ISBN 1-4020-0754-X
20. To be published.
21. W. Sandham and M. Leggett (eds.): *Geophysical Applications of Artificial Neural Networks and Fuzzy Logic.* 2003. ISBN 1-4020-1729-4

KLUWER ACADEMIC PUBLISHERS – DORDRECHT / BOSTON / LONDON

MIX
Papier aus verantwortungsvollen Quellen
Paper from responsible sources
FSC® C105338

If you have any concerns about our products,
you can contact us on
ProductSafety@springernature.com

In case Publisher is established outside the EU,
the EU authorized representative is:
Springer Nature Customer Service Center GmbH
Europaplatz 3, 69115 Heidelberg, Germany

Printed by Libri Plureos GmbH
in Hamburg, Germany